Health, Illness, and Medicine in Canada

Health, Illness, and Medicine in Canada

Second Edition

Juanne Nancarrow Clarke

Toronto   New York   Oxford
Oxford University Press
1996

Oxford University Press
70 Wynford Drive, Don Mills, Ontario M3C 1J9

Oxford   New York
Athens   Aukland   Bangkok   Bombay
Calcutta   Cape town   Dar es Salaam   Delhi
Florence   Hong Kong   Istanbul   Karachi
Kuala Lumpur   Madras   Madrid   Melbourne
Mexico City   Nairobi   Paris   Singapore
Taipei   Tokyo   Toronto

and associated companies in
Berlin   Ibadan

Oxford is a trademark of Oxford University Press

Canadian Cataloguing in Publication Data

Clarke, Juanne N. (Juanne Nancarrow), 1944-
   Health, Illness, and Medicine in Canada

2nd ed.
Includes bibliographical references and index.
ISBN 0-19-541206-0

1. Social medicine – Canada.   I. Title

RA418.3.C3C53 1996   362.1'042'0971   C95-932004-0

Contents

Chapter Fifteen
The Medical-Industrial Complex 363

Preface

AS WE FACE THE twenty-first century, significant changes in the medical care system and in the health of the population loom on the horizon. Perhaps the most basic change confronting Canadians today is the withering away of a universal medical care system. As federal transfer payments continue to decrease and the responsibility for medical care shifts to the provinces, the likelihood of inequality in the availability of services across the nation grows. Universal medical care has never been as threatened as it is today, as Canada shifts to the political right in the context of economic globalization. Unless there is a dramatic alteration in direction, inequality in medical services between provinces, between the rich and the poor, between races and ethnic groups, and between men and women will continue to grow.

Many argue that this crisis in the provision of allopathic medical care will have positive effects on the health of the population. The argument is that as the near monopoly in allopathic medical care is weakened, "complementary" and alternative care strategies will grow in their availability for the population. In fact, already about one-third of the U.S. and Canadian populations use alternatives or complementary health care. To the extent that the current medical care crisis destabilizes the hegemonic status of allopathic medicine and, in turn, other types of health care take up positions in the possible health-care strategies for Canadians, this crisis in medical care must be seen as an opportunity.

Health promotion and disease prevention now predominate in the developing "medical" ideology of the federal government. To the extent that these strategies individualize health problems that result from social-structural inequalities and "blame the victim," they have met with criticism. Focusing on individual "choice" with respect to prevention issues such as seat belt use, smoking cessation, moderation in alcohol consumption, low fat diets, exercise, and the like obscures the social, cultural, and structural constraints within which "choice" operates.

The deleterious effects of a variety of social and economic arrangements appear to be unabated in spite of widespread environment-related policy changes. Changes that the citizenry takes part in on an individual

9

basis, such as paper, metal, glass, and card-board recycling and reducing and reusing whenever possible, have only a minor effect in an industrial and corporate context where environmental controls compete with profitability. Major environmental challenges, such as chlorinated water and the pervasive use of bleach in manufactur-ing, the widespread use of pesticides, her-bicides, and fertilizers in agriculture, and the use of nuclear energy, continue to affect the health of the population and threaten to do so in the future.

Many have noted that sociology is done from four different perspectives: structural-functional, conflict, symbolic interactionist, and feminist. Each of these is based on dif-ferent assumptions, asks different ques-tions, and uses different methods and rules of evidence.

This book is organized into three major sections. The first part, Chapters One and Two, addresses some theoretical and meth-odological issues. Chapter One discusses the theory behind each of the four sociolog-ical perspectives and provides an illustra-tion of each perspective as applied to a concrete study.

Chapter Two examines the methodology employed by each perspective in studying health-related problems, including con-cepts such as causality, objectivity, and reliability; each is illustrated by concrete examples. Chapter Two ends with a discus-sion of the critiques that have been levelled at each methodology.

The second part of the book, Chapters Three to Six, explores the meaning and measurement of health and illness from these four different sociological perspec-tives. Chapter Three examines Canadian morbidity and mortality in historical and cross-cultural context. It examines ques-tions such as: What are the chief causes of death in Canada today? What were the chief causes of death 150 years ago? Why has the difference between male and female mortality rates increased over time?

Chapter Four examines the impact on health of environment, occupational and other health and safety issues, and violence in society. It asks questions such as: What are the major threats to the environment? What do we know about their impact on the health of the population? Are the health effects of environmental degradation spread equally across Canada and through-out the global community? What occupa-tions are safe? Is shiftwork bad for our health? Does violence impact on the health of Canadians?

Chapter Five focuses on the impact of un-equal social status on the health of the pop-ulation. In particular, it looks at the effects of age and gender. Questions such as the fol-lowing are asked: Are older people more likely to be sick or just more likely to be medicalized? Which sex lives a longer and more disabled life? Which sex lives a shorter life span? Do men and women get sick from and die from different or similar diseases?

Chapter Six examines two other aspects of inequity in the social system – class and race/ethnicity. Pertinent questions include: Is there evidence of class differences in morbidity and mortality? Do people of dif-ferent race/ethnic backgrounds experience different health and life chances?

Chapter Seven looks at the social and psychological antecedents to illness, sickness, and disease, and considers such questions as: What is the relationship between stress and illness? Can the loss of a loved one lead to serious illness? Is the inability to express anger sometimes implicated in the onset of cancer? Can social support minimize the effects and shorten the duration of an illness?

Chapter Eight concerns the experience of illness and answers questions such as: How do epilepsy, cancer, or multiple sclerosis affect the everyday life of the patient and family members? Do diseases have social meanings?

The third part of the book deals with the social construction of medical research and practice and the organization of the medical care system. It will address questions such as: Does a universal medical care insurance scheme guarantee universal accessibility? Why has the number of malpractice suits increased? What is the impact of the increasing number of female doctors? What is the "fate of idealism" in medical school?

Chapter Nine discusses the social basis of medical science and medical research. It also examines some strategies of lay resistance to medicine. Questions asked include: Is allopathic medical practice entirely based on the findings of traditional medical science research? Is medical science objective? Are new medical technologies necessarily better? When and why do people sometimes resist medical prescriptions and diagnoses?

Chapter Ten examines the relationship between medicine and religion in cross-cultural and historical context. It considers questions such as the following: Is medicalization increasing? (Medicalization may be defined as a tendency for more and more of human social life to be considered relevant to medicine.) Is medical practice an art or a science? Do physicians act as moral entrepreneurs? What sorts of ethical dilemmas do doctors face as they do their work?

Chapter Eleven examines the history of universal medical care insurance in Canada. What is the impact of universal medical insurance on class-based differences in health status? What is the future of extra-billing? What role did Tommy Douglas play in the foundation of Canada's universal medical care system?

Chapter Twelve examines medical practice as an occupation. Is medicine a profession? What is a profession? How do doctors handle mistakes? What is medical culture?

Chapter Thirteen considers two major critiques of the contemporary medical system – the dominance of one type of medical care system and sexism. It examines questions such as the following: How prevalent is chronic illness? What is the likely impact of the aging of the population on the medical care system? What is sexism? What is patriarchy? Why are most nurses women and most doctors men?

Chapter Fourteen describes other health-care providers – nurses, chiropractors (who focus on spinal alignment), naturopaths (who focus on natural healing), and midwives. Typical questions are: What is the importance of Florence Nightingale for nursing work today? What are the consequences of the bureaucratic work organization of the hospital on the work of the nurse? What is the status of the chiropractor in Canada today? Might naturopathic doctors ever offer traditional medical doctors significant competition for patients? What is the status of the midwife in Canada?

Chapter Fifteen examines the pharmaceutical industry as one component of the medical-industrial complex. It considers questions such as: How does the pharmaceutical industry maintain its position as one of the lowest-risk and highest-profit industries in Canada today? What is the role of doctors and pharmacists in prescribing drugs? Which Canadian people are most likely to use mood-altering drugs?

PART ONE

SOCIOLOGICAL PERSPECTIVES

CHAPTER ONE

Ways of Thinking Sociologically about Health, Illness, and Medicine

ALMOST ALL OF US have been sick at some time in our lives. When do we acknowledge that we are sick? Is it when we stay in bed for a day or two? Perhaps it is when we feel a pain but take a pill and go on with the day as planned? Or perhaps we may not truly claim sickness unless we go to the doctor to find a name for the unusual way or the discomfort we feel? Whatever our view, all of us experience illness in a social context: we recognize it because we have developed a vocabulary that allows us to talk about it with others in our immediate circle or in the larger social world; we learn what to do about it as we interact with others, with friends and family, at times with the formal medical system, and with our society through such media as television and magazines.

Are you aware of the relationship between sickle-cell anemia and ethnicity? Do you know the reason for that relationship? Unemployment has been found to be followed by ill health. What might the sociological explanations of such a finding be? Are most types of morbidity and mortality inversely related to income? That is, is it statistically correct to say that the lower the income, the higher the rates of sickness, disability, and death? These are the sorts of questions asked in the sociology of health and illness, which seeks to describe and explain the social causes and consequences of illness, disease, disability, and death; to show the ways lay people and professionals alike constitute or construct their categories of disease and illness; and to portray the ways that illness affects and is affected by social interaction among various people or institutions.

When we think that we are sick, what do we do? Some of us treat ourselves with our favourite home remedies such as bed rest and tea or chicken soup. Some of us seek advice from friends or family members. Some of us visit our general practitioner, a medical specialist, or a pharmacist. A few head off to the emergency room of the nearest hospital. A few others seek alternative health care such as homeopathy, acupuncture, or Ayurvedic medicine. When you seek the advice of a physician, do you think you would receive better medical care from a physician who works in a fee-for-service setting or from one on a salary paid

Table 1.1   The Sociology of Medicine and of Health and Illness

SOCIOLOGY OF MEDICINE	SOCIOLOGY OF HEALTH AND ILLNESS
The organization of the medical care system	The distribution of disease and death
The profession of medicine (and auxiliary and competing health-care professionals)	Disease and death in socio-historical context
	Socio-demographic explanations for disease and death
Alternative health-care providers	Class, patriarchy, and sexism as explanations for disease and death
The financing of medical care	
The medico-industrial complex	Socio-psychological explanations for disease and death
Class, patriarchy, and sexism and the organization of medical care	Experiencing and talking about disease and death
	Ways people construct or label certain signs as symptoms of disease
The health-promotion industry	
The development and perpetuation of medical discourse and ideology	Environmental conditions and health, occupational health, safety issues, and health consequences

by the state, a corporation, or a clinic? Do practitioners in group practice provide better care than those in practice on their own? Does the Canadian government have adequate drug safety procedures to protect Canadians against another drug disaster such as thalidomide? Why does the universal medical insurance scheme provide guaranteed funds for medical practitioners but only limited funds for chiropractors, naturopaths, and other alternative health-care providers? Did the moratorium on the use of breast silicone implants result from new scientific or medical knowledge or consumer organizing and lobbying?

The sociology of medicine is the study of the ways the institutionalized medical system constructs what it deems to be illness out of what it recognizes as signs and symptoms, and constitutes its response to such "illness" through the treatments it prescribes. This field of sociology examines and offers explanations for such topics as

the varying types of medical practice and medical discourse, the ideology and organization of medicine, different ways of financing medical care, the structure and operation of the hospital, and the occupational worlds of the nurse and the doctor. It also attempts to explain the relationship among the different types of medical care and the importance of the medical care system in the context of the culture and the political economy of the state.

This book provides an overview in Part One of the basic features of a sociology of health and illness and in Part Two of a sociology of medicine in Canadian society. In this chapter we will start by examining the four sociological paradigms that will be applied throughout the book.

Sociologists study the social world from a variety of perspectives. Depending on their perspective, they focus on some aspects of social life and ignore others, and ask different questions and use different ways to answer them. As might be expected, these varied sociological perspectives are manifest in the sociology of health and illness and of medicine: sociologists have approached these fields with different or even contradictory assumptions. At times different sociologists have described or analysed an aspect of illness or medicine from such widely differing points of view that they appear to be discussing different phenomena. There is some agreement that the various perspectives can be distilled into four distinct paradigms or images. It is now conventional to call these perspectives structural-functional theory, conflict theory, symbolic interactionist theory, and feminist theory.

Structural Functionalism

Structural functionalism dominated North American sociology for many years. It has been the reigning paradigm, the "normal" science of the discipline (Kuhn, 1962). Most sociological studies published in North America have adopted this perspective. Auguste Comte, who first gave the name of sociology to the science of society, thought that sociology's goal was to better society so that it might become orderly and progressive. He might be called the godfather of sociology. Emile Durkheim (1858-1917) provided both the theoretical and methodological model for structural functionalism. Durkheim defined sociology as a science of social facts. Social facts, he said, were to be treated and studied as if they were real, external to individuals, and yet capable of constraining and directing human behaviour and thought. The subject matter of sociology was the knowledge of these social facts and their impact on human behaviour. Constrained by the external world, human beings, in Durkheim's view, were predictable and controllable through the power of norms that exist in their own right, aside from their manifestations in individuals.

Sociology in the Durkheimian tradition is often called structural functionalism. It assumes that the proper level of study for the sociologist is the society or the system. The social system is said to be composed of parts, institutions that function to maintain order in the social system. Just as the organs in the human body are inextricably tied to one another and function as interrelated parts, so, too, are the parts or the institutions of society – the family, the economy, the polity, and the educational, welfare, military, and medical care systems. All these institutions operate interdependently to keep the society functioning. It is the problem of maintaining a good working order in society that motivates theorizing and research in this sociological perspective.

Structural-functional theory is often associated with a positivist methodology. Positivists view sociology as a science in the same way that physicists view physics as a science. Positivists assume that social scientists both should and can remain objective and value-free while observing, recording, and measuring external social facts. Just as the natural sciences seek universally true causal explanations of relationships in the natural world, so do positivist sociologists in the social world. As well, since positivists believe that social facts are to be treated as real and external, they tend to rely on data that are assumed to be objective, collected from interviews and questionnaires administered to individuals in survey research, and analysed and organized to reflect the probability of the occurrence of certain behaviours among a certain aggregate or group of individuals.

Five things distinguish structural functionalism from conflict theory, symbolic interactionism, and feminist theory. They are the assumptions (1) that sociology aims to discover and to explain the impact of social facts on human behaviour, attitudes, or feelings; (2) that social facts are to be treated as things that are real and external to human actions, and that determine human behaviour; (3) that social facts can be seen in aspects of the social structure such as the norms that guide behaviour, in social institutions such as the family or the economy, and in social behaviours such as those in relationships between the sexes, in marriage, or at work; (4) that sociology is a science that seeks to describe the world in a series of universal causal laws; and (5) that this science considers that human behaviour is objectively and quantitatively measurable through methods such as experiments and survey research.

One of the most influential contributions to medical sociology from a structural-functional point of view is Talcott Parsons's work on the sick role (Parsons, 1951: 428-79). To understand Parsons's sick role, it is necessary to understand that each individual plays a number of roles in society. Roles arise out of the institutions with which the individual is associated. For example, an individual will likely play some of the following family roles – daughter, son, brother, sister, mother, father, niece, nephew, and a whole series of in-law roles. An individual may also play a variety of work roles, neighbourhood roles, friendship roles, and so on. All roles reflect something of the intermeshing of the individual in society. The idea of role is a pivotal one in conceptualizing the relationship between the individual and society.

Parsons's primary concern was to describe the processes that maintain ongoing societal institutions. His notion of the sick role should be looked at in this context. Sickness could lead to societal breakdown resulting from the inability of the sick to fulfil their necessary social roles, such as parenting, maintaining a home, and working in the paid labour force. Therefore sickness must be managed and must be accorded a special role. However, this legitimation is only temporary and is contingent on the fulfilment of certain obligations by the individual who claims the sick role. There are four components to the sick role. The first two are rights, the second two are duties. Both the rights and duties of the sick role must be fulfilled if the equilibrium of society is to be maintained.

(1) The sick person is exempt from "normal" social roles.

The sick individual has a legitimate excuse for missing an exam or a major presentation at work, for staying in bed all day and neglecting household chores, or for staying home from work. The length of the exemption depends on the severity of the illness. In order to win exemption, the individual may need formal, medical acknowledgement. The sick person may have to obtain an official medical diagnosis and even a medical certificate as proof of illness. Exemptions from examinations, for instance, generally require a formal written note from a physician.

(2) The sick person is not responsible for his or her condition.

The sickness must be the result of an accident or other circumstances beyond the control of the individual if that person is to be accorded the sick role. Thus the individual is not to be blamed or punished. Influenza, a cold, or a broken leg are considered the result of misfortune, not of personal will or desire. Therefore sympathy rather than blame is considered the appropriate reaction of others.

(3) The sick person should try to get well.

The person who is given the legitimacy of the sick role is duty bound to want to get well. Sick role exemption is only temporary. If an individual does not want to get well or does not try to get well then the sick role is no longer legitimate. Thus, if a person has received a diagnosis of pneumonia, he or she must do what the doctor orders. If not, the legitimacy of the sick label deteriorates into the shame of a label such as "foolish," "careless," or "malingerer."

(4) The sick person should seek technically competent help and co-operate with the physician.

The duties associated with the sick role also require that the ill person seek "appropriate" medical attention and comply with the treatment provided. For example, a person with AIDS who refuses both to accept medical care and to change certain sexual or drug-use habits would not be accorded the rights of the sick role but could be subject to legal punishment.

From the viewpoint of Parsons, illness is a form of deviance. It is a potential threat to the social system unless it is managed for the benefit of the social system. Medicine is the institution responsible for providing legitimation and justification and for bringing the sick back to wellness or "normality." Medical institutions can be seen as agents of social control in much the same way that the church or the criminal justice system is.

Parsons's formulation of the sick role was primarily theoretical. It was not based on extensive systematic empirical investigation. There are a number of criticisms of the sick role based on empirical analysis. Some of these criticisms will be examined in the following sections.

(1) The sick person is exempt from "normal" social roles.

The extent to which a person is allowed exemption depends on the nature, the severity, and the longevity of the sickness, and also on the characteristics and normal social roles of the person. A short and

TABLE 1.2 The Four Central Sociological Perspectives

	STRUCTURAL FUNCTIONALISM	CONFLICT THEORY	SYMBOLIC INTERACTIONISM	FEMINIST THEORY
EXEMPLAR Model of subject matter	EMILE DURKHEIM Society is a social system of interlocking and interrelated parts or institutions	KARL MARX Society is a system of classes	MAX WEBER Society is composed of selves who make their social lives meaningful through interaction	DOROTHY SMITH Understanding social organization, structure, power, and knowledge from women's perspective
Model of the subject matter in process	Institutions perform (dys)functions that are both manifest and latent in the interest of the continuation of the social system in equilibrium	Power groups have contradictory purposes, based on their relationship to the basic economic structures	Selves create reality anew from situation to situation in interaction with others	Women's selves are tied to the relations of ruling; feminist research has change as one of its goals; it emphasizes the empowerment of women along with the transformation of a patriarchal social system
Ways of doing sociological analysis	System explainable and predictable through a series of if x . . . then y causal statements; x and y are social facts	Power groups are understandable from a committed stance examining the conflicts in historical context	Selves' world views and symbols arise out of interaction and are made understandable through process of interpretive, empathetic understanding – *verstehen*	All methods of data collection may be used but a collaborative approach (between researcher and subjects of research) is advocated; triangulation is suggested; language is gender appropriate or neutral
Objectivity/subjectivity	Necessary to be objective and to study the social world objectively	Value-committed perspective necessary	Acknowledgement of the inevitability of contextual reflexivity of knower and known	Quite often focuses on/begins with women's experience; impossible to be objective; therefore important to clarify standpoint and acknowledge reflexivity

TABLE 1.2 Cont.

Image of human nature	Human beings believe, think/feel, and do as the result of external constraining forces	Human beings are alienated from self, others, and meaningful work, and need the liberation that would come from revolutionary change	Human beings continually construct reality as they interact with others in their social words	Differences by class, gender/power, sexual orientation, dis/ability limit generalization about human nature

self-limiting burn on the fingers merits only temporary, minimal exemptions from life roles. On the other hand, multiple sclerosis, a chronic and degenerative disease, allows extensive exemptions.

The university student's sick role is mostly informal. Most professors do not take attendance; students can avoid the library for weeks on end without any formal notice being taken; they can stay in bed half the day and stay out half the night. These things are the student's own responsibility. It is only at the time of regularly scheduled deadlines for papers, presentations, and examinations that universities typically take any official notice of the student's actions. At these times the student may need to adopt the sick role formally by obtaining official legitimation from a physician.

(2) The sick person is not responsible for her or his condition.

This belief varies depending on the nature of the condition and the circumstances through which the person is believed to have acquired the condition. The sick person may be held responsible for having a cold, for instance, if he or she stayed out overnight and walked miles in the freezing rain without a jacket. The notion of stress that is prevalent today has an aspect of blame attached to it – that is, people who succumb to disease because they have been overworked or worried may be chastised for having failed to take preventative action. One of the implications, in fact, of the recent emphases on health promotion through lifestyle change (giving up smoking, giving up drinking alcoholic beverages, and so on) is that people who do not change may be more likely to be held responsible for lung cancer or liver cirrhosis, for example, or the person with AIDS may be blamed for his or her sexual habits. In addition, a number of diseases are thought to reflect on the moral and social worth of the individual, and when an individual succumbs to these diseases he or she is blamed by virtue of the stigma attached.

There is considerable evidence that even though the specific causes of particular cancers are not known, there is a way in which the person with the disease is sometimes blamed for succumbing. For instance the person with lung cancer may be held responsible because he/she smoked cigarettes. Several social researchers have noted the way that a person diagnosed with AIDS is often thought to be at fault because of a "nefarious" lifestyle or "immoral" sexual choices (Altman, 1986; Sontag, 1989). Clarke's (1985) study of women with cancer found that respondents sometimes talked of how friends and even some family members seemed to reject them after their cancer diagnosis. Some spoke of people walking on the other side of the street rather than stopping to converse. Some talked of being avoided by their husbands. Some perceived discomfort in their doctor's inability to relate to them once the cancer was discovered (Clarke, 1985: 121). Even daughters experience some stigma and isolation as the result of others' knowledge of their mothers' diagnosis (Clarke, 1995).

Epilepsy, leprosy, and venereal disease are other widely stigmatizing diseases (Schneider and Conrad, 1983). In a sense, people with such discrediting or discreditable diseases are not given the social legitimacy of the sick role (Goffman, 1963). In fact, the negative associations of some diseases have caused discrimination and exile. Such stigma heightens both physical and emotional pain. Even when they are not thought materially responsible, people are sometimes thought morally responsible.

(3) The sick person should want to get well.

There are illnesses from which people cannot or are not expected to recover. People are expected to adjust to such illnesses. A so-called "terminal" illness is a case in point. Patients are not granted legitimacy for wanting to get well once they have been diagnosed with a terminal illness. In fact, if people continue to want to get well when they have been diagnosed as terminal, they are often criticized because they may be said to be denying reality. Similarly, people with a whole range of chronic illnesses are not expected to want to get well but rather to adapt to daily limitations and disabilities.

(4) The sick person should seek technically competent help and co-operate with the physician.

The dominant medical care system is that of allopathic medicine. Allopathic medicine treats disease by trying to create a condition in the body that is opposite to or incompatible with the disease state. While this medical care system still claims a monopoly on the right to provide treatment and thus to legitimate sickness, there are competing medical systems with varying degrees of legitimacy. Some, such as midwifery, have grown in their powers as rightful alternatives (see Burtch, 1994, for an overview of the history and contemporary status of midwifery in Canada). Moreover, allopathic medical practice is the subject of growing critical analysis by consumer interest groups, particularly the women's health movement (see Rachlis and Kushner, 1994, for a popular example of contemporary Canadian critiques of the allopathic medical system). Critical evaluations of such things as unnecessary surgery, side effects from taking prescribed drugs, and unnecessary medical intervention in childbirth raise the possibility that co-operation with physicians may not always be the most efficacious road to good health. The definition of the technically competent profession becomes quite problematic in this situation.

In spite of the critical questions raised above, the sick role concept is important in medical sociology. This was the first analysis explicitly to note that there are ways in which medical practice, its ideology, and its associated medical institutions serve to fulfil social control functions for the society. A number of sociologists since Parsons have examined and critiqued the social control functions of medicine. Szasz (1974), Freidson (1970), Zola (1975), Illich (1976), and Conrad and Schneider (1980) are among those who have expressed concern that medicine is growing in its powers of social control. Behaviours once considered illegal or immoral and thus under the jurisdiction of judicial or religious institutions are more and more likely to be seen as medical problems requiring diagnosis and treatment, and thus come under the jurisdiction of the medical care system. Some argue that medicine has established a jurisdiction far wider than that merited by its demonstrated ability to provide a "cure" (see Freidson, 1970, for a theoretical discussion, and Rachlis and Kushner, 1994, for contemporary Canadian examples).

Although Parsons's sick role concept has, as Freidson says, provided "a penetrating and apt analysis of sickness from a distinctly sociological point of view" (1970: 228), it must be examined critically. It is a generalization, and its empirical application is seriously limited. Criticisms have been levelled at the sick role because of its lack of attention to differences in legitimation given to patients of differing age, gender, class, and racial or ethnic background. Furthermore, different types of illnesses provide different degrees of legitimacy to the sick person. The allopathic doctor described by Parsons is only one of several alternatives available today. Indeed, complementary or alternative medicine is used by a growing number of people in Canada and the United States (Harpur, 1994). Among other factors, doctors themselves differ in their practice according to class, gender, age, and ethnic background. Moreover, numerous alternative healers compete with the allopathic physician.

While positivism is the research methodology most closely associated with the structural-functional perspective, all positivists are not structural functionalists. Parsons's sick role is peripheral to a great deal of medical sociology in the positivist tradition. Contemporary positivists study human health behaviours as they are caused by and as they cause what are called independent and dependent variables. An examination of the impact of a diagnosis, e.g., cystic fibrosis, on the family of the ill person treats human health behaviour as the independent variable. On the other hand, when the impact of income level on the rates of incidence of sickness is studied, human health behaviour becomes the dependent variable. Today, positivists, following Durkheim, assume that the social structure has a constraining impact on individuals. Social structural positions (social facts) determine individual thoughts, behaviours, feelings, and, in this case, health and illness, medical use, professionalization,

FIGURE 1.1 An Example of Model A

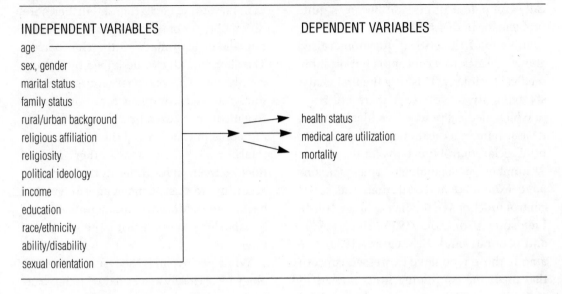

INDEPENDENT VARIABLES

age
sex, gender
marital status
family status
rural/urban background
religious affiliation
religiosity
political ideology
income
education
race/ethnicity
ability/disability
sexual orientation

DEPENDENT VARIABLES

health status
medical care utilization
mortality

and so on. Changes in one institution, such as the family, necessitate changes in other institutions, such as the medical care system. Such changes, according to the positivists, can be described in a series of causal laws of the "if x then y" variety. The underlying assumption of this perspective is that a complete understanding of social facts will explain all that needs to be explained about human beings. Research along these lines examines the effects of such things as gender, class, educational level, family type, marital status, age, rural-urban background, religious affiliation, religiosity, and political ideology on such dependent health-related variables as health experiences, death rates, health lifestyles, and use of medical care. Research that considers health and medical variables as dependent variables is here called Model A.

A prosaic example of the logic of this

research would be a statement such as this: the health status of a certain aggregate, e.g., students at university, is a function of the distribution of such personal characteristics as the age, gender, marital and family status, class, and ethnic background of the students. Thus, for example, one hypothesis might be that female students will have a poorer health status than male students. This might be explained further as follows: female students are more likely to have part-time jobs than male students, and students who have part-time jobs are more likely to be ill.

Other positivists change this causal order and treat the health-related aspects as the independent variables. They examine the impact of the experience of illness on the social structure. They study the ways that such independent variables as health status, death rate, and use of medical care

FIGURE 1.2 An Example of Model B

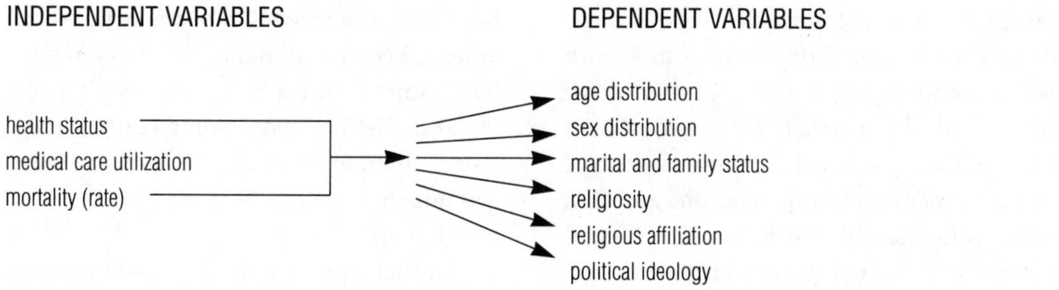

INDEPENDENT VARIABLES

DEPENDENT VARIABLES

health status
medical care utilization
mortality (rate)

age distribution
sex distribution
marital and family status
religiosity
religious affiliation
political ideology

affect social structural conditions such as socio-economic status, educational level, religiosity, marital and family status, and rural/urban background. We call this Model B.

An example of the logic of this research would be a statement such as the following: the age distribution of a given aggregate of people is a function of the mortality rate, the medical care use rate, and the previous health status of the sample.

Summary

Several examples of theory and research have demonstrated the basic principles of structural functionalism. The goal of sociology in this perspective is to discover and explain the place of social facts in human behaviour, attitudes, or feelings. It aims to do this through scientific methods that seek to uncover universal causal laws based on quantitative analysis of "objective" social phenomena. Such empirical analyses are implicitly part of the larger theoretical analysis, which concerns the functions performed by various parts of the social system for the maintenance of that social system.

Conflict Theory

Conflict theory has had a much less dominant role in the development of sociology in North America. It has provided and still provides a radical critique of the more conservative aspects of the mainstream of structural-functional sociology and of the economic and social arrangements found in society. In conflict theory, all social arrangements, all sociological theories, and all sociological methods have political and economic bases and consequences. Conflict theory tends to focus on class/gender power relations and dynamics. Research topics, methodological approaches, and commitment to the use of findings all reflect the political and economic interests of the researcher.

The model of this paradigm is the work of Karl Marx (1818-1883). Marx was directly involved in the analysis of and organization for changes in his society. The author of numerous books, he was the leader of the First Communist International in Europe during the latter half of the nineteenth century and was also a busy and effective investigative journalist. He asserted that human thought and behaviour were the result of socio-economic relations and that both were alterable for human and social betterment. Believing that human beings could change their social order, Marx worked toward human liberation through a social and economic revolution.

Society, according to Marx, has historically been composed of a constantly varying balance of opposite forces that generate change through their ongoing struggle. The motivating force behind this continuous struggle is the way in which people interact with one another as they attempt to obtain their livelihood. Marx described the various modes of production with their corresponding types of social relations that occurred in consecutive historical periods, including primitive communal societies, slave societies, and feudal and capitalist economic systems. He described each of these periods of human history as a period of struggle between classes. The class struggle is related to the means of production, e.g., the land or the factory, because members of one class own the means of production and members of the other class sell their labour for cash. For Marx, an end to conflict was both possible and desirable in a communist state in which all citizens owned the means of production.

In their study of sociology, conflict theorists use information from a variety of sources, but they tend to have an historical and critical focus. As with structural functionalism, the level of analysis is the social system, because ultimately the system must be changed and a new one established. What is distinct about some conflict theorists is that they may also be activists. Some see injustice everywhere, and some try to alleviate it.

Conflict theory can be distinguished from structural functionalism and symbolic interactionism in the following ways: (1) the sociologist's work is to discover and document injustice (and sometimes to attempt to change it); (2) all knowledge is rooted in social, material, and historical context; and (3) sociological research methods must acknowledge social, economic, and historical contexts. When the conflict theorist is particularly influenced by Marx's analysis, the primary subject of study is social classes, because they are thought of as the means to effect change.

Sociology from the perspective of conflict theory is generally thought to involve the documentation of injustice for the purpose of understanding its origins and causes in an historical and socio-political context. The analysis usually focuses on recurrent patterns and the dynamics of power relations between social classes or between the sexes. Social injustice is everywhere, and medical institutions are no exception.

A long tradition of scholarship documents the ways in which health and illness are related to unequal social arrangements. Marx's collaborator, Frederick Engels, in *The Condition of the Working Class in England*

(1845), showed how the working and living conditions that resulted from early capitalist production had negative health effects. Engels described how capitalism introduced mechanization on farms, thus resulting in a mass of unemployed rural workers who were forced to migrate to the cities to make a living. Capitalists in the city, driven to make a profit, kept their labour costs low, and thus the working classes could only afford very cheap shelter and food. The great slums that resulted were the perfect breeding grounds for the epidemic diseases endemic to such living conditions – tuberculosis, typhoid, scrofula, rickets, and other infectious diseases. Thus, ill health was related to the living conditions of the working class and the material conditions of capitalism (Navarro, 1986).

Advanced monopoly capitalism a century and a half later (O'Connor, 1973; Turner, 1987) has generated considerably different but equally troublesome working/living conditions for workers and their families. Standards of living have greatly improved. However, tremendous inequity remains and can be seen in mortality and morbidity statistics. (See "Our Sickening Social and Physical Environment" in Conrad and Kern, 1990, for a discussion of several deleterious consequences of capitalism.) Twentieth-century capitalism is dominated by huge corporations such as General Foods, General Motors, IBM, and Exxon. It is characterized by attempts to increase profit through worker productivity, new time-saving technologies, and expansion into the Third World (both for production and markets). Profits can be increased by getting the workers to produce more in the same time period and by decreasing wages. As a result there is a continuous contradiction between the needs of the workers for a good living wage, good working conditions, adequate time for rest and relaxation, and "meaningful," satisfying work and the needs of the capitalists for expansion and profit.

Vincente Navarro (1976) is one of the foremost of contemporary conflict theorists of medical sociology. He explains the contradictory relationship between capitalism, which is an economic system fuelled by the profit motive, and the health needs of the population. At times the need to make a profit requires that workers labour, live, and eat in unhealthy and unsafe environments. For example, the U.S. National Cancer Institute estimates that 86 per cent of all cancer mortality is the result of (preventable) lifestyle, occupational, and environmental factors (Epstein, 1993: 26; 1994).

The manufacture, sale, and tax level of cigarettes is an example of the way that the need for corporate profits and state tax revenues may outweigh the desire for good health. The tobacco industry has maintained a successful profit margin by opening up new markets in spite of anti-smoking sentiments and some no-smoking legislation. In Canada, for example, the market has expanded into the younger age groups. A 1986 Gallup poll for Health and Welfare Canada reported that the average age for beginning cigarette smoking has gone down over the past twenty years from 16 to 12 years of age. Though there are laws prohibiting the sale of cigarettes to minors, they are rarely enforced, and even when they are enforced the fines are so low that they are virtually useless.

Navarro argues that there are two main goals of contemporary capitalism: the concentration of capital and the growth of the state. He explains that the state intervenes in the health sector to promote capitalist goals. This occurs, first, as the class structures of society are reproduced within the medical sector, so that the distribution of functions and responsibilities of occupational groups within the medical care system mirrors the class, racial, ethnic, and gender hierarchies within the other sectors of capitalist society. Second, the medical system has a bourgeois ideology of medicine that regards both the cause and the cure of illness as the responsibility of the individual. Health itself becomes a commodity with a certain value within the marketplace. In the capitalist view the medical model is politically conservative; it directs attention away from the social-structural causes of ill health such as gender, race, class, occupation, and environmental degradation. Third, the state reinforces alienation because people are not free to choose alternatives to physicians, such as chiropractors, naturopaths, masseuses, or dieticians. The state provides full financial support for only one type of medical service – that provided by the allopathic practitioner.

The state also uses strategies to exclude conflicting ideologies from debate and discussion. One example, cited by Navarro, is the emphasis on the individual causation of disease for research funding. Such viewpoints exclude analysis of the processes through which class origins, environmental pollutants, occupational hazards, and working conditions are significant causes of ill health.

A sociological analysis by Hilary Graham in *Women, Health and the Family* (1984) illustrates research in the conflict perspective. Although Graham's research was carried out in Great Britain, it raises questions for all societies regarding the role of women and the consequences of economic impoverishment for the health of families. It analyses the impact of poverty and the manifold effects of the relative scarcity of resources such as transportation, housing, fuel, food, and health care in the home on the health status of family members.

Health care in the home is composed of four elements. First is the provision of healthy conditions. This involves the maintenance of a warm and clean home with sufficient space for rest and relaxation for all family members, sufficient and adequately nutritious foods, and clean water. Home health care also involves managing social relations and meeting emotional needs for the optimal mental health of family members. The second element is nursing the sick: much of the work of caring for the sick child or adult, and for elderly or disabled persons, falls on the shoulders of women. Furthermore, increasing deinstitutionalization of the mentally, chronically, and acutely ill increases the level of intrafamily responsibility. Nursing the sick is often a very time-consuming and exhausting job. It involves sleepless nights, heavy lifting, preparation of complicated menus, administering medicines, and coping with bandages and the like. A third element is teaching about health, including such things as modelling good health habits and giving instruction on diet, hygiene, and exercise. The fourth aspect of home health care is mediating with outsiders such as doctors or hospitals, making visits to clinics, talking with a social or public health worker, or getting advice from an expert in a health-related area such as nutrition.

Graham documents the existence of class differences in home health care and in mortality and morbidity rates. She notes the consistent inverse relationship between class and some of the most sensitive indicators of a nation's health, such as stillbirths, pre-natal mortality, neo-natal mortality, post-natal mortality, and infant mortality rates. In each case the higher social classes have far lower mortality rates than those in the lower social classes. Babies with low birth weights are much more likely to become sick and die than babies with high birth weights. Women who bear babies with low birth weights generally live in poor households and lack safe and adequate nutrients.

These class differences among infants are mirrored in the morbidity statistics for children and adults. Accidents, the largest single cause of childhood death, are probably one of the best indicators of an unsafe, inadequately supervised environment: the accident rate increases sharply among the lower social classes. Poorer families are also more likely to suffer from other environmental hazards that kill children, such as respiratory diseases. The incidence of infections and parasitic illnesses is also class-related. In contrast, the incidence of childhood diseases without known environmental, nutritional, or other material causes, such as cancer and congenital anomalies, shows no relationship to class.

Samuel S. Epstein, an epidemiologist with an expertise in the occupational and environmental causes of cancer, recently documented the high incidence of preventable cancer deaths and the minimal research investment of the National Cancer Institute in understanding these preventable deaths. He noted that 17 per cent of the $2 billion budget for 1992 of NCI research initiatives goes to research into primary cancer prevention; 1 per cent of the total appropriations are dedicated to research into occupational cancers (Epstein, 1993: 24). The rest of the monies are directed toward diagnosis and treatment. He attributes the bias toward research on diagnosis and treatment to the lack of expertise on occupational and environmental carcinogens on the National Cancer Advisory Board – even though this situation violates the National Cancer Act, which stipulates "that no fewer than 5 members shall be individuals knowledgeable in environmental carcinogens" (ibid.: 19). In addition to the failure to include people with such backgrounds on the board, the NCI is further compromised by institutionalized conflicts of interest. As Epstein says, "for decades the war on cancer has been dominated by powerful groups of interlocking professionals and financial interests, with the highly profitable drug development system at its hub – and a background that helps explain why 'treatment,' not prevention, has been and still is the overwhelming priority" (*ibid.*: 20).

As a particular case, Epstein cites the conflicts among board members of the Memorial Sloan-Kettering Cancer Center in New York. Included among the overseers of this major cancer treatment and research centre are directors, board chairmen, and presidents of pharmaceutical and medical technology corporations. In addition, Epstein documents similarly impressive and powerful directorships and other important ties with various multinational industrial corporations such as Exxon, Philip Morris, Texaco, Nabisco, General Motors, Algoma Steel, and Bethlehem Steel. Even the media, including the New York Times Corp., *Reader's Digest*, Warner Communications, and CBS, are involved (*ibid.*: 22-23). He asks whether people (usually men) with industrial and pharmaceutical interests can reasonably be expected to support research that might criticize and challenge their products. It is no wonder, argues Epstein, that cancer research focuses on diagnosis and treatment to the relative exclusion of prevention.

Summary

Two examples of conflict theory have demonstrated its basic principles. These are that (1) the purpose of sociology is the documentation and analysis of injustice resulting from such factors as class, race, gender, and power; (2) knowledge is never objective but always dependent on its social, material, and historical context; and (3) understanding conflicting social and economic forces is essential for an understanding of all the other conditions of social life.

Symbolic Interactionist Theory

What is the meaning of illness? Does cancer have the same meaningful impact when it happens to an 80-year-old man or woman as when a three-year-old child is diagnosed with cancer? What are the processes through which the slow onset of Alzheimer's comes to be noted by family, friends, and the patient? How do families work through their changing understanding of the uncertainty and then the certainty of death of one of their members? How does the self-identity of the person with AIDS change once he or she has received the diagnosis? How do others change in the ways that they relate to the person with AIDS? These are the sorts of questions asked in the symbolic interaction perspective.

Max Weber provided another definition of sociology: "a science which attempts the interpretive understanding of social action in order thereby to arrive at a causal explanation of its course and its effects" (1947: 88). There are two crucial elements in this statement, each of which exemplifies an aspect of Weber's work. First, social action, as defined by Weber, meant action to which the individual attached subjective meaning. Second, the sociologist, while looking for what Weber calls causal explanation, was actually directed to interpret empathetically the meaning of the situations from the viewpoint of the subject.

Symbolic interactionist sociologists study how the subjective definitions of social reality are constructed and how this reality is experienced and described by the social actors. Human beings create their social worlds. As W.I. Thomas said, if a situation is defined as real, it is real in its consequences (Martindale, 1960: 347-53).

The paradox here is that, just as the subjects who are being studied are busy defining reality for themselves, so, too, are the researchers. Thus, the symbolic interactionist is faced with the problem, when collecting data, of intersubjectivity or reflexivity, that is, that the data are given a subjective slant both by the people being studied and by the researcher.

The sociological researcher must be aware of his or her own processes of attaching subjective meanings just as he/she studies the subjective meanings of the subjects of study. The symbolic interactionist researcher must also face a second problem: that the research act itself creates and changes meanings and processes. In symbolic interactionism it is impossible to gather objective data. All social reality is subjectively defined and experienced and can only be studied through the subjective processes of social researchers.

Empathetic understanding, or what has come to be called, following Weber, *verstehen*, is the desirable methodological stance of the researcher. Generally, sociologists adopting this stance collect data by observing social action in close participation with the subjects or by long, unstructured interviews. The level of analysis is not the system but rather individual interaction

with others, the mind or the self, and meaning. This is micro analysis. The structural functional and conflict theories, because they focus on systems, are macro analyses.

There are three assumptions characteristic of this perspective: (1) sociology is a science whose purpose is to understand the social meanings of human social action and interaction; (2) reflexivity or intersubjectivity, rather than objectivity or critical analysis, characterizes the relationship between the subject and the researcher; and (3) rich, carefully detailed description and analysis of unique social situations from the perspective of the subjects under investigation is typical of symbolic interaction research.

The sociological problem to be understood and explained in the symbolic interaction tradition is the meanings that individuals see in the actions of themselves, of others, of institutions, and so on (Weber, 1968). Analysis of society demands different methods than those used to describe and explain the natural world. It requires methods that attempt to grasp the motives and meanings of social acts. Sociology is a science that must deal with the subjective meanings of events to social actors.

A good contemporary example of work in this perspective is the study of the meaning of the diagnosis of epilepsy within the lives of a sample of people with this disease. *Having Epilepsy* (Schneider and Conrad, 1983) is based on long, semi-structured interviews with a number of people who have been diagnosed with epilepsy. The authors make the point that it is important that sociologists provide an antidote to medical research. It is crucial, they note, to distinguish between disease and illness. Disease is the pathology of the human body; illness is the meaning of the experience associated with a given pathology. In this research the subjects were selected for study because they have epilepsy. However, they live most of the time without the symptoms of the disorder being present. They go about their daily activities, eating, dressing, working, cooking, cleaning, visiting, enjoying leisure and social activities, unstrained and unconstrained by an awareness of their disease. Medicine attempts to understand the nature and cause of disease and to formulate methods for its treatment. One of the tasks of the sociologist is to describe the impact of disease and diagnosis on the individual and on his or her relationships with others. As Schneider and Conrad say: "We cannot understand illness experiences by studying disease alone, for disease refers merely to the undesirable changes in the body. Illness, however, is primarily about social meanings, experiences, relationships, and conduct that exist around putative disease" (1983: 205).

They suggest that one of the most pervasive aspects of epilepsy is the characteristic sense of uncertainty. From the earliest stages of pre-diagnosis and throughout the illness, a sense of uncertainty is a defining quality of this and other chronic conditions. People with epilepsy, like those with cancer, Alzheimer's disease, or multiple sclerosis (just a few of the chronic conditions for which this analysis is relevant), at first wonder what is happening to their bodies or their minds. They wonder whether or not to take this or that small sign of change as a symptom of a disease. They wonder whether it is a symptom of a serious or a minor disease.

Once diagnosed, they wonder how severe their illness will become and whether they will live for long or only for a short time. They ask whether they will be seriously debilitated or only mildly affected. Relationships are altered. Others respond to this sense of uncertainty with their own confusion about the disorder and its likely course. Not only does relating to the self become tinged with ambiguity arising from dealing with change, but so does relating to others become unclear. The lack of easy, honest, open, and straightforward communication is frequently seen as one of the most painful aspects of the disorder. Cancer patients, for instance, have said that the difficulties of communication (born of uncertainty) are frequently even more painful than the disease or its treatment (Dunkel-Schetter and Wortman, 1982).

Chronic illnesses give rise to several sources of uncertainty. First of all, there is ontological uncertainty. Self-identity arises from interaction with the self (identity and body) and others. When a part of this interactive mix is altered, for instance, when the taken-for-granted health of the body is called into question, so is the self. And so people, when confronted with a disease, disability, or accident, ask questions such as: Why me? Why now? Why this disease? Who am I now that I am a cancer patient? Will I still be a husband/wife/lover? Will I be able to continue working? Will my friends continue to be friends?

Another source of uncertainty surrounding some chronic illnesses is the fact that they are poorly understood both by the medical profession and by the patient and loved ones. There is a general lack of knowledge about the probable prognosis of some chronic illnesses. Some chronic diseases receive a considerable amount of press; others receive very little. Some are well understood by the lay public; most are not. Some have been diagnosed for many, many years; others, such as Alzheimer's, are relatively recent diagnoses. For some, the norms of possible remissions, plateaus, and disease exacerbations have long been charted; the short history of others means there are few standards for what to expect.

Epilepsy is a disease with a long history. Like venereal disease and leprosy, and more recently cancer and now AIDS, epilepsy is a disorder that is believed to reflect not just the state of the physical body but also the moral character of the person. At times the person with epilepsy has been considered to be divine, at other times evil. At various times and places seizures have been understood as signs both of prophesy and of madness.

Chronic illness necessitates symptom control. This is particularly necessary when the symptoms can be highly disruptive, as in diabetes, which can exhibit diabetic reaction or coma, or in colitis, which may involve unexpected evacuation, or in epilepsy if there is a grand mal seizure. Symptom control can involve following the doctor's orders. It can also involve non-medical procedures such as biofeedback, hypnosis, diet change, meditation, exercise, relaxation, vitamin therapy, and others. Managing medical regimens does not necessarily mean following the medical rules. Instead, people often manage their medicines according to their own values, habits, activities, relationships, and side effects.

People with epilepsy control their drug use in such a way as to moderate the number and severity of seizures to a level with which they feel comfortable. Doctor's orders are only one of several sources of information upon which people with epilepsy choose to base their use of medication. Drug use patterns develop as an outcome of a complex of self-perceived considerations such as (1) the meaning of the seizure to the subjects, (2) the personal view of the effectiveness of the drug, (3) the personal estimation of the costs of side effects, (4) the desire to test whether the epilepsy is still present, (5) the wish to avoid having others recognize that they have epilepsy, and (6) the need to protect themselves from seizures in particular situations (Schneider and Conrad, 1983).

Relationships within the family and with loved ones are also profoundly affected as people work out how to live with a chronic condition. Interaction with employers and employees may be altered. Recreation patterns change. People may need to reorganize their time in order to manage the disease. Just as interpersonal adjustments are made, so are revisions of the concept of self frequently required as the person copes with illness over time. The work of Schneider and Conrad exemplifies the symbolic interactionist approach to understanding behaviour in health and illness.

Through the presentation of quotations from the subjects intermixed with sociological analysis, the authors have provided a work with relevance to others in similar health circumstances, to their families, to health-care workers who deal with people with such diagnoses, and to the academic sociological community. Research such as this often has an orientation that can be applied to patients and workers in the health-care field.

A similar methodology, this time explaining the world of the health-care provider, is afforded by Strauss (1987) in her study of the caregivers of patients with Alzheimer's disease. It is very difficult to provide a deep, qualitative analysis of the experience of the person who develops this disease. Alzheimer's disease is degenerative and affects the mind. At its onset the individual experiences minor symptoms that may be attributed (both by the person himself or herself and by others) to the natural course of aging, to an emotional upset, or to a physical illness. But as time goes on the person with Alzheimer's may become more and more forgetful, confused, easily angered, irritable, restless, and agitated. Judgement, speech, and concentration are affected. Eventually persons with this disease become totally unable to care for themselves (Health and Welfare Canada, 1984: 4).

As the disease progresses it becomes more and more difficult for the "patient" to provide a reflexive statement of his or her experience. A study based on the world views of Alzheimer's patients would thus be very difficult. On the other hand, talking to the person taking care of the patient with Alzheimer's is useful and important for two reasons. In the first place, these people are in the next best position to describe the life of the person with Alzheimer's because they are most closely involved on a daily basis. In the second place, the description of their own role is an under-researched area. As we have noted (see Graham, 1984), home health care has been largely ignored.

Alzheimer's may be one of the most difficult illnesses to deal with. Its course is uncertain and erratic. Caring for the patient can be emotionally and physically exhausting; at a certain point in the course of the disease, an adult must be cared for in much the same way as one would care for a helpless infant. The pre-diagnosis stage may be the most difficult time of all. At this stage the family members, as well as the future patient, often know that something is intermittently wrong and yet do not know what it might be. One woman explained her experience of the pre-diagnosis stage as follows: "Joe was coming home later from work a lot and I would ask him to be a little more considerate and call the next time and he would just yell back at me and we usually just ended up fighting" (Strauss, 1987: 13).

The early stages are characterized primarily by uncertainty. Sometimes behaviours are interpreted as those of normal aging; other times the same behaviours are thought to indicate a serious problem: "Mom thought dad was just getting miserable and stubborn just like other old people. So it was hard to convince her that dad had a problem and needed help" (ibid.: 17).

Some of the early symptoms of Alzheimer's are quite similar to some of the common stereotypes of old age. They include memory loss, crankiness, and confusion. The early pre-diagnosis stages are difficult. Patients struggle to manage their symptoms and modify and manage their personal habits. As the disease progresses and symptoms persist and increase, medical advice is sought by many.

Diagnosis, however, also involves a difficult and ambiguous process. There are no tests that conclusively prove the presence of Alzheimer's. As a result, many caregivers are left with a sense of uncertainty regarding the diagnosis. One of Strauss's respondents talks about this: "It was a negative diagnosis. . . . she had a thyroid treatment. But it didn't seem to help her mental capacities. Then he sent her to another doctor to see if he could find anything else that could be causing it – but he couldn't. Therefore, it must be Alzheimer's they said" (ibid.: 31). And another caregiver said, "He was never officially diagnosed. After six months of struggling along . . . I said Dr. Jones, is it hardening of the arteries or Alzheimer's? All he said was 'a little of both.' That's as far as any diagnosis went" (ibid.: 32).

Once the diagnosis is made, however tentatively, the family caregivers move into a new stage of adaptation. The confusion involved in dealing with an undiagnosed Alzheimer's patient gives way to the fear of inheriting or passing on the disease. One of Strauss's respondents put this fear as follows: "I've read a lot about Alzheimer's and I feared what lay ahead for him. But my greatest fear was that it was in the genes and it has the tendency to be inherited. But when Jake got it, I became concerned about the grandchildren and the children getting it" (ibid.: 35). For some, and in some ways, the diagnosis was finally a relief: "The diagnosis of Alzheimer's, well, it's like anything. You really don't want to accept it. But at least you do know and you're not hunting anymore. In some ways, it was a relief" (ibid.: 36).

Summary

Two examples of symbolic interaction theory have demonstrated its basic principles. These are that (1) symbolic interactionism is characterized by close attention to the meaningful interaction of social actors; (2) the understandings that the subjects of study have of their own situations become the object of investigation; (3) portrayal of the world views of the respondents in their own language is the desired outcome of such research.

Feminist Theory

Feminist theory and feminist methods have advanced rapidly in the past two decades. A number of journals are now dedicated to feminist studies in many fields of scholarship. The social sciences, in particular, have been challenged and critiqued as having been male-stream in subject matter, research strategies, and theoretical assumptions. A whole new field called Women's Studies has emerged and become institutionalized in universities to the extent of majors, Master's, and Ph.D. degrees (Reinharz, 1992; Richardson and Robinson, 1993; Smith, 1993; Stanley and Wise, 1993). Numerous books have been published on all aspects of social life from this perspective. The work of Canadians, such as Margrit Eichler and Dorothy Smith, has been fundamental to both methodological and theoretical shifts (see, for example, D. Smith, 1987). There is a way in which, following Betty Friedan's *The Feminine Mystique*, the women's health movement can be seen as a major impetus to feminist scholarship and policy. Early rallying of women in the late sixties and seventies for abortion reform, following the liberalization of sexual norms and mores that accompanied the wide prescription of the birth control pill, was a crucial spoke in the whole wheel of the second wave of the women's movement in the twentieth century. The women's health movement led to a radical critique of the patriarchal, allopathic medical care system and practice as exemplified by such monumental publications as *Our Bodies, Our Selves* (Boston Women's Health Collective, 1971).

A major theme in a feminist analysis of health has been the description and critique of the medicalization of women's lives. Much of this critique has focused on the dominance of the medical care system, medical practitioners, and the medical constructions of knowledge with regard to reproductive issues, including birth control and childbirth (Oakley, 1984), PMS (Pirie, 1988), and menopause (McCrea, 1983;

Kaufert and Gilbert, 1987; Walters, 1991). Others have explained women's (poor) health as a result of social structural inequities such as class (Doyal, 1979), labour force participation (Tierney *et al.*, 1990), or familial and domestic roles (Graham, 1984). Nevertheless, as Vivienne Walters (1992) argues, little research had been done on women's views of their own health problems. To address this lack Walters recently interviewed 356 women over twenty-one years old in a community-based survey about their views of their own main health problems and the health-related worries, experiences, and perceptions of women in Canada. The findings of this study stand as a challenge to the widespread understanding and bases of health policy. Whereas key informants for Health and Welfare Canada (now Health Canada) have indicated that women's chief health problems were related to their reproductive system, women themselves considered, when asked an unprompted question, their main problems to be stress (19.7 per cent), arthritis (14.9 per cent), overweight (9.6 per cent), back problems (9.0 per cent), migraines/chronic headaches (8.1 per cent), and blood pressure (8.1 per cent). Only 10.1 per cent of women said that they didn't have any health problems.

Stress, the most important health concern of women, was often associated with family and work responsibilities and worries about money and violence. These issues, while resulting in health concerns, are not best addressed by the medical care system. Nor are health-related research dollars directed toward understanding, minimizing, or eliminating these concerns.

This example of feminist health research demonstrates several of its basic principles. It begins with the experience of women and asks them to articulate their own views of the issues. It relies on a combination of qualitative and quantitative data collection methods. It challenges the definitions provided by the powerful (in this case, major state health policy and funding bodies).

Summary

Feminist theory and research can be distinguished by several fundamental principles. (1) By virtue of gender, men and women occupy different places in the social structure and live in distinct yet overlapping cultures. (2) Men tend to dominate in all institutions in society. They tend to have more power, more money, and more access to all types of resources. (3) Sociology, including the sociology of health, illness, and medicine, has historically reflected male dominance with respect to subject matter and styles of theorizing and research. (4) Feminist researchers theorize, problematize, describe, and explain the social world so that women and gender are always central. More recently, feminists have also realized their own myopia and have argued for the necessity of paying attention to class, race/ethnicity, ability/disability, and sexual orientation as fundamental social categories along with gender.

Sociology of Health in Canada

Much of the literature referred to in this book is based on Canadian populations; much is also based on populations of other countries, particularly the United Kingdom and the United States. A recent article in the founding edition of a new multidisciplinary journal concerned with health and medical sociology, *Health and Canadian Society*, reviewed Canadian literature in this field and drew some interesting conclusions. The authors (Coburn and Eakin, 1993) argue that this subdiscipline has, to a large extent, mirrored the overall disciplinary trends. On the one hand, the emphasis on applied work is parallel to the overall disciplinary focus, particularly in the United States. On the other hand, the growing emphasis on theoretical analysis and development tends to be a more important aspect of Canadian work. American work has tended to focus on socio-psychological variables such as stress, health locus of control, health behaviour models, and social support and has tended to use survey research methods. American medical sociology has also tended to have a more applied approach. Mechanic (1993) evaluated the field in the U.S. and drew the conclusion that there is a lack of structural and critical analysis. Canadian sociology, as Brym and Fox (1989) have argued, has moved from a focus on culture to a focus on power. Moreover, it has often taken a critical or political economy approach. These concerns with gender, class, race, and so on, while not yet theorized adequately, are increasingly well informed by feminist and other developments in sociological theory (Coburn and Eakin, 1993).

Summary

(1) Illness is experienced in a social context: we learn to think and talk about it with a vocabulary that others share and we learn what to do about it through interactions with family and friends, the formal medical system, and the media.

(2) The sociology of health and illness describes and explains the social causes and consequences of illness, disease, disability, and death. The sociology of medicine is the study of the institutionalized medical recognition of and response to illness.

(3) Sociologists use four main perspectives to study the social world: structural functional theory, conflict theory, symbolic interaction theory, and feminist theory. Each of these paradigms has different assumptions about the social world and therefore different ways of understanding it.

(4) Structural functional theory was first discussed by Emile Durkheim. Its goal is to understand the social causes of social facts; it does this by studying the causal relationships among institutions. The parts of society are inextricably bound together to form a harmonious system. Human beings are constrained by the external world and are therefore predictable and controllable through the power of social facts.

(5) The origin of the conflict perspective is attributed to Karl Marx. Conflict theorists study competing groups within societies through history. The basic competing forces are class and gender. Conflict theorists are committed to the description and documentation of injustice through the understanding of economic arrangements and their impact on other conditions of social life. In the conflict perspective of the sociology of health and illness, health and illness are related to the unequal social arrangements found in capitalist, patriarchal societies.

(6) Symbolic interaction theory is based on the definition of sociology given by Max Weber. Symbolic interactionists attempt to understand the subjective meanings and causes that social actors attribute to events. The meaning social actors give to their disease affects their self-concepts and their relationships with others.

(7) The feminist perspective provides a thoroughgoing critique of sociology to date to the extent that it has omitted and neglected women or presented women's lives in biased and narrow perspective.

CHAPTER TWO

Ways of Studying Health, Illness, and Medicine Sociologically

HOW DO WE COME to know what we know? What do we understand to be the most important sociological questions concerning health or medicine? In fact, how do we conceptualize health, illness, and medicine? What assumptions do we make about the nature and essential characteristics of proof in any sociological study? Is sociology a science? Is the scientific approach to knowledge always the best approach? Each of the four perspectives discussed in Chapter One encompasses a different picture of the relevant subject matter for the field, a different set of assumptions about the nature of proof, a different strategy for the analysis of the data.

The purposes of this chapter are to describe and critically analyse the methodologies used in the construction of knowledge in the four theoretical paradigms: structural functionalism, conflict theory, symbolic interactionism, and feminist theory. Just as the previous chapter described and illustrated each perspective, this chapter will discuss the ways of knowing and the methods of proof adopted by each perspective. It will clarify the limitations and strengths, the presuppositions, and the consequences of each perspective's methodological strategies.

Positivism

The methodology most often associated with structural functionalism is positivism. Positivism is distinguished by three fundamental presuppositions: (1) that sociology is a science that seeks to describe the social world in a series of universal causal laws; (2) that this science sees human behaviour as objectively measurable through such methods as survey research and experimental designs; and (3) that social facts are to be treated as things because they determine human social behaviour and attitudes through the norms that regulate human behaviour.

Much of the sociology of medicine falls within this paradigm, including studies of (1) who seeks medical services (from doctors, hospitals, etc.) and how frequently; (2) what role is played by such social factors as social norms for defining mental and physical illness; (3) how important social support from family and friends is in preventing, minimizing, or helping to adjust to illness; (4) what role the quality of life – work, recreation, community, physical activity, occupational conditions – plays in the health and well-being of people. Essentially, positivist studies analyse the relationships between social facts and various sorts of health-related variables.

The two fundamental models of analysis in this paradigm were described in Chapter One. In the first, the one we have called Model A, health-related variables are dependent. For example, there is a well-documented finding that women are more likely to be ill than men; gender is considered the independent variable and illness is considered the dependent variable – the one that is being explained. In the second model, Model B, the health-related variable is the independent variable and the study would be concerned with its impact on a social factor.

An example of Model B research might be studies that examine the impact of changes in health status, such as serious illness, on income level. Any of the following socio-psychological and demographic variables may be selected for study: age, sex, gender role, marital and family status, ethnicity, race, social class, occupation, education, social support, and beliefs about the disease and its curability. It must be emphasized that the preceding list is by no means inclusive.

In these models of research, a variable can be causal in one study and caused in another. However, before causal connections can be determined and verified, at least three conditions must be met. They are: (1) an association between the variables; (2) evidence that one variable precedes the other in a time sequence; (3) the elimination of other, potentially intervening variables. Furthermore, causal relationships may involve more than two variables at a time. Such complex modelling of the relationship between independent and dependent variables is called multi-variate analysis; when causal ordering is tested, it is called path modelling. Path modelling involves assumptions about the time sequence of a number of variables that either affect the dependent variable directly or affect other variables and through them the dependent variable. Figure 2.1 illustrates the theoretical logic of path modelling.

In the illustration each arrow is meant to indicate a direct causal connection (e.g., gender causes or has an effect on quality of housing). The model in Figure 2.1 suggests that illness can be conceptualized as the end result of a chain of variables that operate together. There are both direct relationships and indirect relationships. An example of a direct relationship is the one between gender and working conditions. For instance, it might be hypothesized that women are paid less than men for full-time work. An example of an indirect relationship is one between class and stress management: for example, it might be hypothesized that members of the working class tend to have longer working hours and therefore less time available to learn or use stress-management techniques.

FIGURE 2.1 Hypothetical Path Model

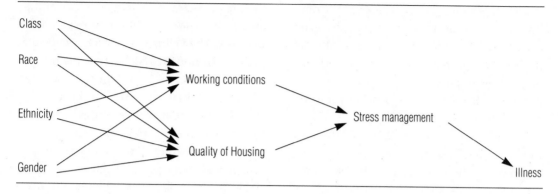

Epidemiology

Epidemiology, the study of the causes and distribution of disease, is one example of a positivist methodology. Some argue that epidemiology is not a type of sociology. To the extent that the purpose of epidemiology is to understand disease patterns in order to minimize or alleviate their effects, it is not sociology. However, to the extent that social labelling of disease, social relations, and social/structural positions of people are considered relevant subjects for the analysis of disease incidence (the number of new cases of a disease in a given period of time) and prevalence (the number of cases of disease in a population), then it can be considered to be an example of one use of sociology.

Epidemiological investigation began in the last century when the prevalent diseases seemed to be contagious and infectious. Patterns of human contact were obviously relevant in studying the incidence of these diseases. The discovery of bacteria and their connection with disease

raised hopes that clear and observable cause and effect relationships would be found in all diseases. There are several interesting, classic examples of early "epidemiological" work. Sir John Snow is said to have done the first epidemiological research when he observed and noted that a good proportion of the people who became ill with cholera had drunk water from the same water pump. Later, Sir Percival Pott discovered that scrotal cancer was prevalent among chimney sweeps. Further investigation led him to the conclusion that the extensive bodily exposure to soot, which was a necessary part of the work of the chimney sweep, was implicated in the development of cancer in the scrotum.

One of the most important contemporary examples of epidemiological research concerns AIDS – auto-immune deficiency syndrome. The following is a brief illustration of some of the methodologies involved in the epidemiological investigation of this disease. In this case, AIDS is the dependent variable. The causes of AIDS, the independent variables, are being sought. The first issue in the epidemiological analysis was the definition of the strange new disease syndrome that was later called AIDS. In all positivistic research, one of the first and one of the most important steps is the accurate, precise, valid, and reliable description of each of the relevant variables.

In the late spring of 1981, several doctors in Los Angeles noticed what seemed to be a surprising medical mystery. In the previous six months they had between them treated five young men with pneumodystis carinii pneumonia (PCP). This was a very rare infection. Even more unusual was the finding that all these men had other "opportunistic infections," which had normally only been seen in organ-transplant patients whose immune systems had been intentionally depressed to aid the body's adoption of the new organ. All five men were homosexual. The doctors reported these unusual circumstances in the June 5, 1981, issue of the *Morbidity and Mortality Weekly Report* (*MMWR*) published by the Centers for Disease Control (CDC) in Atlanta, Georgia. At about the same time, a New York doctor called the CDC about an unusual number of cases of Karposi's sarcoma (KS). The July 3, 1981, issue of *MMWR* reported that 26 cases of Karposi's sarcoma had been diagnosed in New York and California in the previous 30 months. All 26 men were homosexual, their ages ranging from 26 to 51. Four also had PCP. Eight died within two years of the diagnosis of KS.

In reaction to these remarkable findings, the CDC established an epidemiological research program designed to investigate the prevalence and incidence of PCP and KS and to look for any possible explanatory patterns. They chose to investigate hospital records, tumour registries, and the records of medical doctors in certain cities selected because of their differing proportions of homosexuals. Next they interviewed as many patients as possible, using fairly wide-ranging interview schedules, during which they gathered information about the patients' lifestyles, sexual behaviour, drug use, and medical history. It was quickly discovered that most of the patients used "poppers," such as amyl and butyl nitrates, as sexual stimulants.

The researchers designed a case-control study to compare diseased and healthy homosexuals in order to identify risk factors. With 20-page questionnaires as guides, the investigators studied 50 homosexual men with the illness(es) and 120 healthy homosexual men. The 60-90 minute interviews covered medical histories, drug use, sexual behaviour, occupation, travel, family history, and other lifestyle issues. In addition, the researchers collected various types of biological specimens for comparison.

After months of detailed analysis, these findings were noted: the diseased men (1) had more sexual partners; (2) were more likely to use bath houses; (3) had histories of syphilis and hepatitis; and (4) were more frequent users of marijuana and cocaine. "Poppers" seemed equally popular in both groups. The major conclusions of the study were that (1) the disease could be transmitted through the blood or semen; (2) patients tended to engage in frequent, anonymous sex; and (3) these sex practices often produced abrasions that exposed them to small amounts of blood, semen, and feces.

A year after the first questions about this new disease were raised, 216 cases had been reported to the CDC: 84 per cent were male homosexuals, 9 per cent were intravenous drug users, 2 per cent were Haitians, and 5 per cent were women. Eighty-eight of the patients had died. The common cause had yet to be discovered. Some thought that it was a new virus, others believed illicit drug use to be the cause, and yet others favoured a theory of overload to the immune system. The Haitian background of some of the diseased did not fit into any theory. Later the syndrome was found in hemophiliacs and in a 20-month-old boy who had received a blood transfusion from an AIDS patient. Still later, female sexual partners of AIDS patients and children of at-risk group members were found with the disease. As knowledge of the types of people found to suffer from AIDS grew, so, too, did the incidence grow – astonishingly quickly.

Looking for the causes of AIDS also involved the search for an accurate, precise, working definition of the disease. The operating definition that was developed included three criteria: (1) the patient must be under 60 years old, (2) have a specific diagnosed disease, such as PCP or KS, suggesting an underlying cellular immune deficiency, and (3) the disease had to occur without the presence of an immune deficiency that could be ascribed to another factor. Epidemiological research verified that there was no incidence of a disease

fitting these characteristics until 1978 (DeVita *et al.*, 1985).

The process of tracing the development of AIDS and isolating causal factors is still continuing. Researchers today are concerned not only with documenting the spread of the disease but also with predicting it. Increasingly, complicated statistical models are being developed to help predict the future of AIDS. However, a model containing all the variables now known to be relevant would overtax even the most advanced computers.

In addition, numerous methodological problems limit the potential predictive power of any model, no matter how complex. First, there seem to be a number of somewhat different AIDS epidemics happening at the same time, in different places, with somewhat different spread patterns, with different specific associated diagnoses, and so on. Second, there appear to be two periods when an individual is highly infectious; one shortly after the disease is contracted, and another when he or she begins to show symptoms. Third, the length of time a person can have the disease before symptoms are observed is unknown. Fourth, reporting rates vary tremendously among different cultural groups, medical systems, public health regions, and so on. Fifth, changes in the spread of the disease, and thus the risk of getting it, are dependent on choices, most especially about sexual behaviour, that individuals make today, tomorrow, and thereafter. All these are influenced by a myriad of social structures such as: economic, political, family practices, and gender. All may change in response to changes in any of these social structures. Epidemiological research into AIDS, therefore, has been and continues to be a very complicated and multifaceted procedure.

The above case study provides an illustration of epidemiology, one type of positivistic research. Recent research on the spread of AIDS demonstrates a more theoretically based "historical materialist epidemiology." The focus of this research is compatible with conflict theory. This approach assumes the primacy of economic conditions and practices in the explanation of health outcomes. In an attempt to understand the spread of AIDS in Africa, Hunt (1989) related its incidence to the ways in which people in eastern, southern, and central Africa make a living. In particular, he investigated the impact of the migratory labour system, which is accompanied by long absences of workers from their families, increased family breakdown, and an increased number of sexual partners. These patterns have historically resulted in recurrent epidemics of heterosexually transmitted diseases. Now these labour force demands are associated with a higher incidence and prevalence of HIV and AIDS. Here AIDS is primarily a heterosexually based disease and has been called Type II AIDS. Type II AIDS, characterized by a 1:1 ratio (male to female), is the predominant "type" in Third World countries. By contrast, in the developed world Type I AIDS, with a 16:1 sex ratio (male to female), predominates. Hunt (1989) argues that policies that maintain the economic dependence of the Third World on the First World are fundamental to the epidemic spread of Type II AIDS in the Third World.

Table 2.1 summarizes the assumptions of positivistic science.

Conflict Theory

Have you ever been ill because you lacked nutritious food, a warm place to live, or clean, comfortable, and adequate clothing? Have you ever developed a cold because your family couldn't afford boots to keep your feet warm and dry in the snow and cold of the winter? Have you ever developed emphysema because you worked with asbestos in a factory producing fire-retardant fabrics? Have you ever had a lung disease because you smoked? Have you ever known someone who was killed in a car accident because of excessive alcohol consumption? Have you ever had an allergic reaction to a prescribed drug such as penicillin? Do you know anyone who has suffered from the effects of thalidomide, DES (the drug diethylstilbestrol), or Valium? Have you ever had unnecessary surgery? Have you ever eaten contaminated food and suffered food poisoning? Have you ever suffered ill health because you lacked clean drinking water?

All these health problems are related to unequal social arrangements in North America and in other parts of the world. Social conflict theory predicates behavioural patterns on power differences in society resulting from class and gender: the poor, women, Native people, etc. have unequal access to rewards and to health-promoting

TABLE 2.1 Methodological Assumptions of Positivist Social Science

Objectivity	Social science can be as objective as physical science and should be modelled on the physical sciences.
Generalizability	One of the most important goals of social science is to generalize and thereby to describe the world in a series of "if x then y" causal laws.
Validity: Construct	It is possible to design measures that accurately and briefly describe sociological concepts.
Validity: Internal	It is possible in any social scientific research design to say with a degree of surety that "x" is the probable cause of "y."
Validity: External	It is possible to select a sample so that generalization from the sample to the total population is accurate with a known but limited amount of error.
Reliability	It is possible for the same research to be completed in different settings and by different researchers and with essentially the same findings.
Adequacy	The data collected adequately describe and explain the phenomenon under investigation.
Causality	It is possible to demonstrate probable causal relationships between social science variables.
Data Collection Strategies	The usual data collection strategy involves survey research with either a questionnaire or an interview, either administered in person or over the telephone. Experimental laboratory research is sometimes done.
Quantification	The incidence of sociological phenomena can be quantified in statistical data.
Probability	Analysis is based on assumptions of probability, not determinism, i.e., hypotheses are put forward as probabilities.

resources. The poor are more likely to be ill than the rich. They have a shorter life expectancy than those with more resources. Third World peoples are more likely to suffer the effects of malnutrition, of impure drinking water, and of fatal infectious diseases than people in developed countries. Women are more often prescribed addictive drugs such as Valium and other mood-altering pharmaceuticals for their problems than are men. Men are still more likely to smoke cigarettes and to suffer such repercussions as a higher mortality rate from lung cancer than are women, though the incidence of women who smoke has been increasing. Women are more likely than men to fall ill from any number of causes throughout their lives. Men are more likely than women to die from homicide, suicide, and motor vehicle accidents. People who stand in different positions in the social structure have correspondingly different levels of health and different rates of health-care use.

What methods are used to demonstrate and explain findings such as those above? What responsibility does the researcher have to attempt to change the sorts of injustice he/she is committed to discovering? The methods used by conflict theorists can be distinguished in the following ways. Conflict theorists tend to believe that: (1) sociological research, indeed, all knowledge, is limited by the perspective derived from the place in the social system of those who develop knowledge and those who receive and interpret it; (2) research should be comparative and historical in scope; (3) it is impossible for research to be objective; and (4) understanding inequalities arising from class, gender, and other differences is the foremost purpose of such research.

Several examples of research within this perspective will be presented in illustration. One is the general analysis of medicalization: the tendency for more and more areas of people's behaviour to be subject to medical intervention as an outgrowth of industrialization. This analysis also involves a critique of the growing power of the medical care system and its institutions to affect the lives of individuals, groups, and societies.

Ivan Illich has offered an influential critique of medicalization in *Medical Nemesis: The Expropriation of Health* (1976). His argument is that contemporary medical practice is iatrogenic, that is, it creates disease and illness even as it provides medical assistance. Three sorts of iatrogenesis are isolated and explained.

[It is] clinical, when pain, sickness, and death result from the provision of medical care; it is social, when health policies reinforce an industrial organization which generates dependency and ill health; and it is structural, when medically sponsored behaviour and delusions restrict the vital autonomy of people by undermining their competence in growing up, caring for each other and aging. (Illich, 1976: 165)

Clinical iatrogenesis, that is, injury and/or disability that result directly from the work of the doctor in the hospital or in the clinic, is the first problem addressed by Illich. Addictions to prescribed drugs, the side effects of prescribed drugs, harmful drug interactions, and suicide resulting from prescribed medication are specific examples of clinical iatrogenesis. Thalidomide, prescribed in the 1940s in West Germany, Canada, and elsewhere to women with a history of miscarriage, resulted in untold tragedy when numerous children were born without limbs. DES (diethylstilbestrol), again prescribed (in the 1950s) to women with obstetric problems, has been found to result in thousands of cases of ovarian and cervical cancer in the daughters of those who used DES, and in other cancers in their sons. Unnecessary hysterectomies have caused extensive emotional, marital, and other social problems for the women who have undergone this procedure, as well as for their partners and families. Breast silicone implants have been found to be associated with numerous and various deleterious health outcomes (Rachlis and Kushner, 1993). Today, millions receive chemotherapy, radiation, and/or surgery as treatment for a variety of cancers. However, such treatments are often felt by the patients to be more painful than the disease itself. Those for whom a cure is effective may feel that the cost is worth the pain. On the other hand, those who are not cured, and whose lives can only be extended for a limited period, may regret having submitted to such treatments as chemotherapy. These are just a few of the troubles that occur in an over-medicalized clinical practice.

In the Third World, sanitation, unsafe drinking water, malnutrition, and insufficient and/or dangerous birth-control measures are the major health problems. However, large parts of the health budgets of poor nations are spent on drugs that do little to alleviate any of these problems. Substantial expenditures for pharmaceuticals by Third World nations minimize and prevent expenditures for clean water, the development of a good agricultural base, and the promotion of safe and inexpensive birth-control devices and practices.

Social iatrogenesis is evident in the impact of medicine on life spans. Medical and technological intervention begins at birth and ends with the care of the aging and dying. There are medical specialties to deal with pregnancy and childbirth (obstetrics), childhood (pediatrics), women (gynecology), the elderly (geriatrics), and the dying (palliative care). The presence of medical specialists to deal with various normal stages of the human life span is symbolic of the trend toward the medicalization of life and the increasing addiction of modern people to medical institutions.

The most onerous example of medicalization is the growing dependence on care in Western industrialized societies. The huge growth in spending on medical treatment, on hospitalization, and on pharmaceuticals is just one example of this. The fact is that many physicians say that most of their patients come to them with self-limiting (minor) illness or with social problems that would be better handled elsewhere. In repeated studies (Cartwright, 1967; Cartwright and Anderson, 1981), general practitioners have claimed that approximately one-quarter of their consultations were "trivial, unnecessary or inappropriate."

Considerable evidence suggests that pharmaceuticals are often prescribed to people for social and psychological problems. Women consistently receive more prescriptions for tranquilizers than men do. Moreover, women in the middle and older age groups are at highest risk. Yet research has found that the majority of those subjects who used tranquilizers explained that they needed them because of a variety of societal, familial, and occupational demands and expectations rather than physical need (Cooperstock and Lennard, 1979). Even women's body shape and appearance have become medically diagnosable by plastic surgeons who recently instituted a new "disease" category – small breast syndrome. Naomi Wolfe in *The Beauty Myth* (1990) says that medical discourse tells women that beauty is, in fact, the equivalent to good health.

By structural iatrogenesis Illich means the loss of individual autonomy and the creation of dependency. The responsibility for good health has been wrested from the individual as a result of the imposition of the medical model – the prevalence of medical institutions and medical practitioners. Pain, suffering, disease, and death are important experiences for all human beings. They can encourage the development of service to others, compassion, and connectedness with others. But medical bureaucracy and technology minimize the possibilities for the fertile development of family and community-based models of care. The medical model and its institutions usurp individual initiative and responsibility and thus attenuate humane responses and spiritual development. In sum, in Illich's view, we rely excessively on medical care and this overdependence has many destructive consequences for people and their communities. To correct this problem he advocates the deprofessionalization and debureaucratization of medical practice, and the maximization of individual responsibility. Self-care, autonomy, and self-development should be the guiding principles.

Navarro, a leading Marxist critic of medicalization, takes issue with Illich's explanation. Whereas Illich's main foe appears to be the bureaucratic organization of medical practitioners and medicine-related industry, Navarro (1976) argues that medicine is a mere pawn in the hands of a much greater power – the power of the state directed through the dominant class. The health industry in the United States, he claims, is administered by, but not controlled by, medical professionals. Members of the corporate class (the owners and managers of financial capital) dominate in health and other important spheres of the economy. The upper middle class (executive and corporate representatives of middle-sized enterprises and professionals, primarily corporate lawyers and financiers) have major influence in the health delivery sector of the economy. Together these groups comprise a minority of the health-care system, yet they control most of the health institutions. The majority of the population has little or no control over either the production or consumption of health services.

The writings of both Illich and Navarro illustrate conflict methodology. A position is taken with regard to injustice, and then documentation of the injustice is found in both historical and other available evidence. Another type of conflict research uses a more positivistic methodology, including questionnaires and interviews as well as available statistics, to document injustice. The explanations given in studies of class and illness described in Chapter Four illustrate this aspect of conflict methodology.

A persistent theme in the criticism of health and medical care is that sickness has societal origins. Some social arrangements generate the potential for disease and death. Certain social structural positions determine the likelihood of the type of illness, disease, and the incidence of death. Social class, age, gender, race, ethnicity, and region are all correlated with particular health and illness profiles. People who differ in social background have differing rates of various types of illness, of illness in general, and of mortality. When the explanation for this inequity focuses on capitalism, it is usually Marxist; when it focuses on gender relations, the explanation is usually feminist. The focus on inequity and the type of explanation offered make this research an example of conflict theory rather than of structural functionalism.

TABLE 2.2 Methodological Assumptions of Conflict Theory

Value Commitment	Rather than seeking objectivity, the conflict theorist believes that sociologists must discover, document, and record recurrent patterns and dynamics of power/class/gender relations both because they have no choice (members of a society are committed to the ongoing action of the society) and because they believe that this is the morally correct position.
Historical Specificity	Rather than looking for generalizations, conflict theorists assert the necessity of understanding the unique features of the particular situation in its socio-historical context as an example of these recurrent patterns of power/class/gender relations.
Validity: Construct	Formal tests of validity are considered irrelevant. Researchers, it is assumed, of necessity study what they claim they are studying.
Validity: Internal	Formal statistical tests of causal relationships are not always necessary. Rather, logical meaningfulness may be the relevant criterion of causality.
Validity: External	The conflict theorist assumes that inequities based on power/class/gender relations are ubiquitous, yet analyses the components separately in each historical situation.
Ethical Concerns	The primary importance in the research of the conflict theorist is the commitment to such ethical and humanitarian principles as justice and equality.
Data Collection	The usual sources of data are historical documents. As well, the conflict theorist may use other data collection methods, including surveys, statistical data, and methods such as unstructured interviews and participant observation that provide subjective and descriptive data.
Objectivity	Objectivity is not possible. Knowledge cannot be separated from the power/class/gender relations of the researcher and the subjects.
Quantification	While numerical data may be used to document an argument, they are not always considered necessary.

The common theme in conflict research, as outlined in Table 2.2, is that it is impossible to do objective research, and that historical analysis of power/class/gender relations provides an appropriate scheme for understanding the true dynamics of social relations.

Symbolic Interactionism

Symbolic interactionist sociology can be thought of as sociology from the inside. Its focus is on the world views and the meanings given to reality by the subjects of study. Definitions and understandings of social circumstances made by social actors are the topic of analysis. Several characteristics distinguish the methods of symbolic interactionist sociology. All are based on the assumption that a science of human subjects must take into account the particular character of unique social actors embedded in discrete situations. Therefore symbolic interactionism considers that: (1) sociology is a science whose purpose is to understand the social meanings of human social action and interaction; (2) reflexivity or intersubjectivity, rather than objectivity, characterizes the relationship between the subject and the researcher because human subjects construct meanings out of social contexts or, to some extent, create each situation anew; (3) rich, carefully rendered, intimate detail, descriptive and analytic of unique situations, is the ultimate goal of study within this paradigm.

Research within this perspective is often based on intensive participant observation of a medical setting (for classic empirical examples, see Scully, 1980; Sudnow, 1967; Fisher, 1986; Goffman, 1963), long, unstructured interviews (see, for instance, Schneider and Conrad, 1983), or close analysis of language in context (Raffel, 1979) or of documents. What is common to these methods is the focus on the frame of reference, the world views of the subjects of study. When the method is participant observation, a number of possible roles can be taken by the researcher, depending on whether she/he gives precedence to the observer role or the participation itself. These are called: (1) complete observer, (2) complete participant, (3) observer as participant, and (4) participant as observer. When a hypothesis is to be tested by participant observation (or other inductive research strategies), the conditions of proof are very exacting. Called negative case analysis, this mode of proof requires that every single piece of evidence must support the hypothesis (Kidder, 1986). As the data are collected the researcher must make an effort to find a negative example, and if one is found the hypothesis must be revised accordingly. This form of proof to confirm a hypothesis uses deterministic logic based on the assumption that a variable must always cause or be associated with another variable. Deterministic logic contrasts with the probabilistic logic of positivism, which merely asserts the probability that the two variables are associated.

The data collected are usually qualitative rather than quantitative. Generally, positivism is very exacting in its use of numerical data because these are used in sophisticated statistical analyses. However, symbolic interactionism tends to use numerical data for descriptive rather than analytic purposes. But data more usually consist of exact descriptions of events and lengthy quotations, frequently from the subject of study. The following is one example of participant observation that involved long, unstructured interviews and qualitative data analysis. *The Unkindest Cut* (Millman, 1977) is a field study based on several years of participant observation in three different hospitals in the United States. Millman's method of research involved attaching herself to a particular physician or resident for a period of time and following him or her through daily rounds. As she says,

> For example, on the days I followed a surgical resident or a team of residents, I would arrive at the hospital at six-thirty in the morning, change into surgical dress and laboratory coat, and accompany my residents on the morning rounds, then to breakfast, and afterward into the operating room where they would spend a large part of the day. During surgery, I usually stationed myself next to the anesthesiologist, by

the patient's shoulders, so that I could watch the operation and listen to the conversations of the surgeons and staff as they worked. Between operations I joined the residents and nurses in the operating room [or] staff lounge, so that I could learn what was happening in other operating suites. Since I was free to wander, I could then circulate to whatever operating rooms were the scenes of interesting incidents on any particular day. (Millman, 1977: 14)

As she indicated, Millman was given relative freedom to move throughout the hospital, both its public and private areas, into operating rooms, examining rooms, waiting rooms, emergency rooms, and into the rooms of individual patients, generally in the company of a physician or other hospital staff. In addition, she attended all the medical mortality review conferences held during the period of her research. She was introduced as Dr. Millman and came, simply, to be regarded as one of the team. Millman also conducted more formal interviews with various hospital administrators, individual department chiefs, attending physicians, and other members of staff. To all of them she disclosed the purposes of her research: the conflicts among various hospital groupings and the management of medical mistakes.

As often happens in such field research, Millman became especially close to some of the staff and was somewhat alienated from some others. The research milieu was extended when Millman became a team member and was asked to take part in house parties, dinners, and casual conversations over coffee. Yet there was a way in which Millman maintained a distance from the ongoing interaction. The research demanded that she both empathize with the subjects of study (mainly the doctors) and maintain the identity of the objective outside observer, the researcher.

Such marginality – the experience of belonging entirely in the shoes of neither the researcher nor the subjects – is an inevitable dilemma for the field researcher (Shaffir et al., 1980). "Going native," or becoming a member of the social group that one is studying, is always a potential threat in fieldwork in which the researcher becomes deeply involved. On the other hand, a certain degree of closeness is necessary for the researcher to be able to describe the world as it looks to the subjects of research and as it is acted upon by those subjects. Gaining access or getting close (Shaffir et al., 1980) to subjects and to their social world is also a significant challenge in fieldwork. Whether to introduce oneself as a social researcher, an historian, a social worker, or in some other role is a question that must be addressed and answered before the research begins. Learning how to spend time with the subjects; determining the desired depth or intimacy of the involvement; becoming aware of the language, habits, and locations of the subjects; and deciding which section of the community or institution to contact first: these are among the issues faced by field researchers. When "leaving the scene," participant-observers must consider ethical questions such as: What are the responsibilities of

TABLE 2.3 Methodological Assumptions of Symbolic Interactionist Social Science

Reflexivity	Social researchers interpret the sayings and behaviour of their subjects from the subject's own perspectives and within the context of the researcher's own perspective. The researchers are affected by the needs and expectations of their subjects and the subjects' knowledge of the data collection process, and at the same time, they change the subjects' understanding. Thus it is impossible to measure human social behaviour objectively.
Ethical Concerns	Just as it is impossible to study human social action as if it were the action of so many atoms, molecules, neutrons, and protons, so it is impossible not to change the social situation that is the subject of the analysis.
Generalizability	While the method of analytic induction, one of the operating logics of this perspective, claims universality, most research in this paradigm is based on the specificity of human social action.
Causality	Causality is recognized in this perspective as a subject of study, e.g., "I believe I have cancer because I sinned against God," rather than the "if x . . . then y" causality of positivism.
Proof	The most stringent criterion of proof is sometimes required within this perspective – negative case analysis.
Validity	Validity is always hampered by intersubjectivity, but the depth and detail of the description of the data and their "meanings" are considered important criteria.
Reliability	This is considered less important than validity because it is assumed that different researchers would be researching different situations and would therefore have (at least somewhat) different findings.
Scope of Analysis	The symbolic interactionist sociologist is generally content to describe the social world of a small population of people in rich complexity and detail.
Advocacy	Some symbolic interactionists view their work as advocacy (e.g., Millman, 1977; Scully, 1977). Others fervently argue for the value of knowledge for its own sake.

the researcher to the community and the people in it? How can the anonymity of the subjects be guaranteed when the results of the study are reported? How can the sociologist be assured that the research will not cause any harm to any of the participants? Does the researcher have a responsibility to try to improve the circumstances of the subjects? (Shaffir *et al.*, 1980).

Symbolic interactionism leads to qualitative methods based on the assumption that the meanings people attach to social actions must be the subject matter of the discipline. Table 2.3 highlights the basic features of this methodology.

Feminist Theory/Methodology

Are the worlds of men and women different? In what ways? Do men and women relate to their bodies differently? Do their doctors? Are women more likely to be involved in the domestic sphere? Do they spend more hours in household-based work, domestic and family caretaking? And are such differences complicated by race, ethnicity, class, and power? Do black women, while having a lower morbidity rate from breast cancer, have a higher mortality rate from breast cancer? Do poorer people live shorter lives? Do those with less power suffer more illness? These are the types of questions that are addressed within this perspective. Feminist methodologies and their logical corollaries in relation to race, ethnicity, class, and power are built on the observation of two features of gender (race/class/power) organization – differentiation and inequity. Feminists argue that the institution of medicine operates to maintain the subordinate position of women in a patriarchal society. Indeed, medicine's dominance, in this view, maintains and constrains women's experience by providing a male medical conceptualization of and vocabulary for women's bodies and their functioning.

Feminist methods transform positivist methods by critiquing such issues as objectivity, the purposes for research, and the like. They are based on a number of challenges to positivism (Clarke, 1992). The first is from the work of sociologists in the interactionist and ethnomethodological traditions. This work has emphasized the ways in which individuals as social actors continually create social reality as they interact with other social actors. From this perspective it became clear that the "meaning" of an event varied depending on the viewpoint (structural and cultural positions) of the interactant. The second challenge to positivism was from the work of Thomas Kuhn (1962), who demonstrated that science is not best represented as the continual accumulation of truths in the search for the ultimate causal explanations of the social and physical world. Rather, Kuhn represents the history of science as a history of growth and development in one hegemonic tradition followed by a challenge to

this tradition. Revolts against findings, methods, and other aspects of scientific convention rather than the accumulation of truths is thus a more perceptive description of the development of science. Third, the findings of quantum physics in the twentieth century emphasized the impossibility of an objective outside observer because of the profound universal interconnection of the subject and the object. Fourth is the challenge of the second wave of the women's movement in the second half of the twentieth century. The medical profession was among the first of the institutions challenged by the new feminism. Childbirth reforms, including more "natural" childbirth, home birth, midwifery, and abortion availability, are among the issues fought for by the new feminist movement as women seek to control their own bodies. Current legislation and public consciousness about (and in opposition to) violence against women, sexual assault, and harassment are some of the more current struggles in the "fight" of women to reclaim their bodies. Vivienne Walters has undertaken and published research on women's views of their main health problems. As discussed in the previous chapter, the problems mentioned by women are not those that are successfully treated or diagnosed by the medical profession. Nor are they "diseases" that have received high priority for research funding (Walters, 1992: 372).

When asked whether they worried about or experienced a list of sixty-seven health and social problems, Walters's respondents said that they worried most about car accidents, breast cancer, being overweight, stress, and arthritis. The pattern was different with regard to the health problems they had experienced. Here they mentioned stress, being overweight, tiredness, disturbed sleep, difficulty finding time for themselves, and anxiety. Over 20 per cent indicated that they were dissatisfied with the quality of medical care, often mentioning doctors' reluctance to acknowledge women's problems. Large numbers of women also mentioned social problems, including money problems, various forms of violence, and worries about caring for an elderly or sick relative.

This research adopts some of the tenets of feminist methodology but not others. Most importantly, in keeping with the fundamental principles elucidated above, the work focuses on gender as a significant social category worthy of extensive analysis. Equally important, the research asks women to describe their own experiences and their own viewpoints regarding health issues. Currie (1988) takes issue with an exclusive focus on women's personal experience at the expense of social structural explanations of behaviour that has characterized some versions of feminist research. Instead of promoting a solitary focus on women's experience, she argues that feminist research should follow the course of inductive theorizing and theory-testing based on women's views of their own experiences. In her work on reproductive decision-making, Currie noted that when women discussed their decision-making regarding whether they would have children or not, their explanations could at first be considered to be private troubles or personal issues. For instance, when a woman says that she doesn't want children because they're too expensive she is really making a "private" statement. However, these "private" troubles can easily be understood as the result of social structural limitations due to, for example, workplace and family organization, support structures, and pay scales. Bridging the gap between structural and personal issues allows social scientists to see how solutions to opportunities and constraints are individually negotiated.

Pirie critiques current theory in medical sociology that emphasizes the patriarchal nature of the medical system and its dominance of women's experience of their bodies. She asks how it is that some women more than others, and women at some times but not others, may adopt or reject a medical definition of reality. Noting illnesses about which there is at least some dispute, such as premenstrual syndrome, she questions the processes whereby some diseases at some times acquire medical legitimacy, while others may not. To compensate for the over-emphasis on structural influences on women's medicalization, Pirie argues for further analysis of "(1) the productional activities of dominant groups with commercial and/or political self-interests in medical labelling; (2) the productional activities of those adopting the label; and (3) the cultural pathways or determinants which predispose the collective adoption of some illness categories, and not others" (Pirie, 1988: 629). In other words, first Pirie advocates sociological research on strategies used by doctors and medical device and pharmaceutical industries to

TABLE 2.4 Methodological Assumptions of Feminist Social Science

Objectivity	It is impossible to be objective in social research. Therefore it is important to be as clear as possible about the biases that are brought to any research study. The continual necessity of reflexivity in research is acknowledged.
Generalizability	Class, gender, race, and power differences limit generalizability.
Subjectivity	Often focuses on women's experiences and/or their own viewpoints.
Subject Matter	Gender is always an important component of the investigations.
Language	Uses gender neutral language where appropriate and specifies actual gender when relevant.
Data Collection Methods	All methods are used but a collaborative approach between researcher and subjects of research is advocated. Triangulation is suggested.
Purpose	Feminist research has change as one of its goals. It emphasizes the empowerment of women along with the transformation of patriarchal social structure.

persuade the laity of the existence of certain diseases and of appropriate treatments for these diseases. Second, she suggests a new line of sociological research on how women come to adopt, with varying degrees of scepticism and acceptance, medical definitions of reality. Third, she argues for studies of cultural forces, such as the mass media, and how they work to promote the legitimization of some illnesses but not others.

Research done by feminists shares aspects of all of the previous research traditions. Those who have added to our understanding of women's health issues have used research questions to guide the research in all traditions described. Those whose method is explicitly feminist often prefer a combination of conflict and symbolic interaction types of methods. Following Dorothy Smith (1987), they try to understand social structure and culture from the standpoint of women. More recently, feminists have added race/class/sexual orientation/ability/disability as problematics to be raised in all feminist research.

Table 2.4 lists the assumptions of Feminist Social Science.

Theoretical/Methodological Critiques

There are several important critiques of the way the sociology of medicine and sociology of health and illness have been practised. Many of these are relevant to greater or lesser degrees to the three paradigms discussed above. The first critique is that the discipline has been medico-centric or what Strauss (1957) called "sociology in medicine" rather than "sociology of medicine," and what Roth (1962) called "management bias."

This critique suggests that sociological research has, to an extent, been dwarfed by the great power and wealth of medical institutions and practitioners. Sociological studies have been considered fundable and legitimate to the extent that they have asked and answered questions of direct relevance to medical practice. Studies of compliance with doctors' orders are one example. As a result of focusing on questions of interest to medical practice, studies have adopted the point of view of medical practice, which tends to be individualizing and curative in its approach to disease and treatment (Davis, 1979). This has served to reinforce the power of the medical profession and to enhance the trend toward medicalization (Phlanz, 1975a). Indeed, there are those who have argued that medical sociology's primary purpose is to serve medicine; they consider its contribution to medicine to be as important as such fields as physiology, endocrinology, and biology (McKinlay, 1971).

Johnson (1975) and Frankenberg (1974) have both noted that medical sociology has emphasized conservative theoretical models such as structural functionalism. Such models have reinforced the power of allopathic medicine. They have described illness as a deviance in need of the powers of social control of the established medical practice. As Carpenter (1980: 104) says, "the obsessive focus on institutionalized medicine has meant a failure, for the most part, to transcend the medical concern with individual pathology which, arguably, is one of the most fundamental ways in which medicine's activities do not profoundly disturb established interests in industry and society."

In contrast to medico-centrism, some argue that sociology should stand outside the medical concerns and focus on health in a much broader context. Davis (1979) noted that medical sociology has neglected areas of study such as prevention, illness resulting from differences in social-structural (class/gender) access to resources, occupational health and safety, and environmental regulation. Twaddle (1982) suggests that the discipline is in the process of becoming less medico-centric. Twaddle's argument is that medical sociologists have already corrected whatever medico-centrism they might be seen to have had. His claim is that what had been called medical sociology ought now be called the sociology of health (see Table 2.5). As he explains it, the shift is a change in focus from medicine as the key health-related institution to health in a more general sense, including characteristics of social, psychological, and biological well-being, with or without a concern with medicine *per se*. He suggests that the following trends are occurring:

(1) from an individual or group level of analysis to a social, structural, or societal level;

(2) from the physicians and other medical professionals as the key healers to politicians, nutritionists, educators, lay healers, health promoters, practitioners, and physician substitutes;

(3) from the main means of healing – pharmaceuticals and surgery – to an emphasis on healthy lifestyles and environmental control;

(4) from goals of the care and cure of the individual to health, well-being, and reduction in mortality and morbidity in the population.

TABLE 2.5 Key Characteristics of the Shift from Medical Model of Sociology of Medicine to a Sociology of Health and Illness

	SOCIOLOGY OF MEDICINE	SOCIOLOGY OF HEALTH AND ILLNESS
Purpose of Sociology	To serve medicine and through this to serve the health- and illness-related needs of the population	To pursue and accumulate sociological knowledge for its own sake
Level of Analysis	Aggregated individual data	Social system, social group, social class, gender, group, and individual
Metaphor	Illness as breakdown of machine	Illness as socially defined, caused, and experienced
Methods	Positivist methods; survey, experimental design	Positivist, critical, ethnomethodological, symbolic interactionist
Powerholders	Medical professionals, pharmaceutical supply companies, hospital and clinic systems, and governmental bodies	Funding sources for sociology of medicine, universities, colleges, and, increasingly, governmental bodies
Illustration	Intervention research designed to increase patient compliance	Analysis of the history and role of pharmaceutical industry in the Canadian economy

Turning the social scientific critique of encroaching medicalization on its ear is a paper by P.M. Strong, "Sociological Imperialism and the Profession of Medicine: A Critical Examination of the Thesis of Medical Imperialism" (1979). His argument is that just as the medical profession can be seen to have been pursuing dominance in the medical division of labour, so have sociologists pursued dominance and legitimacy in academia. Strong is not alone in his critique. Alvin Gouldner (1970) portrayed sociology as a discipline in the service of the bureaucratic state – applying sociological research methods to questions asked in the interest of state administration. But Strong goes further: sociology, he claims, is not simply a discipline in service of the powerful; rather, it has actively pursued its own legitimacy as a distinct, and important, academic discipline. It, too, moved into new fields of human endeavour – into, for example, a study of medicine. It, too, made claims about its relevance to almost the whole of human thought and activity. As Strong says, "In criticizing the imperialism of other professions, sociologists also advance their own empire and do so under exactly the same bases as other professions – the service of humanity; a creed which provides the only public rationale for its existence" (1979: 202). Thus, just as medicine can be charged with promoting medicalization, sociology can be charged with a parallel phenomenon – "sociologization."

Sexism

Another critique of medical sociology comes not from the examination of this particular substantive field of study but rather from the more widespread critique of sociology: sexism. Here it has been useful to consider three different dimensions of sexism: theoretical assumptions, methodological decisions, and the organization of the profession (Wilson, 1982). The "founding fathers" of sociology were, as Julia and Herman Schwendinger (1971) have shown, "sexist to a man." Their social theories, informed as they were by prevailing definitions of women's nature and role, ignored or stereotyped women. Similar critiques about modern sociology are widespread. Modern theorists have done little to redress the balance. Thus women are excluded from some types of sociological analysis, such as the study of politics, because they do not participate in the same ways men

do. In other types of analysis, for example, the study of the family, women have traditionally been defined in a stereotypical way, for instance, as the expressive or emotional leaders in the family.

An early review of sexism in medical sociology is Lorber's "Women and Medical Sociology: Invisible Professionals and Ubiquitous Patients." Lorber (1975: 98) notes that the best-known studies of the professionals, except for nurses, who have been assumed always to be female, have excluded women. We know very little about the processes whereby women came to be doctors and how they work (Lorber, 1984, is an exception to this). On the other hand, there is an overemphasis on women as patients in the sociological investigation of illness, and there is a tendency to generalize from one woman patient to all women. Women are often seen as health-care consumers; they are seldom studied as health-care professionals.

Just as medical sociologists have been constrained to describe health and illness through the eyes of physicians, so has the experience of women been distorted because of the dominance of the male sociologist in the discipline of medical sociology. Women are seen as more ill than men as a result of their reproductive functions, as more neurotic than men, and as more likely to complain of psychogenic illnesses (Ehrenreich and English, 1973b; Scully, 1980). Women are more likely to suffer unnecessary surgery – consider, for instance, rates of hysterectomy (Scully, 1980) at the hands of male doctors. Thus women are more often included in rates of illness. Sexism in medical sociology may reflect two important features. The relatively greater number of male than female medical sociologists and the male-dominated medical interests in the specification of subjects suitable for research (Clarke, 1983).

In their critique of sexism in sociology, feminists have advocated placing women at the centre of the research, both as subjects of study and as researchers. The focus on women's lives shifts the lens from the public and visible to the invisible, informal, and private activities of everyday life. This method/theory has produced books such as Hilary Graham's *Women, Health and the Family*.

Some feminist scholars have gone beyond a concentration on women to an analysis of the gender bases of all social and cultural life. The feminist critique of social science has noted the historical emphasis on white, middle-class, Euro-American women as a basis for generalizing to all women. A more inclusive approach, it is argued, would include race, class, and sexual preference, as well as gender. Methodologically speaking, feminist research advocates a reflexivity that asks the scholar to consider her effect on her subjects and their reciprocal effect on her. It does not necessarily claim objectivity. Sometimes, in fact, feminists argue that social researchers must take an active role in changing sexist social arrangements.

Summary

(1) For the most part, structural functionalism uses positivism as its methodology. The social world is described using a series of universal causal laws: human behaviour is seen as objectively measurable, and social facts are treated as things. Most of medical sociology falls within this paradigm.

(2) An example of positivist methodology is epidemiology – the study of the causes and distribution of disease. A clear and observable cause and effect relationship is looked for in order that the disease be defined and the spread of the disease documented and predicted. Much of AIDS research is carried out in this way.

(3) A positivist methodology claims to be objective and describes the world with a series of causal laws. It assumes that accurate measures can be designed to describe sociological concepts and that it is possible to say with certainty that "x" is the cause of "y." It assumes that it is possible to generalize from a sample of the population and that the same research can be completed in different settings with different researchers. It also assumes that the data collected will adequately describe and explain the phenomenon under investigation, and that it is possible to demonstrate causal relationships between social science variables. Questionnaires and interviews are typically used for data collection.

(4) Conflict theorists do not attempt to be objective or to generalize. Formal tests of validity are considered to be irrelevant, and formal tests of causal relationships superfluous. Components of injustice are analysed uniquely in each situation, and conflict theorists maintain a commitment to ethical and humanitarian principles. Historical documents are the usual source of

data, but many other methods of data collection are used.

(5) The symbolic interactionist perspective focuses on the meanings that the subjects of study attach to reality. Data collection methods traditionally used are participant observation, long, unstructured interviews, and qualitative content analysis.

(6) A difficulty that many researchers have as participant observers is that they must empathize with their subjects yet remain objective. "Going native" can create more problems than it solves for those involved in fieldwork. Ethical considerations are also a concern.

(7) Researchers studying women's health and medical care have used all of the research traditions explicated. Those whose work adopts an explicitly feminist methodological perspective are distinguished by explicit attention to a primary focus on the standpoint of women in society and culture.

(8) Recently feminists have argued for viewing as problematic not only gendered standpoints but those pertaining to race, class, sexual orientation, and ability/disability.

(9) Several critiques of the way that the sociology of medicine and the sociology of health and illness have been practised are relevant to all three sociological paradigms. One is that its focus has been primarily on allopathic medicine and on questions of interest to medical practice. There appears, however, to be a shift away from this focus.

(10) Another criticism concerns sexism. Social theories of the "founding fathers" of sociology ignored or stereotyped women, and there have not been many attempts to change this. Women are most often studied as health-care consumers, while their roles as providers of health care are ignored. These trends appear to be slowly changing.

PART TWO

SOCIOLOGY OF HEALTH AND ILLNESS

Introduction

The next six chapters illustrate each of the theoretical perspectives in the sociology of health and illness. Chapter Three describes some of the changes in mortality and morbidity in Canada over the past century and a half and some reasons for these changes. It also explores the importance of such things as nutrition, clean water, birth control, and immunization for the health of a population. It focuses on early Canada and the Third World. It discusses various "causes" of contemporary mortality and morbidity. Chapter Four describes environmental and occupational health and disease in the context of Canadian society as a whole. The analysis in these chapters is not unified by a single theoretical framework. These two chapters examine such variables as food availability and quality (and thus, implicitly, the institutions that provide food and oversee occupational health and safety). Therefore, the analysis can be considered an example of descriptive analysis in the positivist tradition. Such analysis, which uses the structural-functional perspective without explaining how institutions function together, is typical of the contemporary practice of much of structural functionalism.

Chapter Five discusses age and gender. Chapter Six examines social class, race, and ethnicity. These chapters examine each of these in turn as a cause of differing life expectancy among different groups. When the analysis deals with these variables, considers aspects of the social system and the ways that they relate to human behaviour,

and uses a variety of explanations, it would be considered structural-functional. When the analysis focuses on the single prime economic determinant of capitalism, as it does in the final few pages of Chapter Four, it can be seen as an example of conflict theory. Chapter Seven, while it deals with the functions of the social system in a positivist way and is thus compatible with structural functionalism, focuses on the social-psychological behaviour of the individual and its relationship to illness and death. Chapter Eight is symbolic interactionist in that its focus is on the description of the illness experience from the perspective of the subjects, whether these are the people who are ill or other people associated with them.

The following table provides an overview of many of the factors that need to be considered in explaining the health of a population. Disease and death are seldom the result of isolated conditions or incidents. Death rates, and most deaths (perhaps with exceptions such as sudden death because of car accidents), are the result of complex causes, including the direct cause of death (which may be, in the case of cancer, for instance, starvation); the underlying cause, which in this case might be the growth of malignancies; the bridging cause, which might be the malfunctioning of the stomach resulting from malignant growth so that food cannot be absorbed; contributing causes, which might include smoking and excess alcohol ingestion; predisposing conditions, which might be air pollution resulting from a certain industrial process; and generating conditions, which might be

TABLE PART II.1 Conditions Affecting Life Expectancy

PREDISPOSING CONDITIONS	GENERATING CONDITIONS	
HISTORY (A) wars famine epidemics	SOCIAL-STRUCTURAL POSITION WITHIN A SOCIETY (B) age sex marital and family status	SOCIO-PSYCHOLOGICAL CONDITIONS (C) stress experience and stress management type A or B behaviour
ECOLOGICAL/GEOGRAPHICAL CONDITIONS natural disaster (earthquake, tornado, heat wave)	class education level occupation rural/urban location religion	sense of coherence gender role expectations LIFESTYLE CONDITIONS (D) smoking habits
ENVIRONMENTAL quality and quantity of water quality of air quality and quantity of foodstuffs (nutrition) safety: roads, airways, waterways, transportation vehicles, workplace, home, tools, equipment birth control	religiosity region ethnicity	seat belt use alcohol consumption rate sexual behaviour drug use and abuse EXISTENTIAL FACTORS (E) the meaning of the illness experience of illness
MEDICAL immunization antibiotics other chemotherapy surgery, radiation		
SOCIETAL STRUCTURE political-economic system cultural values		

the economic position of the worker who had no choice but to work in an asbestos mine.

This table lists a number of different groups of causes of death. It also considers causes at the individual level under the heading of existential factors. The next chapters will examine some of these causes in detail. Others are not dealt with because of space limitations and/or because of lack of research. Generally speaking, Chapters Three and Four focus on (A) and, to a limited extent, (B); Chapters Five and Six focus on (B) and, to a limited extent, (A); Chapter Seven focuses on (C) and (D), and Chapter Eight focuses on (E).

Definitions of illness and health are complex and at times equivocal. Disease is determined in a number of ways: self-report, which includes asking respondents to describe their own state of health; clinical records, which include the records of physicians as well as hospital statistics; and physical measurements, which include such things as blood pressure readings and tests of tissue pathology. Each of these may record a disease state at a different level of development or potential acknowledgement (e.g., a Pap smear may indicate evidence of pre-cancerous or cancerous tissue before either the physician or the patient would be able to notice its occurrence). None of these three can be considered true or objective ways to determine illness, but in combination they may point to true disease (or some other condition).

Not only is disease hard to define, but it is also difficult to determine the actual cause of disease. As in the case of death, the determination of causality is complex both because of the operation of several variables at one time and because of the possibility of a variety of causes over time. In addition to understanding disease causation, students of sociology are interested in questions regarding the meaning of and social construction of disease and the differentiation among disease, sickness, and illness (see Chapter Eight for further discussion of this issue). Finally, sociologists are critical of the "validity" of official statistics regarding both disease and death.

CHAPTER THREE

Disease and Death:
Canada in International and Historical Context

WHAT ARE THE major causes of death in Canada today? What were the major causes of death in Canada a century ago? What is the average life expectancy in Canada today? How long did Canadians live on average when your grandparents were children? A century ago? How important has medicine been to the increase in life expectancy over the last century or so? You have probably heard that penicillin is a wonder drug. What was the importance of penicillin in the overall improvement in the health of the nation? What about other "wonder drugs"? How important have immunizations been in the extension of life for the average Canadian?

Life Expectancy

The average life expectancy for men and women has varied considerably through the millennia and in many different types of social and economic structures. For example, the average life expectancy of late Ice Age hunter-gatherers about 11,000 years ago has been estimated to have been approximately 38 years (Eyer, 1984). According to available records and estimates, the average life expectancy in Europe varied between 20 and 40 years from the thirteenth to the seventeenth centuries (Goldscheider, 1971). In Canada, too, there have been wide variations in life expectancy. In 1831 the average for Canadians is estimated to have been 39.0 years. Females born in 1991 can expect to live for almost 81 years, while males can expect to live for 74.6 years (see Table 3.1). This is an increase of 19.6 years for males and 22.6 years for females since 1920-22 (*Canada Year Book*, 1994: 31). These figures are among the highest in the world.

Against this optimistic picture must be set the fact that the gap between the life expectancy for men and women in Canada remains high. Canadian women live an average of about 6.4 years longer than men (see Table 3.2). Furthermore, the life expectancy gap between men and women in Canada widened over the sixty-year period beginning in 1931 from 2.1 years to about 6.4 years. In effect, mortality declines over this period have benefited females more than males.

TABLE 3.1 Life Expectancy at Birth by Sex, Canada, 1831-1991

PERIOD	MALE (years)	FEMALE (years)	BOTH SEXES (years)
Around 1831	38.3	39.8	39.0
Around 1841	39.4	41.3	40.3
Around 1851	40.0	42.1	41.0
Around 1861	40.6	42.7	41.6
Around 1871	41.4	43.8	42.6
Around 1881	43.5	46.0	44.7
Around 1891	43.9	46.5	45.2
Around 1901	47.2	50.2	48.7
Around 1911	50.9	54.2	52.5
Around 1921	55.0	58.4	56.7
1930-32	60.0	62.1	61.0
1940-42	63.0	66.3	64.6
1950-52	66.3	70.8	68.5
1955-57	67.6	72.9	70.2
1960-62	68.4	74.2	71.2
1965-67	68.8	75.2	71.9
1970-72	69.3	76.4	72.8
1975-77	70.2	77.5	73.8
1980-82	71.9	79.0	75.3
1983-85	72.9	79.8	76.4
1985-87	73.0	80.0	76.5
1991	74.6	81.0	77.8

Sources: Prior to 1930-32: Bourbeau and Légaré, pp. 77-86; 1930-32: Statistics Canada, *Vital Statistics* 1977, vol. III: Deaths, Catalogue 84-206, p. 2; 1980-82: Statistics Canada, *Life Tables, Canada and Provinces*; 1980-82: Catalogue 84-532, May, 1984, pp. 16, 18; 1983-87: *Canada Year Book*, 1988, Catalogue 11-402. Ottawa: Statistics Canada, 1989, p. 3-18; *Canada Year Book*, 1994; 1991: *Report on the Demographic Situation in Canada 1994*, Catalogue 91-209E (Ottawa: Minister of Industry, Science and Technology, 1994), p. 50. Reproduced with permission of the Minister of Supply and Services Canada, 1989.

Such dramatic changes in life expectancy as have occurred in the past 100-150 years in Canada (and elsewhere in the developed world) can be explained by a whole host of factors. One description of the process is epidemiologic transition (Omran, 1979), which is based on the theory of demographic transition. Simply put, this idea suggests that as the economy changes from low to high per capita income, there is a corresponding transition from high mortality and high fertility to low mortality and low fertility.

TABLE 3.2 Life Expectancy at Birth, Selected Countries

COUNTRY	YEAR	MALES (M) (years)	FEMALES (F) (years)	DIFFERENCE F – M
Japan	1991	76.1	82.1	6.0
Netherlands	1991	73.7	79.8	6.1
Denmark	1990	72.0	77.7	5.7
France	1991	73.0	81.1	8.1
Canada	1991	74.6	81.0	6.4
Spain	1990	73.4	80.1	6.7
Australia	1991	73.9	80.0	6.1
United Kingdom	1990	72.9	78.5	5.6
Italy	1988	73.2	79.7	6.5
Portugal	1990	70.2	77.3	7.1
Mexico	1991	66.5	73.1	6.6
Ireland	1990	71.9	77.4	6.5
New Zealand	1991	71.9	78.0	7.1
Belgium	1990	72.7	79.4	6.7

Source: *Canada Year Book*, 1994, Cat. no. 11-402E/1994 (Ottawa: Minister of Industry, Science and Technology, 1993), Table 3.1, p. 111.

Changes in the patterns of disease occur in three distinct stages: the Age of Pestilence and Famine, the Age of Receding Pandemics, and the Age of Degenerative and Man-Made Diseases (Omran, 1979).

The *Age of Pestilence and Famine* is characterized by socio-economic conditions in which communities are traditional, economically underdeveloped with a low per capita income, and generally agrarian. Women usually have low status, the family is extended, and illiteracy is high. The high mortality rate is largely attributable to famine and infectious diseases. The *Age of Receding Pandemics* is characterized by a decrease in epidemics and famine and a consequent decline in the mortality rate. At this point the fertility rate continues to be high, resulting in a "population explosion." The fertility rate then begins to decline as people begin to live longer and to die of emerging industrial and degenerative diseases such as cancer, stroke, and heart disease. This characterizes the *Age of Degenerative and Man-Made Diseases* (these terms are Omran's). This description provides an outline of stages, but not an explanation.

The question remains, then, what sorts of factors are responsible for the overall decline in mortality rate and the increase in life expectancy? McKeown (1976) has offered an explanation based on studies of the decline in mortality in Britain (and supported by findings for Sweden, France,

Ireland, and Hungary) over the last few hundred years. First, the decline in the mortality rate was almost entirely due to a decline in infectious disease. Second, the decline in infectious disease was the result of three basic changes: (1) improvements in nutrition, (2) improvements in hygiene, and (3) increasing control of disease-causing micro-organisms. Improvements in birth control were also a factor. Infant and early childhood mortality are usually one of the most sensitive indicators of the health of a nation. Table 3.3 shows the tremendous discrepancies among different countries in this mortality rate. In Canada for 1951 to 1988 the infant mortality rate dropped 82 per cent, largely as a result of better nutrition and living standards for the mother and baby, coupled with pre- and post-natal medical care (*Canada Year Book*, 1994: 131).

A second study on the twentieth century (1901-71 is the period actually considered) showed that this increase in overall life expectancy was the result of very similar conditions (McKeown and Record, 1975): improved nutrition accounted for about half the increase, and better hygiene, resulting in fewer water- and food-borne diseases, accounted for about one-sixth of the decline in mortality. Immunization and medical therapy were together responsible for about one-tenth of this decline. Recent research, based on 1988 data, evaluated the contemporary importance of medicine, as compared with various socio-economic resources, for infant mortality rates in 117 industrialized, developing, and underdeveloped countries (Kim and Moody, 1992). Using GNP, energy consumption,

daily caloric supply per capita, percentage of population enrolled in secondary education, urbanization, and safe water supplies as indicators of socio-economic status, and comparing their effects to those of health resources (i.e., population per physician, nurse, hospital bed) on infant mortality rates, Kim and Moody noted the continuance of the pattern described above. Namely, health resources make only a small contribution to the health of the population when compared to socio-economic resources. The following sections will discuss the ways that some socio-economic resources affect health and longevity. Table 3.4 illustrates this by detailing the present causes of childhood deaths world-wide and the measures that could readily prevent these unnecessary deaths. Such measures would, of course, eliminate a great deal of suffering over and above these particular deaths. The guilt-stricken families and communities, as well as those children who do not die but suffer from all of the same diseases, are not counted in the statistical picture presented.

Poverty

Both the overall level of income and the relative income of a people have major impacts on health. Income level is the context for all of the other elements of daily life, including work, education, food, shelter, water, hygiene, and sanitation. Poverty is also often associated with political powerlessness and marginalization. While the economic growth of a whole country is not

TABLE 3.3 Number of Children Dying before Age Five per 1,000 Births

COUNTRY	1975 VALUE	1987 VALUE	COUNTRY	1975 VALUE	1987 VALUE
Mali	358	296	Peru	167	126
Sierra Leone	358	270	Guatemala	162	103
Niger	281	232	Kenya	161	116
Chad	281	227	Papua New Guinea	152	85
Burkina Faso	275	237	Zimbabwe	151	116
Senegal	275	220	Nicaragua	150	99
Mauritania	266	223	El Salvador	137	87
Ethiopia	262	261	Botswana	136	95
Liberia	262	150	Ecuador	136	89
Somalia	262	225	Madagascar	135	187
Benin	255	188	Dominican Republic	132	84
Nigeria	250	177	Brazil	125	87
Sudan	245	184	Syria	125	67
Bolivia	244	176	Jordan	116	60
Cote D'Ivoire	240	145	Colombia	102	69
Egypt	240	129	Philippines	101	75
Nepal	240	200	Mexico	100	70
Rwanda	237	209	Thailand	91	51
Haiti	232	174	Paraguay	82	63
Bangladesh	228	192	Chile	79	26
Burundi	228	191	Sri Lanka	79	45
Pakistan	226	169	Mauritius	77	30
Gabon	223	172	Panama	68	35
India	218	152	Venezuela	67	45
Togo	204	156	Costa Rica	64	23
Algeria	200	111	Malaysia	62	33
Morocco	190	123	Republic of Korea	61	34
Turkey	184	97	Yugoslavia	53	28
Ghana	183	149	Uruguay	52	32
Lesotho	180	139	Jamaica	48	23
Tunisia	180	86	Trinidad and Tobago	37	24
Indonesia	173	120			
Honduras	171	111	Mean	173	125*

*By comparison, the U.S. and Japan had an under-5 mortality rate of 21 and 15 (respectively) in 1975, and a rate of 13 and 8 in 1987.

Source: *Social Forces*, 71, 3 (March, 1993), pp. 632-33.

TABLE 3.4 The Five Leading Preventable Diseases that Kill Children, Late 1980s

	RANK ORDER OF PREVENTABLE CHILD DEATHS	AVERAGE ANNUAL NUMBER OF CHILD DEATHS
Diarrhea	First	5 million
Acute respiratory diseases	Second	4 million
Measles	Third	2 million
Tetanus	Fourth	1.8 million
Malaria	Fifth	1 million

Source: Adapted from UNICEF, 1990.

necessarily associated with better health for all, economic decline usually affects the standard of living and, consequently, the health of many. "This is because the costs of an economic recession tend to fall most heavily on those who were least well off to start with" (*Beyond Adjustment*, 1993). Public policies regarding income security are also associated with infant mortality rates. On the basis of international comparative data, Wennemo (1993) concluded that: (1) relative income inequality within a country seemed more important than overall level of economic development in the country with respect to infant mortality rates; and (2) the level of unemployment and the availability of unemployment and family and social security benefits were related to infant mortality rates. Other comparative and contemporary analyses reinforce the primacy of socio-economic conditions to health and life expectancy (Wnuk-Lipinski and Illsley, 1990). Navarro (1992) surveyed the health conditions of the world's population, continent by continent, and showed how, contrary to current Western beliefs, socialism

and socialist forces have been, for the most part, better able to improve health conditions than capitalism. This finding tends to be true in both the developed and the developing world.

Food Security

The availability of an adequate amount of nutritious food has an enormous effect on the health of a population. Experimental field studies have demonstrated that improvements in nutrition, particularly in the quantity and quality of the protein component, have a much larger impact on morbidity and mortality than any other public health or medical measure (see Eyer, 1984).

Nutrient deficiency in women has been an even more prevalent and significant problem than in men. In a global context, women work almost 65 per cent of the total hours worked and in return earn 10 per cent of world income and own 1 per cent of the world's property. Because women have less income, they have less access to adequate

Box 3.1 The Debt Crisis and Childhood Illness

Every day 40,000 children die from malnutrition and other preventable causes, including particularly diarrheal diseases, measles, tetanus, whooping cough, and pneumonia. Table 3.4 portrays the percentage of children who die from the five leading causes of childhood death. It also describes means necessary to prevent these deaths. Not only are these causes of death simple to prevent, they are relatively inexpensive. Also, the death rates for children in the Third World are considerably higher than those in the developed North. UNICEF has estimated that full immunization would cost $1.50 per child. Oral rehydration kits that could prevent diarrheal deaths cost approximately 10¢ each. Antibiotics, to combat pneumonia, cost about $1.00 per child. A co-ordinated program dedicated to combating childhood malnutrition and death would cost only about $2.5 billion a year world-wide.

High rates of childhood death persist in the Third World although they have diminished somewhat. There are a number of explanations for this situation:

urbanization, fertility rates, development itself, foreign investment, foreign control, and foreign exploitation of the environment and people for financing by the developed world. Today one of the chief causes of the persistence of childhood disease is the global debt crisis. The developing nations owe $1.3 trillion to governments and financial institutions in the developed world. In consequence, the international financial community now requires that borrowers implement severe austerity measures (including government spending cuts and wage freezes) before they can obtain new loans on restrictive outstanding debts. Domestic austerity measures, also known as "structural adjustments," are intended to increase efficiency and save resources to repay foreign debt. The result is the allocation of fewer resources for immunization, general health maintenance, prenatal care, nutrition, urban development, and other programs with direct impacts on children's health (Bradshaw et al., 1993).

FIGURE 3.1 The Hunger Cycle

| pregnant mother with nutritional inadequacies more likely to have low birth-weight baby | → | low birth-weight baby is more likely to die or to succumb to one of the many nutrition-based diseases | → | nutritional inadequacies result in physical and mental "stunting," listlessness. Limits stimulation and later leads to poor school attendance, concentration, and performance, all of which contribute to a cycle of hunger and poverty |

Source: Adapted from *Women, Health and Development Kit* (Consultant: Eva Zabolai-Csekme), undated.

nutrients (Zabolai-Csekme, 1983). However, because women bear and breast-feed babies, they have a greater need for calories

(Figure 3.1). The low standard of nutrition among women is pivotal, too, for the health of their offspring. In brief, underweight

TABLE 3.5 Some Nutrition-Related Diseases of Development

DISEASE	CAUSE	SYMPTOMS	PREVENTION
Kwashiorkor	Severe protein-calorie malnutrition in young children, due to insufficient and inadequate diet; infections; and other causes related to underdevelopment	Swelling, loss of weight, apathy, skin and hair changes	Better nutrition of mothers, breastfeeding, enough and adequate solid food, prevention of infections
Nutritional marasmus	Severe protein-calorie malnutrition in infants, due to insufficient and inadequate diet; infections; and other causes related to underdevelopment	Wasting, fat and weight loss, arrested growth	As for kwashiorkor
Vitamin A deficiency	Lack of vitamin A	Loss of appetite and weight, eye damage, blindness	Breastfeeding, dark green leafy vegetables, mangoes, papayas, liver
Nutritional anemia	Lack of one or more essential nutrients	Too little hemoglobin or too few red blood cells, pallor, easily fatigued, breathlessness	Improved diet to supply missing nutrients, prevention of nutrient loss due to illness
Endemic goitre	Lack of iodine	Enlarged thyroid gland in the neck; physical and mental retardation in children of affected mothers	Iodine added to salt or given by injection

Source: *Women, Health and Development Kit* (Consultant Eva Zabolai-Csekme), undated.

babies are born to women who suffer nutritional inadequacies. Underweight babies are more likely to die as infants and to suffer from any one of a number of diseases (see Table 3.5).

The Physical and Social Environment

The availability of a sufficient amount of clean drinking water is another crucial factor in health. Available statistics indicate that three out of five people (and an even greater proportion among the poor) in developing countries do not have access to safe drinking water, and numerous fatal and debilitating chronic illnesses are spread by unsanitary water. Two of the most prevalent are cholera and dysentery. Diarrheal disease is the single greatest killer of children in the developing world (*Beyond Adjustment*, 1993). Its prevention depends on changes in water supply, hygiene, and sanitation. Yet the infrastructure developments necessary for adequate improvements are extensive. They include drainage systems for the disposal of human and animal wastes, access to potable water, and water for irrigation. Environmental problems such as air, water, and soil pollution and forest destruction continue to have catastrophic health impacts.

The devastating effects of unclean water have nowhere been as dramatic as in death rates among babies in the Third World who were fed by "instant" infant formula that had to be mixed with water (for an overview, see Box 3.2). The export of infant formula to the developing world was responsible for a widespread increase in infant mortality from a specific source. Not only were babies made sick by being given contaminated water, but also, lacking adequate information (either because of illiteracy or because mixing directions were not available) or lacking sufficient income to purchase adequate amounts of formula, mothers were often diluting the formula so extensively that babies were dying of starvation.

Safety

Personal safety is of great concern, particularly in times of heightened national tensions. Intra- and international warfare and violence in communities, workplaces, and the home are all threats to fundamental safety. As inequities grow, so, too, can violence and war. As a consequence death, disability, and disease can be expected to increase.

The Centrality of Women

For a number of reasons the position of women in a society has a significant impact on the health of the people. In a world-wide context women's health is considerably poorer than that of men. This is particularly true in the developing world. The lifetime chance of maternal death is 1:6,366 in North America and 1:121 in Africa; 99 per cent of the half-million who die annually in childbirth are from developing countries. Almost half of the 1.1 billion women in developing countries are malnourished, 250 million suffer from iodine deficiency, and almost 2 million are blind because of vitamin deficiency. Poverty, malnutrition, illiteracy, the level of education, and access to medical care all contribute significantly to women's health status world-wide. Women's reproductive and sexual health is also often seriously compromised by the lack of availability of contraception or abortion, cultural inhibitions, and absent, distant, and/or male doctors. Particularly in sub-Saharan Africa, discrimination results in feeding boys more and better than girls and the greater proportion of female babies suffering infanticide. Violence, both sexual and non-sexual assault, is also a major issue for women's health world-wide (Koblinsky et al., 1993). Interestingly, one of the most important interventions possible for the improvement of a nation's health is the education of women.

Box 3.2 Infant Formula Feeding: The Crisis in the Developing World

In the early 1980s a crisis occurred in the developing world over the seemingly innocuous question of how best to feed infants. Starting in 1969, a new market for infant feeding formula was developed in Third World countries. It was regarded as a wise and humane move that would extend the lives of many of the millions of children in the Third World who died annually as a result of malnutrition. Advertising and promotion for infant feeding formula quickly became successful. Free samples donated by the companies manufacturing the formula were given to women who had just given birth. In the Third World, feeding babies formula rather than breast milk was very quickly taken to be a symbol of mother love and responsibility, because formula was associated in the minds of Third World peoples with the successful middle and upper classes in the Western world. A number of unexpected negative consequences resulted.

(1) When women used the free samples given to them at the birth of their babies, their own milk supply would dry up and breastfeeding would become impossible.

(2) When women went home with their babies, they were often ill-prepared to continue with the infant feeding formula for a variety of reasons.

   (a) Formula was frequently unavailable in the small villages and communities, and as women's breast milk had dried up, the babies starved to death.

   (b) When the formula was available, there were often no instructions on how to use it, the mothers were illiterate, or the instructions were in a language the mothers could not read. Thus many women mixed the formula with too much water, so that the nutrients in the mixture were inadequate for the baby's growth.

   (c) Generally, the formula was sold in powdered form, to be mixed with water. Frequently, the water source was polluted, resulting in unnecessary illnesses and death.

   (d) The bottles used for feeding the babies should have been sterilized. Many Third World mothers were unable to sterilize the bottles because they lacked clean water, a heat source, or a chemical sterilizing agent.

   (e) Aside from the problems enumerated above, one other fact stands out: breast milk is actually better for babies because it passes important immunities from mother to offspring. In the developing world, babies breastfed for less than six months are five to ten times more likely to die than those breast-fed for a longer period.

When the dangers of infant formula became clear to the Western world, the companies manufacturing infant formula were boycotted, especially Nestle's, which held more than 50 per cent of the world's market. A World Health Organization conference called in 1979 initiated the adoption of a code to govern infant formula sales. Its aim was:

"the provision of safe and adequate nutrition for infants; by the protection and promotion of breast feeding, and by ensuring the proper use of breast milk substitutes . . . on the basis of adequate information through appropriate marketing and distribution."

The infant formula companies agreed to support the code. While the extent of the problem declined, there were, and continue to be, numerous violations of the code. To a degree, then, this is an ongoing problem.

"A Boycott over Infant Formula," *Business Week*, April 12, 1979, pp. 137-40; Eva Zabolai-Csekme, *Women, Health and Development* (Geneva: World Health Organization, 1983); *International Code* (Geneva: World Health Organization, May, 1981), Article 1.

Box 3.3 The Impact of Free Trade on Health: A Few Incidents

A number of documented health and safety hazards, particularly at the Mexico/U.S. border, resulted from the introduction of free trade zones or Maquiladoras just across the border in Mexico. The North American Free Trade Agreement among Canada, the U.S., and Mexico will not alleviate these problems. Northern companies (from the U.S. and Canada) can make more profit in Mexico. Labour costs are much lower, unionization is scarce, and environmental, occupational, and safety standards are far weaker. The budget amount dedicated to environmental protection and rehabilitation in Mexico is one-sixteenth of the budget of just one of the United States – Texas. Such inequities have led to profound consequences for people.

For instance, in an area on the Texas-Mexico border there have been 50 cases of anencephalic babies – four times the national average and the largest cluster ever documented. A Brownsville, Texas, pediatrician found that all mothers who gave birth to these babies lived within two miles of the Rio Grande River, where wastes from U.S.-owned companies were being dumped. The levels of chemical disposal of xylene ranged from 6,000 times (by General Motors) to 53,000 times (by the Stephen Chemical Company) the U.S. standard. Researchers continue to try to ascertain the link between industrial pollution and these health tragedies.

The costs of NAFTA are not only close to the U.S./Mexico border. The quality of food supplies is jeopardized in all three nations involved in NAFTA by what has been called a circle of poisons. In this situation one country, for example, Canada, might export into Mexico pesticides that have been banned for use in Canada. Later, Canadians and Americans may then import the produce grown with the pesticides. Checks on the quality of imported food, already minimal (only 2 per cent of food imported from the U.S. is inspected), have declined even further since NAFTA (*Beyond Adjustment*, 1993: 13).

Birth Control

Effective birth control is an important cause of the decline in the mortality rate. In developing countries such as India, Indonesia, Mexico, Brazil, Nigeria, and Morocco, the average woman has seven to eight live births and approximately five additional pregnancies that end in miscarriage (Hammer, 1981). Too many pregnancies or pregnancies spaced too closely together are a threat to the health of the mother and the child (*Health and the Status of Women*, 1980) for a number of reasons. In the first place, when the pregnancy is unwanted, women may seek illegal abortions, which are extremely dangerous. In Latin America, for instance, illegal abortions are said to account for between one-fifth and one-half of all maternal deaths. Second, because of malnutrition, women's bodies may be undernourished and small, their pelvises misshapen, and they may experience fatigue during pregnancy and a difficult delivery. Third, pregnancy itself takes a toll on a woman's body because nutrients are needed for the baby as well as the mother. During pregnancy large increases in calories, vitamins, and minerals are required. More iron, vitamin B_{12}, and folic acid are also needed, especially during the last trimester of pregnancy. Fourth, because

Box 3.4 Violence against Canadian Women

A recent survey (1993) based on a telephone interview with a random sample of 12,300 Canadian women 18 years and older found that 51 per cent of the respondents had experienced violence (this was limited to violence after 16 years of age). Twenty-five per cent of women reported they had experienced violence at the hands of a marital partner (including common-law); 20 per cent of all reported experiences of violence resulted in physical injury.

Several features of this study are important to note. First, this is the first national study of its kind based on a random selection of respondents. Most studies to date have been non-representative and/or based on samples drawn from women's shelters or police records. Second, the definitions of violence used in the study were those included under the Criminal Code as physical and sexual assaults ranging from threat of imminent attack, attack with serious injury, sexual assault including unwanted sexual touching, up to, and including, violent sexual attacks with severe injury to the victim. Under these definitions, 34 per cent of the women reported actual physical assault, 6 per cent experienced threats only, 39 per cent reported sexual assault, and 15 per cent reported unwanted sexual touching only. This study reinforces what other research has suggested: that women face the greatest risk of violence from men they know – 23 per cent experienced violence from a stranger; 17 per cent experienced violence from both strangers and men they knew.

The rate of actual violence is the background for pervasive fear among women. Over 83 per cent of women who use parking garages stated that they were very or somewhat worried when walking alone to their car; 76 per cent were worried when they walked or used public transportation alone after dark; 60 per cent of women who walk alone after dark in their neighbourhood were worried about their personal safety while walking; and 39 per cent of women indicated that they were worried when they were home alone in the evening. As people age they generally reported less fear. Those in small communities tended to be less fearful. Women were also asked about actions they take for their own personal safety. Seventeen per cent always or usually carry something to defend themselves or to alert others; 31 per cent avoid walking past teenage or young men; 60 per cent who drive check the back seat before they get in; 67 per cent lock the car doors when they drive alone; and 11 per cent have taken a personal safety course (Statistics Canada, 1993a).

---

energy is used up during pregnancy, rest, especially in the last trimester, is important. Most women in developing nations do not, however, have the leisure to take the necessary rest (*ibid.*). Fifth, childbirth itself, because of the lack of sanitation, lack of prenatal care, or lack of emergency medical services, is responsible for a much higher rate of maternal mortality among the developing nations than in the developed nations. Some of the most prevalent causes of child birth-related deaths are postpartum hemorrhage, which occurs when a woman has anemia, and sepsis (infection), which occurs because of inadequate sanitation or because of hypertensive disorders of pregnancy.

Infectious and parasitic diseases result from nutritional inadequacies and polluted

Box 3.5 The Global Health Burden of Rape

This review of numerous studies from all around the globe documents the prevalence of rape and the correspondingly widespread health effects. One way of conceptualizing the health effects is in terms of DALY, or disability-adjusted life years. According to the *World Development Report – 1993: Investing in Health* (World Bank, 1993), rape and domestic violence are major causes of disability and death, especially among women in their reproductive years. The report estimates that one in every five years lost (either by death or disability) results from gender-based victimization in the developed economies. In the developing world the health burden resulting from rape and domestic violence is about the same but because of the overall disease profile the percentage attributable to gender-based victimization is smaller. On a global basis the health burden of gender-based victimization (9.5 million DALY) is comparable to HIV (10.6 million DALY), tuberculosis (10.9 million DALY), sepsis during childbirth (10 million DALY), all cancers (9.0 million DALY), and cardiovascular disease (10.5 million DALY).

The health consequences discussed include psychological distress, sociocultural impacts, and somatic consequences. Psychological distress may last throughout the lifetime of the woman who has been raped. The symptoms can be diverse and extensive. In North America they have been conceptualized by the psychiatric profession as a type of PTSD – post-traumatic stress disorder. Survivors of rape and violence are "more likely to have received several psychiatric diagnoses during their adult life including major depression, alcohol abuse/dependence, drug abuse/dependence, generalized anxiety, and obsessive compulsive disorder."

Sociocultural effects are the effects that spread beyond the suffering of the individual woman and lead non-victimized women to change their behaviour and restrict their movements out of fear. Surveys done in countries around the world indicate that many women consider the fear of rape a major stress in their lives. In some countries women who have been raped may be doubly abused. For instance, in parts of Asia and the Middle East, consequences of rape may include being divorced by one's husband, ostracized by one's family, or even killed by family members to cleanse the family honour. Rape can lead the victim to commit suicide or to be the victim of an "honourable" murder. Physical illnesses that are disproportionally diagnosed among women who have been raped include chronic pelvic pain, arthritis, gastro-intestinal disorders, headaches, chronic pain, psychogenic seizures, premenstrual symptoms, and substance abuse. For example, women in the United States who have been victims of rape (or other crimes) report more symptoms of illness across virtually all bodily systems and perceive their health less favourably than non-victimized women. There are also numerous reproduction-related issues that may result from rape: pregnancy, sexually transmitted diseases, including AIDS, future high-risk sexual behaviours, and loss of self-esteem.

Mary P. Koss, Lori Heise, and Nancy F. Russo, "The Global Health Burden of Rape," *Psychology of Women Quarterly*, 18 (1994), pp. 509-37.

water. Vulnerability to such diseases is increased by a high birth rate. Much of the dramatic decline in mortality in the last century or so in the developed world, including Canada, is the result of the drop in death rates from infectious or parasitic diseases. This drop is chiefly the result of rising standards of nutrition and improvements in sanitation and birth control, resulting in a "favourable trend in the relationship between some micro-organisms and the human host" (McKeown and Record, 1975: 391). As has been discussed, the Third World is far behind in standards of nutrients, clean drinking water, sanitation, and birth control.

Comprehensive Health Care

Comprehensive health care, while relatively powerless without such fundamentals as adequate food and clean water, is also an important factor in good health. Policy decisions about the relative amount of primary, secondary, and tertiary health care to provide in a society are crucial, but these depend on a national health-care program, which is often lacking. Primary health care emphasizes equitably distributed prevention through community development and education. Environmental issues may also be addressed. Secondary health care is directed toward disease treatment in hospital and community via various (usually Western-style) medical practitioners. Tertiary care especially occurs in a teaching hospital attached to a university and has a side emphasis on health promotion.

Immunization

Immunization affects the health of a population, although its importance appears to have been overemphasized. McKinlay and McKinlay (1977), using American data from about 1900, show that most of the decline in mortality from the infectious diseases prevalent in 1900 (about 40 per cent of the deaths in 1900 were attributable to infectious diseases) was the result of public health measures such as water purification and of improvements in nutrition and birth control. Data for tuberculosis, typhoid, measles, scarlet fever, polio, whooping cough, influenza, diphtheria, and pneumonia demonstrate that the significant decline in mortality for each disease came before the introduction of the vaccine or drug to treat it. In detailing their findings, they present a series of graphs that show that the significant declines in mortality preceded medical treatment of each disease (Figure 3.2).

Infectious and bacterial diseases have all but disappeared in the developed world (with the notable exception of AIDS). They persist in the developing world. But while immunization and medical measures may have made a relatively minor contribution to the decline of mortality in the developed world, they do, along with improvements in public health, nutrition, and sanitation, have a role to play in the developing world, where only 10 per cent of all children receive protection from measles, tuberculosis, whooping cough, polio, tetanus, and diphtheria. Five million children still die from these diseases, and an additional five million are disabled annually.

FIGURE 3.2 The Fall in the Standardized Death Rate (per 1,000 Population) for Nine Common Infectious Diseases in Relation to Specific Medical Measures, United States, 1900-1973

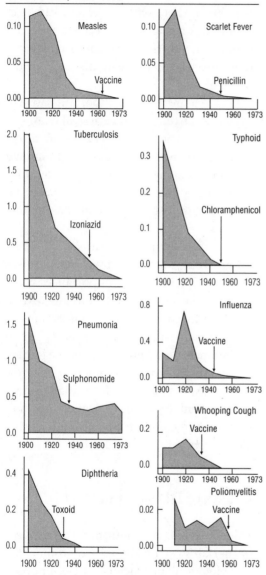

Source: John M. McKinlay and Sonja M. McKinlay, "Medical Measures and the Decline of Mortality," in Howard D. Schwartz, ed., *Dominant Issues in Medical Sociology*, Second Edition (New York: Random House, 1987), p. 699.

Death, Disease, and Disability in Canadian Society

The chief causes of death in Canada today are heart diseases, cancer, and accidents. In 1990, cardiovascular diseases, including heart disease and stroke, accounted for 39 per cent of Canadian deaths; 27 per cent of the deaths were attributable to cancer, 8 per cent to respiratory disease, and 7 per cent to accidents and violence. Accidental and violent deaths are particularly prevalent among 15-24-year-old men. Contrary to commonly held beliefs, however, the rates of violent and accidental death are decreasing – 27 per cent from 1980 to 1990. As well, the proportion of deaths from heart disease has been decreasing over the past twenty years or so. Over that time there was a 30 per cent reduction in death by cardiovascular disease (*Canada Year Book*, 1994). On the other hand, the proportion of deaths due to cancer has been increasing (Angus and Manga, 1985). While the rate of cancer deaths for men has been evening out, the rate of cancer deaths for women continues to rise (*Canada Year Book*, 1994). Heart diseases, cancer, and accidents are called the diseases of civilization or diseases of affluence, or what Omran (1979) has called "man-made" diseases. Their causes are different from those of the diseases of development. Food security and lack of clean water and birth control are no longer problems for most people in most of the developed world. Rather, a combination of lifestyle, environmental, work-related, and other factors are important in explaining today's mortality rate. As mentioned previously, the

FIGURE 3.3 Life Expectancy at Birth and Infant Mortality

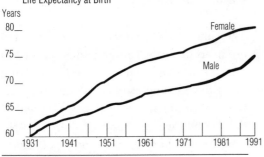

Life Expectancy at Birth

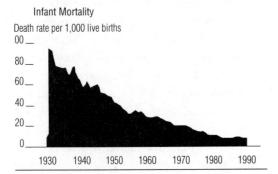

Infant Mortality

Source: *Canada Year Book*, 1994, p. 131.

infant mortality rate is one of the most sensitive measures of the overall health of a people. Figure 3.3 illustrates the downward trend in infant mortality in Canada.

The mortality statistic, called PYLL or potential years of life lost, sheds a particularly useful light on mortality in contemporary society (Table 3.6). Figure 3.4 shows the specific causes of death for men and women of all ages in one year. Assuming life expectancy to be a minimum of 75 years, PYLL shows the average number of potential years of life lost among all Canadians who die before seventy-five years of age.

Potential years of life lost is a valuable statistic for examining "premature death." It gives heavier weight to deaths that occur at an early age than to deaths that occur late in life. Several contrasts are apparent when we compare PYLL with the leading causes of overall death (Figure 3.3). While malignant neoplasms are the most important PYLL, they are second to heart diseases as the cause of overall death. Accidents and suicides are a much more important cause of PYLL (23 per cent) than of overall death (11 per cent). Suicides and accidents are also a much more important cause of PYLL among men than among women. They account for three and a half times as many potential years of life lost among males as among females.

The most important causes of death differ from the most important causes of PYLL. Of significant importance in the major causes of death are smoking, excessive alcohol consumption, diets rich in fat, sodium, and sugar, and stress. Of great importance in potential years of life lost are occupational and environmental hazards, lack of health and safety legislation or non-enforcement of existing legislation, and risk-taking behaviour, including drinking alcoholic beverages and driving.

Causes of Disease and Death

Lalonde (1974) distinguished three causes of mortality: self-imposed, environmental, and biological host factors. By self-imposed factors Lalonde meant such things as (1) excessive alcohol consumption, (2) smoking, (3) drug abuse, (4) nutritional inadequacies

TABLE 3.6 Potential Years of Life Lost (PYLL) by Selected Causes and Sex, 1985

CAUSE OF DEATH	PYLL BETWEEN 0 AND 75 YEARS				DEATHS BETWEEN 0 AND 75 YEARS	
	MALES	FEMALES	TOTAL	%	#	%
All malignant neoplasms	222,862	194,178	417,040	24.2	30,592	31.8
Diseases of the heart	220,537	75,422	295,959	17.2	26,582	27.7
Motor vehicle accidents	112,275	45,667	167,942	9.8	3,999	4.2
All other accidents	97,012	26,679	123,691	7.2	3,656	3.8
Suicide	87,827	21,316	109,143	6.3	3,113	3.2
Congenital anomalies	47,794	43,108	90,902	5.3	1,369	1.4
Causes of perinatal mortality (excluding stillbirths)	51,930	38,505	90,435	5.3	1,214	1.3
Respiratory disease	37,462	22,583	60,045	3.5	5,115	5.3
Cerebrovascular	26,450	22,855	49,305	2.9	4,370	4.5
All other causes	198,096	117,892	315,988	18.3	16,098	16.8
Total	1,112,245	608,205	1,720,450	100.0	96,108	100.0

Source: *Canada Year Book*, 1988, p. 3-20, Table 3.6, Catalogue 11-402E/1987. Reproduced with permission of the Minister of Supply and Services of Canada, 1989.

such as over-consumption of sugar or fat, (5) lack of exercise or recreation, and over-work, (6) careless driving and failure to wear seat belts, and (7) promiscuity and sexual carelessness. Environmental factors include physical factors such as contaminated water, acid rain, and air pollution, and social-economic factors such as urbanization and working conditions, including inadequate health and safety measures on the job, which contribute to death, disability, and disease. Finally, host factors are the result of individual biological heritage or genetics. The following sections will examine self-imposed factors.

FIGURE 3.4 Causes of Death

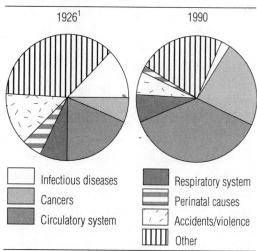

[1] Excludes Newfoundland, the Yukon, and the Northwest Territories in 1926.
Source: *Canada Year Book*, 1994, p. 132.

Box 3.6 Occupational Mortality among Bartenders and Waiters

Dimich-Ward *et al.* (1988) studied death registrations in British Columbia to determine the strength of the association between the exposure of waiters and bartenders to smoking and alcohol consumption, and their risk of death from lung cancer and cirrhosis of the liver. Records on 254,920 males and 165,912 females, representative of deaths in British Columbia between 1950 and 1978, were selected through the Division of Vital Statistics. Of these deaths, 1,280 men and 436 women had their occupation recorded as bartender or waiter.

It was found that alcohol- and tobacco-related causes of death were predominant among male bartenders and waiters. Measured in terms of proportional mortality ratios (PMR), their death rates from cancer of the mouth, esophagus, larynx, and lung, and from bronchitis and emphysema, cirrhosis of the liver, accidental poisoning due to drugs or alcohol, and homicides were higher than average. Elevated risks for female bartenders or waitresses included esophageal cancer, lung cancer, cirrhosis of the liver, and accidental death.

There are many potential explanations for these rates. A survey of smoking habits of U.S. workers found that bartenders and waiters are among the highest percentages of cigarette smokers. It is, however, not only active smoking that is a contributing factor. Pollutants in the air of bars include carbon monoxide, nicotine, particulates, and aromatic hydrocarbons.

If 400 cigarettes are smoked per hour within a poorly ventilated tavern, the benzo(a)pyrene content would be the equivalent of 36 cigarettes smoked in an eight-hour period. Cooking fumes also contribute to benzo(a)pyrene levels. Benzo(a)pyrene has carcinogenic properties and is therefore likely to be involved in the etiology of respiratory cancers. It is said that ease of access to alcoholic beverages contributes to the elevated rates of cirrhosis of the liver for those whose occupation is bartender or waiter. These findings are important in identifying ways to cut the risks involved in such occupations.

The Impact of Alcohol

Alcohol ingestion appears to affect health in paradoxical ways. On the one hand, excess consumption is known to be associated with morbidity and mortality through alcoholism, cirrhosis of the liver, malnutrition, accidents, obesity, suicide, and homicide. On the other hand, moderate drinking appears to have a beneficial impact on health (*Health and Social Support* 1985). According to Angus and Manga (1985), alcohol consumption was estimated to be the direct cause of 2,520 deaths and the indirect cause of 5,668 others in 1978, and alcohol played a part in 10,142 other deaths in Canada. Approximately 11 per cent of all deaths that occurred in 1978 were related to alcohol. While the data indicate that there has been a decline in the amount of alcohol consumed on average among Canadians, by the late 1970s approximately one in ten persons was said to be suffering from an alcohol-related limitation. About 8 per cent of all Canadians (over 15 years old) reported drinking 15 or more drinks per

FIGURE 3.5 Proportion of Current Drinkers Aged 15 and Over, by Weekly Volume, Canada, 1978, 1985, and 1991

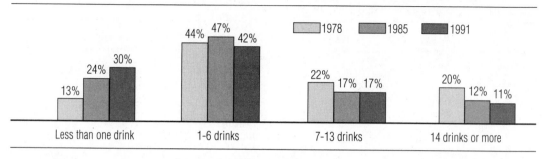

Note: Proportions have been recalculated to exclude the unknown category from the total.

Source: Wayne Millar, "A Trend to a Healthier Lifestyle," *Canadian Social Trends* (Spring, 1992), Cat. no. 11-008E, no. 24.

week (Canada's Health Promotion Survey, 1988). In 1982, 3,063 deaths were directly attributed to alcohol consumption; 78 per cent of these were the result of either liver disease or cirrhosis. In 1981, there were 397,000 males and 189,000 females suffering from alcohol dependence syndrome. The consumption of alcohol was involved in 60 per cent of all traffic fatalities investigated in 1982 (*Canada Year Book*, 1988).

By 1978 alcohol consumption was double the rate in 1950. Yet, as Figure 3.5 shows, alcohol use has been declining since 1978. In 1991, 55 per cent of Canadians over 15 were "current drinkers" (drank alcohol at least once per month). This is down from 65 per cent in 1978. In all age groups men are more likely than women to drink, and to drink more alcohol when they do drink. This gender gap increases as age increases; it is, however, almost non-existent for the youngest age group. The highest proportion of the heaviest drinkers were men who were single, separated, or divorced between 20 and 24 or over 65 years of age. Ten per

cent of the separated and/or divorced men had 28 or more drinks in the week preceding the survey (Canada's Health Promotion Survey). Alcohol consumption increases with income for both men and women. Drinking rates are highest for both men and women of 20-24 years of age and decline with advancing age. In 1991, 80 per cent of men and 58 per cent of women aged 20-24 were current drinkers. By age 65 and up, 49 per cent of men and 29 per cent of women were current drinkers. Although Canadians drank more moderately in 1991, the proportion of people who never drank declined (*Canadian Social Trends*, 1992). Table 3.7 shows that, with two exceptions, the health risks in a number of particular categories of disease are increased by heavy drinking; when heavy drinking is combined with smoking, the risks are even greater.

Impact of Cigarette Smoking

As Figure 3.6 shows, the decline in teenage smoking is the mirror image of the increase

TABLE 3.7 Age-Standardized Odds Ratios for Selected Risk Factor Comparisons by Selected Health Measures, Population 15 Years of Age and Over, Canada, 1985*

	REGULAR SMOKERS VERSUS NEVER SMOKED	HEAVY DRINKERS VERSUS MODERATE DRINKERS	CURRENT SMOKERS AND HEAVY DRINKERS VERSUS NEVER-SMOKED AND MODERATE DRINKERS	OBESE** VERSUS ACCEPT-ABLE WEIGHT	UNDER-WEIGHT** VERSUS ACCEPT-ABLE WEIGHT	SEDENTARY VERSUS MODERATE ACTIVE
HEALTH VARIABLES						
Fair or Poor Self-Rated Health Status	1.55	1.16	1.89	1.65	1.46	1.40
Ever Diagnosed with High Blood Pressure	.92	1.21	1.26	2.49	91	1.09
Ever Prescribed Treatment for High Blood Pressure	.81	1.13	.99	2.68	.83	1.07
Ever Had Trouble with Heart	1.09	.96	1.11	1.20	1.32	1.00
Ever Diagnosed with Diabetes	.56	1.42	.14	3.33	.44	1.12
Has Respiratory Problems	1.90	1.04	2.09	1.38	1.43	1.10
Has Arthritis/Rheumatism	1.04	1.03	.88	1.70	.96	.94
Had One or More Bed-Days in Past Two Weeks	1.26	1.17	1.81	1.30	1.37	1.12
Had One or More Nights in Hospital in Past Year	1.18	.74	.86	1.25	1.48	1.25
Have Long-Term Activity Limitation	1.25	1.16	1.51	1.63	1.24	1.22
Somewhat or Very Dissatisfied with Health	1.60	1.35	2.18	2.14	1.80	1.54
Somewhat or Very Dissatisfied with Life	1.81	1.24	1.70	1.25	1.84	1.23
Somewhat or Very Unhappy	1.38	1.17	1.33	1.28	1.97	1.58

*The odds ratios were constituted by dividing the percentage of the population in a risk group with that particular health condition by the percentage of the population not in the risk group who also had that particular health condition. A ratio of greater than 1.0 therefore indicates that the likelihood of having a condition is greater when the risk factor is present than when it is absent.

**Age 20 + for obese and underweight.

Source: *General Social Survey Analysis Series, Health and Social Support,* Catalogue 11-612E, no. 1 (Ottawa: Statistics Canada, 1985), p. 170. Reproduced with permission of the Minister of Supply and Services, Canada, 1989.

FIGURE 3.6 Real Cigarette Prices and Cigarette Smoking among Canadians, Ages 15 to 19

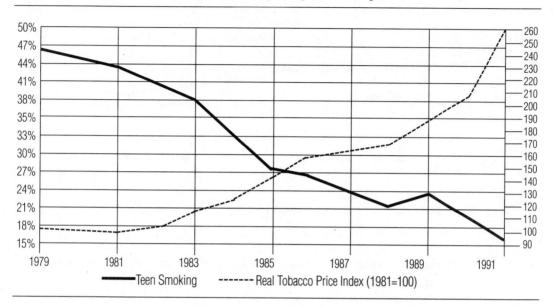

Sources: *Canadians and Smoking: An Update* (Ottawa: Health and Welfare Canada, 1991); Statistics Canada, *Labour Force Survey*, 1991.

in tax levels. Over the past 25 years the rate of cigarette smoking has continued to decline (see Figure 3.7). In 1991, about 26 per cent of the population (15 and over) smoked daily, as compared to 41 per cent in 1966. The decline was more dramatic for men (from 54 per cent in 1966 to 26 per cent in 1991) than for women (from 28 per cent to 26 per cent). There is a significant change in the smoking trends for men and women. In the age range of 15-19, 20 per cent of females as compared to 12 per cent of males smoked cigarettes. The rate of cigarette smoking for women in this age group has actually risen since 1966, while it has declined for men in this age group. The decline in smoking has been associated with public health campaigns, warnings on cigarette packages, banning of cigarette advertising, and increased tax

levels. In some jurisdictions there have been concerted efforts to charge store owners who sell cigarettes to underaged people. On the other hand, the increase in smoking for adolescent women is difficult to understand and explain.

The lowest rate of smoking is among those with a university degree (18.3 per cent). Among those with secondary and some secondary education, 35.5 per cent were smokers. Of those males with little education (0-8 years), 38.0 per cent were smokers. Further, the tendency to smoke low-tar cigarettes is positively related to years of education (*Health and Social Support 1985*). While these figures might be somewhat different in 1991 the class and education correlates of smoking remain. Current and former drinkers are slightly

FIGURE 3.7 Age-Adjusted Smoking Rates for Population Aged 15 and Over, By Sex, Canada, 1966-1991

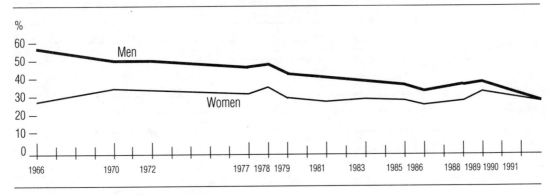

Note: Data are for regular smokers, and are adjusted for differences in age distribution across time.

Source: *Canadian Social Trends* (Spring, 1992), Cat. no. 11-008E, no. 24.

more likely than the population in general to be regular cigarette smokers. Further, those who have never drunk alcohol are twice as likely never to have smoked (*ibid.*).

Some interesting patterns emerge from recent data pertaining to the percentage of Canadian males and females who smoke. When the population is categorized according to level of income, of those who earn under $40,000 nearly 40 per cent of males and 35 per cent of females are smokers. However, of males who earn $40,000 or more, only about 28 per cent are smokers, and of women in the higher income category 25 per cent are smokers (Valpy, 1994).

Valpy reports that among blue-collar workers 40 per cent of males and 39 per cent of females are smokers. Among white-collar workers, 33 per cent of males and 35 per cent of females are smokers. In the professional/managerial category, 25 per cent of males and 28 per cent of females smoke. These data show little difference between men and women.

A significant decrease in smoking is noted as the level of education increases. For those who completed elementary school only, 38 per cent of males and 31 per cent of females were reported as smokers; for secondary school graduates, 30 per cent of males and 31 per cent of females smoke; among university graduates, only 19 per cent of males and 16 per cent of females smoke.

Smoking has harmful effects on health. As Table 3.8 shows, those who are regular smokers have a greater risk of many of the health problems considered. People who smoke are twice as likely to suffer from emphysema, for instance. There is a direct parallel between the rise of lung cancer deaths and tobacco consumption.

The majority of Canadians report that they have experienced unpleasant side effects from the smoking of others (Jossa, 1985). In 1985 more than 50 per cent of the Canadian population either smoked regularly themselves or were exposed to second-hand smoke in their households.

TABLE 3.8 Population 15 Years of Age and Over, by Regular Smokers and Non-Smokers, Prevalence of Selected Health Problems, Canada, 1991 (nos. in thousands)

	REGULAR SMOKERS (total: 5,434)		NEVER SMOKED (total: 9,422)	
Hypertension	775	14%	1,417	15%
Heart Trouble	309	6%	532	6%
Diabetes	182	3%	272	3%
Arthritis/Rheumatism	1,067	20%	1,677	18%
Asthma	315	6%	584	6%
Emphysema	626	12%	530	6%
Stomach Ulcer/ Other Digestive Problems	759	14%	997	10%
Migraines	614	11%	815	9%
High Blood Cholesterol	426	8%	703	7%

Source: Statistics Canada, *Health Status of Canadians*, Cat. 11-612E, no. 8 (Ottawa, 1994), Table 9-6, p. 154

These figures are down somewhat, but a significant proportion of the population still suffers the effects of second-hand smoke. Evidence of deleterious effects of second-hand smoke has been growing (*ibid.*). The United States Surgeon General has concluded from available evidence that parental smoking can cause respiratory problems in children and that the chance of contracting lung cancer is heightened by the experience of being subjected to second-hand smoke.

Smokers are known to miss more days of work, and to suffer more disability days, than non-smokers. Current smokers of age 15 and up experienced 13.2 days of disability (male) and 22.1 days of disability (female) per year (Collishaw, 1982: 62). Smoking contributes to certain oral and respiratory cancers, heart and blood vessel diseases, chronic bronchitis, and emphysema (see Health and Welfare Canada and Statistics Canada, 1981: 48). Pipe and cigar smokers run increased risks of lip and throat cancers. In addition, there is considerable evidence that smoking and drinking (which are likely to be found together) along with other health risks (such as those risks associated with working with asbestos) multiply the negative impact on health of each one of these factors.

Physical Activity

Canadians of all ages have increased their levels of physical activity recently. In 1991, 32 per cent of Canadians over 15 years of age were "physically very active" as compared to 27 per cent in 1985. Physical activity rates are highest among the younger age groups of Canadians. In 1991, 55 per cent of Canadians between 15-24 and 36 per cent between 24-44 were physically active; 21 per cent of men and 12 per cent of women over 65 were physically active. Men are

Box 3.7 NAFTA Supports Tobacco Companies' Opposition to Plain Packaging

In order to diminish the appeal of cigarettes the Canadian government has introduced tax increases and, among other things, banned advertising. A recent strategy before a House of Commons committee was to require that cigarettes all be packaged in plain wrappings. Two multinational giants – Reynolds and Philip Morris – reacted forcefully against this proposal. They responded by sending Julius Katz, deputy U.S. chief negotiator of NAFTA, to the Commons committee. He argued that such a move would violate the free trade agreement. Katz's appearance was followed by a letter from the president of Philip Morris warning that plain packaging would be seen as a threat to all of the company's business interests in Canada, including Kraft General Foods, which has eleven plants and 4,700 employees, and its 20 per cent ownership of Molson.

more likely to be physically active than women, although the rates of activity are increasing for both men and women (Millar, 1992).

The Impact of Weight

In spite of the increased levels of physical activity, the number of overweight Canadians has increased: 23 per cent of those aged 20-64 were overweight in 1991, 17 per cent in 1985. Overweight is more common among older Canadians. Men are more likely to be overweight than women, but women are more likely to be underweight. Despite women's tendency to be underweight (especially when young) more than one-third of Canadian women with normal weights believed that they weighed too much (*Canada Year Book*, 1994).

Overweight and obesity, too, have deleterious effects on health. As Table 3.9 shows, men and women who are obese are about three times more likely to suffer from hypertension than those of average weight. What this table does not show, but what was evident in a government-sponsored survey of a representative sample of the Canadian population – the General Social Survey (1987) – is that those who are obese, as compared to those whose weight is "acceptable," are three times as likely to have been diagnosed with high blood pressure and three times as likely to have been diagnosed with diabetes.

AIDS

A discussion of the health of Canadians would not be complete without a brief description of the newest infectious and contagious disease – AIDS. Discovered in the late 1970s in the United States, its presence was not officially noted until 1981. The first case recorded in Canada was in 1982 (Weston and Jeffrey, 1994). It is estimated that by 1991 some 50,000 Canadians were infected with the HIV virus, the precursor of AIDS, and by 1993 approximately 7,000 people would have progressed to AIDS (*ibid.*: 722). While the earliest reported cases were among homosexual men, the disease

TABLE 3.9 Population 20 Years of Age and Over, by Body Mass Index, Sex, and Prevalence of Selected Health Problems, Canada, 1985

ALL AGE GROUPS

	ACCEPTABLE		OVERWEIGHT		OBESE	
MALE	4,406		3,052		557	
Hypertension	542	12%	621	20%	204	37%
Heart Trouble	293	7%	260	9%	56	10%
Diabetes	74	2%	84	3%	34	6%
Respiratory Problems	405	9%	344	11%	81	15%
Arthritis/Rheumatism	684	16%	689	23%	153	28%
FEMALE	4,926		2,424		852	
Hypertension	642	13%	578	24%	347	41%
Heart Trouble	267	5%	234	10%	93	11%
Diabetes	91	2%	76	3%	75	9%
Respiratory Problems	507	10%	310	13%	140	16%
Arthritis/Rheumatism	1,081	22%	822	34%	449	53%

Source: Statistics Canada, *General Social Survey Analysis Series, Health and Social Support*, 1985, Table 54, p. 161. Catalogue #11-612E. No 1. Reproduced with permission of the Minister of Supply and Services, Canada, 1989.

has spread beyond this group of people. According to the Canada Communicable Disease Report (1992), the risk factors in Canada today appear to encompass all of the following categories (Weston and Jeffrey, 1994): homosexual/bisexual activity only (78 per cent), injection drug use only (2 per cent), both of the above (4 per cent), recipient of blood or blood products (5 per cent), heterosexual activity (origin in country defined by the World Health Organization (WHO) as having a high rate of HIV infection) (4 per cent), sexual contact with a person at risk (4 per cent), and no identified risk factor (4 per cent). In 1992 the WHO estimated 1.7 million cases of AIDS world-wide with 69 per cent in Africa, 16 per cent in the U.S., 9 per cent in the Americas (excluding the U.S.), 6 per cent in Europe, and 1 per cent in Asia and Oceania. By the year 2000,

WHO estimates 30 million people around the world could be infected with AIDS (90 per cent of carriers would be in the Third World and 50 per cent of those carriers would be female) (Colombo, 1993). Because it is inevitably fatal and because of its pattern of contagion and spread, AIDS has been called an epidemic. There is no known cure as of yet. Presently, the disease stands as a major threat to the health of the population of the world – especially, and more immediately, the health of Africans.

Summary

(1) Over the past 150 years there have been dramatic changes in life expectancy rates in Canada and other developed nations. Many explanations have been suggested

for this, such as improvements in public health, nutrition, hygiene, and birth control. Immunization is also an important component in the health of a population.

(2) The chief causes of death in Canada today are heart diseases, cancer, and accidents. These can be referred to as diseases of civilization and affluence as they are typical of developed nations. Causes of death are related to lifestyle, environmental, and work-related factors. Self-imposed contributions to mortality include: smoking, drug abuse, excessive alcohol consumption, nutritional inadequacies, lack of exercise or recreation, overwork, reckless driving, and sexual carelessness. The diseases that result from smoking and alcohol consumption and the social characteristics of their users are discussed.

(3) Physical activity and obesity levels are related to the health of Canadians.

(4) AIDS is an important new disease with both spreading and devastating effects in populations.

CHAPTER FOUR

Environmental and Occupational Health and Illness

The Major Environmental Issues

Have you heard of the twentieth-century disease? Do you know anyone diagnosed with allergies or asthma? Is housework safer than factory work? Are occupational hazards greater for men or for women? Is environmental degradation a threat to health? What are the specific relationships between environments, occupations, and health? What is the significance, for health, of Chernobyl, Bhopal, Love Canal, the PCBs in the Great Lakes, the depletion of the Newfoundland fisheries, or the Westray mine disaster? The purpose of this chapter is to provide an overview of some of the major environmental and occupational health hazards, their effects on health, and the available sociological explanations.

There are three fundamental components of the environment: air, land, and water. They effect our health both directly (e.g., through the air we breathe and water we drink) and indirectly (e.g., through the food we eat). Environmental hazards in air, land, and water have increased tremendously in the twentieth century. It has been estimated that 60 to 90 per cent of all cancers are environmentally caused. Moreover, environmental hazards may affect our health in ways that are unseen in the short term and only evident in the long term in the health of our future generations. Many other diseases of major organ systems such as the lungs, heart, liver, kidneys, as well as reproductive problems, birth defects, and behavioural disorders seem to be associated with environmental factors. There are between 50,000 and 70,000 chemical substances in commercial use in farming, manufacturing, and forestry industries. Every year about 1,000 new chemicals are introduced in North America, the majority of which have not been tested for potential ill effects. Radioactive waste with a half-life of 250 centuries, uranium mine tailings, and low-level radiation leakage from routinely functioning nuclear power plants and weapons facilities are taken for granted as an inevitable part of the environment by most Canadians. Nuclear accidents are almost "normal" events. Yet they have the ongoing potential for massive death and destruction.

Environmental risks are now ubiquitous and increasing. The whole world is a united ecosystem. Changes in one nation-state's

Box 4.1 Exxon Valdez Oil Spills

On March 24, 1989, the oil supertanker *Exxon Valdez* ran into a reef off northern Alaska and spilled 260,000 barrels, or 11 million gallons, of crude oil. The oil contaminated thousands of square kilometres of water and hundreds of kilometres of shoreline. It has been estimated that 30,000 birds, including sea ducks, loons, cormorants, and bald eagles, and between 3,500 and 5,000 sea otters were killed. The most massive oil spill in history was intentional, when 250 million gallons of crude oil were deliberately dumped into the Persian Gulf during the Persian Gulf War in 1991. This spill set oil wells on fire and covered large parts of the desert and shoreline. These disasters are a critical problem. However, the day-to-day dumping of oil by home mechanics is a much more significant problem. It has been estimated that in the U.S. alone, the same amount of oil that was spilled in the *Exxon Valdez* disaster is spilled by home and auto mechanics every two and a half weeks; 240 million gallons of oil are improperly discarded by American drivers every year, yet it takes only one quart of used motor oil to contaminate 250,000 gallons of drinking water (Raven *et al.*, 1993: 178–205).

environmental policies and procedures, in the amounts of air, water, and land pollution, for example, have the capacity to affect aspects of the ecology of the rest of the world. Even the snows of the remote, virtually uninhabited Antarctica contain residues of PCBs, DDT, and lead, which have emanated most directly from the industries in North America and the former Soviet Union. Water, air, and soil have all been infiltrated with various types and degrees of toxic chemicals. As other health threats, environmental hazards are unequally distributed. Poorer people are less likely to be able to move away from a toxic waste dump, to drink bottled water, to buy organically grown foodstuffs, and so on. A recent U.S. study noted that visible minorities (particularly Aboriginal, black, and Hispanic peoples) are more likely to live near uncontrolled waste sites (Lee, 1987).

The poorer, Third World countries, too, are unequally subject to the destructive effects of environmental degradation when they, for instance, cut down precious rain forests to provide timber for furniture, housing, or other purposes for the developed world or to create pasture for cattle. Moreover, cash-strapped economies of the developing world, lacking alternatives, are more likely to allow the dumping of wastes within their borders in return for cash payments. Thus, there are ways that the environment of the Third World, in spite of a relative lack of industrialization, is more vulnerable than that of the developed world.

A number of environmental issues threaten the everyday health and safety of all of the people on the planet. The most critical environmental issues arguably include: (1) carbon dioxide and climate change (the greenhouse effect); (2) acid precipitation; (3) the depletion of the ozone layer; (4) chemicals; and (5) the disposal of hazardous wastes (*Economic and Ecological Interdependence*, 1982).

Box 4.2 Bhopal

The discussion of one example of the affects of the political economy on the health of Third World peoples brings to light a number of issues. In 1984, in one of the worst industrial disasters ever, an explosion in a Union Carbide pesticide factory in Bhopal, India, spewed more than forty tons of lethal methyl isocyanate gas into the slums immediately surrounding the plant. According to recent government figures, more than 6,600 people were killed and well over 70,000 were blinded, disabled, or otherwise injured. Death and injury claims filed by citizens are much higher, exceeding 16,000 and 600,000 respectively (Karliner, 1994). Although specific figures are not available, an estimated two people still die every day as a direct result of this exposure more than a decade ago (*Scientific American*, 1995).

The accident was absolutely avoidable; that it happened is the result of a number of important factors common in industries imported from the West to the Third World. First, the Union Carbide plant in Bhopal was relatively unprofitable as compared to other divisions of Union Carbide elsewhere in the world. At the time of the accident, the Bhopal plant was for sale. It lacked top-level interest or support. A number of parts of the plant had been closed down. Personnel had been let go and not replaced. Thus, it was operating with only a partial complement of workers and with equipment that was ill-repaired. It was also, however, involved in the manufacture of dangerous chemicals such as methyl isocyanate. In spite of the objections of the municipal authorities, the central and state governments allowed it to continue operating without adequate safety precautions and regulations. There were no adequate plans for dealing with a major accident and the company personnel did not really understand the potentially lethal effects of the chemical they were producing (methyl isocyanate gas). The external regulation was extremely weak. In the interest of fostering the importation and investment of capital to their countries, governments in the Third World frequently ignore or are unaware of even minimal health and safety standards. The disaster at Bhopal is just one example of the potential for widespread industrial-based devastation in the Third World.

Carbon dioxide is produced by the burning of fossil fuels used to provide heat for residential purposes. Fossil fuels are one of the causes of increasing amounts of carbon dioxide in our atmosphere. Carbon dioxide reflects additional quantities of the sun's radiant energy back to earth, causing a warming trend that will affect, among other things, the growing of crops and the probability of flooding from melting of the glaciers. The depletion of the ozone layer, which is already happening rapidly as the result of the use of aerosol spray containers, among other things, has increased this warming trend and has led to and will continue to cause an increased incidence of skin cancer and extensive damage to animal and plant life.

The primary agents in the production of acid precipitation are sulfur and nitrogen. They are released as gases from ore smelters, coal-fired generating stations, automobile exhausts, and ore and gas refineries. Sulfur and nitrogen combine with water vapour in the air to form acidic solutions – primarily sulfur dioxide and several nitrous oxides.

Box 4.3 Environmental Illness

An increasing number of North Americans are suffering from what they claim to be the results of low-level exposures to synthetic chemicals (Ashford and Miller, 1991) and what has been called by a variety of names, including "multiple chemical sensitivity," "chemically induced hypersusceptibility," "immune system dysregulation" ("environmental illness" or EI), and "twentieth-century disease." These are similar to other new diseases such as AIDS (see Sontag, 1989: 104), in which the sufferer is subject to a whole range and variety of signs and symptoms, of varying severity, some of which vary considerably from day to day. Among the various symptoms often reported are the following: headaches, rashes, depression, shortness of breath, muscle and joint inflammation, fatigue, nausea, and other gastro-intestinal and nervous system disorders. Those with EI are often sensitive to a wide range of synthetic environmental contaminants, and equally troubling, perhaps, because of its delegitimizing character, is the fact that symptoms often cannot be detected by standard methods. One consequence of the lack of reliable diagnostic markers is that most allopathically trained medical doctors do not recognize this as a real disease. In 1965 a small group of physicians, scientists, and health professionals founded the Society for Clinical Ecology to address the issues being brought forward by sufferers. Renamed the American Academy of Environmental Medicine in 1984, it now has about 600 physician members (Oberg, 1990: 6).

In 1987 the Environmental Research Foundation and the National Academy of Scientists in the United States suggested that between 15 and 20 per cent of the American population may have allergic sensitivity to chemicals in the environment. The Environmental Protection Agency has also stated that health problems result, for some people, at levels of exposure that are considered below regulatory concern. People with environmental sensitivities have to be careful about where they live, go to school, and play, and what they eat, drink, and smell. Their whole lives may be affected by their sensitivities. Some have even established a protected community, in Wimberly, Texas, an area with little industry or farming (Belkin, 1990).

These acids then fall as precipitation and enter the soil and surface water, ultimately affecting plant and animal life. There is growing evidence that acid precipitation can lead to gastroenteritis and respiratory damage (Weller, 1986). Air pollution may result from the more than 70,000 chemicals on the commercial market, many of which are released into the environment; little or nothing is known of their potential long-range effects. Finally, the problems of disposing of hazardous wastes particularly threaten those less economically able to resist "dumping" in their own backyards.

Pollutants pose significant threats to health. Polluted water and air are known to affect rates of dysentery, typhoid, various bacterial and infectious diseases, lung disease, various respiratory problems, and cardiovascular diseases. Pesticides are known to kill or induce disease in fish and wildlife. They may also have an adverse effect on people when they enter the food chain. Solid wastes (such as plastics, rubber, glass, and metals) that do not break down except over a very long period of time

Box 4.4 Environmental Illness: An Anomaly for Allopathic Medicine

Characteristics of Allopathic Medicine	*Nature of Environmental Illness*
Doctrine of Specific Etiology (specific disease due to a specific cause)	• non-acute, chronic sensitivity to low-level exposure • symptoms can involve any bodily system or several systems at once • symptoms may change over time in location, severity, number • "total body load"
Doctrine of Diagnostic Technology (the definition and diagnosis of a particular disease depend on the availability and accessibility of measurement tools and techniques)	• patients' experience of symptoms not measurable • levels of exposure for risk variable and low (often too low to be measurable)
Allopathy and Heroic Cures (allopathic medicine's strength lies in its capacity to cure acute problems, often dramatically)	• sufferers are likely to subscribe to treatment that emphasizes avoidance rather than consumption • treatments vary and may include most or all features of lifestyle (including diet, exercise, and drink) and explicitly exclude pharmaceuticals
The Sick Role	• EI politicizes illness by pointing to the contradictions of contemporary society: chemically contaminated environments, the workplace, home, and the atmosphere

That these are basic assumptions of the biomedical professions has been explained by Gordon, 1988; Kirkmayer, 1988; Freund and McGuire, 1991.

pollute the environment and take up space that could be used by living things. Noise pollution is known to be responsible for loss of hearing, accidents, stress, cardiovascular disease, and disturbed sleep. These five critical environmental issues threaten the quality of the air, the water, and the land. Some specific contemporary concerns with respect to air, water, and land will be examined in turn.

Indoor Air Pollution and Human Health

Both indoor and outdoor air contribute at least low levels of pollutants, including ozone, sulfur oxides, nitrogen oxides, carbon monoxide, and other particulates that may irritate eyes and inflame the respiratory tract. There is evidence that long-term exposure may have a negative effect on the immune system and be implicated in

Box 4.5 Last Gasp

Asthma is often seen as an elusive disease. Although it tends to run in families, it also develops in those who have no history of asthma among their kin. This disease may strike at any age, but children, with their small breathing tubes, are particularly vulnerable.

Approximately 1.2 million Canadians suffer from asthma, the sometimes lethal inflammation of the airways of the lungs. In asthma sufferers, the bronchial tubes are extremely sensitive to a variety of triggers unique to each patient. These triggers include those in modern office buildings, such as the more than 900 chemical and biological agents in the air, including chromium dust, acrylates, and epoxy resins. In factories and shops there are more unseen dangers, such as fluorocarbon propellants breathed in by beauticians, sulfur dioxide fumes inhaled by brewery workers, and chlorine gas encountered by petrochemical workers.

Other triggers lie in wait at home. Central heating and wall-to-wall carpeting have become the breeding grounds for dust mites, microscopic animals that churn out a potent allergen, Der p 1, in their dung. The vapours of common household products like cleaning solvents and paint thinners and the fumes from such personal products as spray deodorants and scented cosmetics can also set off an attack. Some individuals display fewer symptoms while others face life-threatening attacks. Some physicians confuse this disease with respiratory infections, especially in children.

For all of its elusiveness, asthma can be a deadly disease. In 1985, asthma was the seventh leading cause of hospitalization among men and the thirteenth leading cause among women, and in 1987-88 hospitalization for asthma represented an estimated $120 million worth of Canadian hospital bills. It is also the most common medical emergency in children. The rate of fatal asthma attacks, where patients are unable to pump oxygen through their airways and therefore suffocate to death, has escalated. For patients between the ages of 15 and 34, the asthma mortality rate has risen 50 per cent over the last ten years. Today, ten Canadians each week die of asthma.

Asthma has been documented as being the only common chronic disease in the Western world with an escalating death rate. Why is this the case? In attempting to find the answers, researchers are uncovering clues that may end up benefiting patients. Researchers are uncovering new strategies to control the disease. Classic experiments on air pollutants are providing important insights into asthma triggers, and there is beginning to be growing appreciation of the subtle connection between the body and the environment. Some feel that the increasing rate of mortality to asthma could be a result of the increasing amounts of pollution entering the indoor and outdoor air. Since the 1950s, more than 60,000 new chemicals have been introduced into the commercial market. This is another source of struggle for the asthmatic.

Beta agonists, an adrenaline-like drug, are often prescribed to asthmatics as a form of treatment. Beta agonists, inhaled as a fine spray, work to dilate airway passages by relaxing the bronchial muscles. The common dosage is eight puffs a day. Concern has been raised regarding the prolonged use of these beta agonists. Some believe they become addictive, that one has to take the drug regularly or be faced with an attack. Others believe that the common dosage of eight puffs a day somehow aggravates asthma. Some studies conclude that the chance of death rises with greater use of beta agonists.

It is still uncertain as to how these popular drugs induce fatal attacks. Studies now contend that beta agonists oversensitize the body's response to foreign substances entering the lungs. It is believed that

Box 4.5 continued

whenever a pollutant or allergen enters the lungs, the body's defence system reacts by releasing chemicals that inflame the airways, resulting in scar tissue build-up. Using beta agonists interferes with the body's ability to regulate its chemicals and the resulting scar tissue quickly obstructs the flow of mucus and eventually cuts off any chance of breathing properly. Most physicians recommend avoiding asthmatic triggers as much as possible and reducing beta agonist use gradually.

"Last Gasp," *Equinox* (May/June, 1992), pp. 85-98.

long-term respiratory problems such as emphysema and chronic bronchitis, cancer, asthma, cardiovascular disease, chronic obstructive pulmonary disease, and various respiratory infections (Raven *et al.*, 1993: 435). An extreme case of indoor pollution is the "sick building syndrome" in which the presence of air pollution inside tightly sealed office buildings can lead to a whole variety of illnesses. As well, total environmental sensitivity, in which a person is allergic to myriad components in the modern environment, has forced some people to live in a totally sterile environment.

The most seriously harmful indoor pollutant may be radon, a tasteless, odourless gas that forms naturally during the radioactive decay of uranium in the earth's crust. Radon seeps through the earth and through basements. It greatly increases the deleterious effects of smoking on the lungs. Another indoor pollutant with serious health costs is asbestos. Often used as insulation because it does not conduct heat or electricity, asbestos also can break down into almost invisible fibres that can be inhaled. When inhaled it irritates the lungs and is known to be related to lung cancer and mesothelioma, a rare and almost always fatal cancer of the body's internal linings. In the last several decades various levels of government have acknowledged the dangers of asbestos, which have been known since the 1920s, and have introduced laws requiring the removal of asbestos insulation from public buildings such as schools, offices, and homes and eliminating its use in new buildings. Unfortunately, some research has shown that removing asbestos can release fibres that would otherwise remain stable and thus it is sometimes safer to seal asbestos in place (*ibid.*).

Second-hand Smoke

Second-hand smoke is both an environmental and occupational health issue. It is an environmental issue because smoking may affect others in homes, on the streets, and in public buildings, public transportation, restaurants, stores, and so on. It is a workplace issue because workers may be involuntarily exposed to the second-hand smoke of their colleagues. Not only is smoking a direct cause of lung cancer but so is breathing in the second-hand smoke of others. Researchers at Canada's Laboratory Centre for Disease Control have estimated that as

TABLE 4.1 Principal Air Pollutants, Their Sources, and Their Respiratory Effects

POLLUTANT	SOURCES	HEALTH EFFECTS
Sulfur oxides, particulates	Coal and oil power plants Oil refineries, smelters Kerosene stoves	Bronchoconstriction Chronic bronchitis Chronic obstructive lung disease
Carbon monoxide	Motor vehicle emissions	Asphyxia leading to heart and nervous system damage, death
Oxides of nitrogen (NO_x)	Automobile emissions Fossil fuel power plants Oil refineries	Airway injury Pulmonary edema Impaired lung defences
Ozone (O_3)	Automobile emissions Ozone generators Aircraft cabins	Same as NO_x
Polycyclic aromatic hydrocarbons	Diesel exhaust Cigarette smoke Stove smoke	Lung cancer
Radon	Natural	Lung cancer
Asbestos	Asbestos mines and mills Insulation Building materials	Mesothelioma Lung cancer Asbestosis
Arsenic	Copper smelters Cigarette smoke	Lung cancer
Allergens	Pollen Animal dander House dust	Asthma, rhinitis

Source: from H.A. Boushey and D. Sheppard, "Air pollution," in J.F. Murray and J.A. Nadel, eds., *Textbook of Respiratory Medicine* (Toronto: Saunders, 1988).

many as 330 non-smoking Canadians may die yearly from lung cancer due to regular exposure to second-hand smoke (Canadian Cancer Society, 1994). There are two sources of second-hand smoke, sidestream smoke (given off by the burning tip of a cigarette, pipe, or cigar) and exhaled smoke (puffed out by the smoker). Among the toxic chemicals released into the air in these ways are nicotine, tar, carbon monoxide, formaldehyde, hydrogen cyanide, ammonia, and nitrogen oxide. Among the health effects

Box 4.6 Sustainability

The sustainability of the planet is threatened. Environmental protection is a fundamental part of world development and the world community is at a crossroads. Between 1990 and 2030 the world population is expected to grow by 3.7 billion people. Food production will need to double. Industrial output and energy use are expected to triple in a world-wide context and increase fivefold in developing countries. Meanwhile, already one-third of the world's population has inadequate sanitation, one billion people lack safe water, and 1.3 billion people suffer from unsafe levels of soot and smoke. Between 300-700 million women and children suffer from severe indoor air pollution and cooking fires. If development and the planet are to continue, there is an urgent need for policies to sustain development. The World Development Report advocated a number of important policies. Among them are the following:

- removing subsidies that encourage excessive use of fossil fuels, irrigation, pesticides, and excessive logging;
- clarifying the rights of ownership of land, forests, and fisheries;
- increasing the development and provision of sanitation and clean water;
- education, family planning, and agricultural extension, credit, and research;
- empowering, educating, and involving farmers, local communities, indigenous people, and women. (World Development Report, 1992)

are lung cancer, bronchitis, emphysema, asthma, hay fever, cystic fibrosis, headaches, coughs, throat irritation, heart and circulatory diseases, pregnancy complications, and low birth-weight babies. Table 4.1 lists the major air pollutants, their sources, and their health consequences.

Greenhouse Effect

The greenhouse effect, which has received considerable media attention in recent years, can be thought of as a rise in Earth's average temperature due to increasing concentrations of carbon dioxide (and other similar gases) in the environment. This global warming has both direct and indirect effects on human health. The indirect effects operate through the many changes in the physical environment such as drought on the Prairies, decline of water supplies in southern Canada, soil degradation, erosion, and flooding of coastal regions (*Canada's Green Plan*, 1994: 99). These changes result in changes in quality and availability of primary requisites for human life – food and water. Temperature increases may also directly cause certain health problems, especially affecting the cardiovascular, cerebrovascular, and respiratory systems. For instance, one U.S. study noted increased rates of death and stroke at about 25° C (Chivian *et al.*, 1993). As the concentration of carbon dioxide increases the number of deaths can be expected to increase. For example, during the summer heat waves in Los Angeles, when temperatures averaged about 41° C,

FIGURE 4.1 Relationship of Temperature to Heart Disease and Stroke Mortality

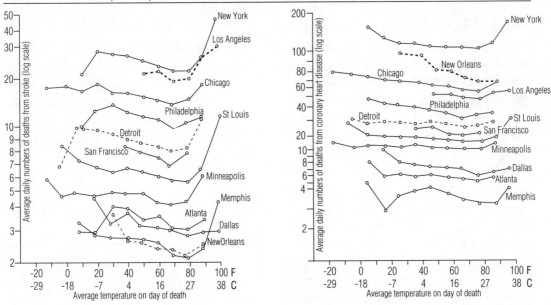

Source: A. Haines, "The implications for health," in S. Leggett, ed., *Global Warming: The Greenpeace Report* (New York: Oxford University Press, 1990); reproduced with permission of Oxford University Press and *American Journal of Epidemiology*.

the peak mortality was between 172 per cent and 445 per cent higher than would be expected (at all ages). Among people over 85 the peak mortality was affected even more: it ranged from 257 per cent to 810 per cent of expected mortality levels (Chivian *et al.*, 1993).

Prolonged high temperatures in urban areas may cause heat exhaustion and even death. Figure 4.1 illustrates the striking correlation between temperature and stroke in twelve American cities.

Medical Pollution

The burning of medical wastes is another serious, yet relatively unknown source of air pollution, particularly in urban areas. On average, each hospitalized patient contributes 13 pounds of waste on a daily basis. The wastes emitted include soiled bandages and bedding, replaceable syringes and other surgical/medical tools, contaminated plastics, and pathogenic remains, such as blood and body parts. The problem is compounded by the fact that most medical incinerators do not meet adequate standards of waste disposal. Waste is burned incompletely, thereby emitting acidic gases, heavy metals, toxic organics, and dioxins that can be from 10 to 100 times higher than waste from municipal incinerators (Rabe, 1992).

The Great Lakes

The Great Lakes comprise one-fifth of the world's fresh surface water. Their destruction is both a Canadian and an international disaster. The Great Lakes are both a source of drinking water for about 40 million people on both sides of the Canada-U.S. border and a garbage dump for industrial and domestic waste. Already, over 1,000 chemical and metal pollutants have been observed in the Great Lakes (Harding, 1994: 653). In the Golden Triangle between Oshawa and Hamilton in southern Ontario there are more than 50 sources of industrial pollution and more than 30 sources of municipal sewage. Moreover, for many years, untreated industrial and human wastes have been dumped directly into the lakes from both Canada and the United States (*ibid.*).

The overuse of water is another environmental threat. Canada already has one of the highest per capita consumption rates of any country. Ninety per cent of water is used for industrial and agricultural consumption. The Canada-U.S. Free Trade Agreement and the pressure from the U.S. to divert some of Canada's water to the States potentially comprise a great threat to Canadian health via the availability of clean water for Canadian consumption.

More recently, overfishing and bottom dredging by foreign and Canadian ships have contributed to the exhaustion of the cod stocks off Newfoundland. In response, the Canadian government has declared a moratorium on cod-fishing. This is a disaster for the province as cod have long been the staple bedrock to the variable economy of the island.

Waste Disposal

One of the most contested of contemporary issues is what to do with solid waste, whether it be domestic, manufacturing, hospital, radioactive, or from any other source. Debates about waste disposal have even spawned a new acronym, NIMBY (not in my backyard). Without doubt, however, the most serious environmental problem concerns the effect on humankind of the thousands of by-products and wastes of our industrial society, ranging from slightly annoying products such as mild caustics to deadly toxins and chemicals.

Increasingly, literature on hazardous waste disposal in Canada and the United States points to a growing problem with seemingly fewer and fewer solutions (Rabe, 1992). Many facilities, unable to meet tightening regulatory standards, have closed. Others have been planned but have been prevented from opening because of local opposition. Deciding the location for the hundreds of millions of metric tons of hazardous wastes is one of the most important political and policy issues of the day. Alberta, Manitoba, and Quebec appear to be among the most effective jurisdictions involved in providing sites in North America. Success in these jurisdictions has taken place via a procedure that has rejected top-down planning in favour of extensive public participation, creative types of community compensation, and

Box 4.7 Why Is It Difficult To Demonstrate Effects of the Environment on Health?

For a number of reasons, it is difficult to assess the effect of the environment on health. The two main reasons can be categorized as methodological and ideological.

(1) There is a time lag between exposure to carcinogen and health/illness consequence.

(2) It is difficult to measure the environmental impact for the following reasons:

(a) the environment is complex: differentiating among different parts of the environment and their independent effects is practically impossible;

(b) since new chemicals are released into the environment daily, noting and then measuring the particular amount of the chemical in the changing environment is exceedingly difficult;

(c) the ratio of the potential contaminant to the environment is usually so extremely small that instruments capable of measuring such minuscule amounts are not readily available;

(d) double-blind studies with human subjects are considered unethical, yet the amount of the potential contaminant necessary to account for the time lag and the low weight of the typical laboratory mammal (rat) is enormous;

(e) synergistic relationships are inevitable in the environment, yet because all of the elements in the synergistic relationship are not known the effects cannot be duplicated;

(f) there is a variable latency period between exposure and the onset of disease; both this variability in latency and its length make identifying causal connections between toxic substance and illness difficult;

(g) few physicians are trained in environmental and occupational health;

(h) the structure of medical examination (i.e., doctors working for companies) means that there is a tendency for doctors and policies to favour a diagnosis to causal connection;

(i) a definite cause-effect link demonstration is required as proof.

In consequence, only a small number of environmentally based illnesses have yet been noted and a fraction of occupationally caused illnesses compensated (one out of 17 occupationally induced cancers are estimated to be compensated by Workers' Compensation in Ontario) (Makdessian, 1987). The ideological/financial issues include the following:

(a) frequently, sponsors of research are pharmaceutical and medical device companies with a vested interest in research that involves their products;

(b) when research is funded by interest groups, independent research is difficult at best;

(c) medical journals, the major legitimate purveyors of new scientific findings, are also often funded by major pharmaceutical companies;

(d) certain types of basic research have dominated to date and because of the peer review system (a good old boys' network, some would say) basic research has a greater likelihood of support than other types of research.

Box 4.8 Critical Epidemiology

Critical epidemiology (Jones and Moon, 1987) is a particular type of epidemiological investigation that examines the ways in which social and economic organization is implicated in disease causation. The following two cases illustrate potential lethal effects of industrial production and agricultural productivity.

A 1988 study (Shannon *et al.*) investigated the link between air pollution and lung cancer rates. When the researchers controlled for age, smoking, and occupational histories the elevation in mortality for men decreased from 100 per cent to 15 per cent in a comparison between the least and most polluted areas. There was no relationship between air pollution and lung cancer for women. Godon *et al.* (1989) found a statistically significant relationship among source of drinking water (well as compared to river), pesticide use, and leukemia and brain cancer deaths among men in an area of rural Quebec.

There has been some research on communities that are or consider themselves to be contaminated. Such research has looked at risk perception, political and social organizing, and the health effects of living in the vicinity of garbage dumps or solid waste facilities (Edelstein, 1988; Elliott *et al.*, 1993). The dramatic effects of the Hagersville (southwestern Ontario) tire fire have been extensively documented in the media and in scholarly reports (Eyles *et al.*, 1990; Baxter *et al.*, 1992). Among the health, economic, and personal concerns were the following: stress, uncertainty and fear among community members, conflict in families and divisions in neighbourhoods, concerns about property values, livelihoods, and marketing of produce from the rural community's farmland, ongoing doubts about water quality, concerns about the trustworthiness, usefulness, and responsibilities of the government.

---

solid partnerships among public and private organizations and local and provincial governments (*ibid.*).

As the deleterious effects of hazardous waste disposal become more widely known, "dumping wastes" becomes a complicated legal and political issue. Chapter Fifteen discusses drug dumping in the Third World. Hazardous waste dumping in the Third World is a similar problem. When some industries in the richer developed world have needed to get rid of hazardous wastes they have shipped them to countries in the developing world where people may have agreed to receive – or been bribed into taking – hazardous wastes without adequate information or a safe destination. At

a UN-sponsored conference in 1989, 105 nations tentatively agreed to some controls on international shipments of wastes such as those from hospitals and pharmaceutical companies, PCBs, mercury, lead, and other chemicals that are known to be harmful.

Biodiversity

All of these threats to the air, water, and land have another profound implication for the future of life on the planet – the decline in biodiversity. For instance, although the rain forests comprise only 7 per cent of the earth's surface, they are home to almost half of the living species of the planet.

TABLE 4.2 Number of Accepted Time-Loss Injuries, by Sex, 1987,1991-1993

SEX	YEAR	INJURIES
Male	1987	473,741
	1991	387,403
	1992	335,546
	1993	311,854
Female	1987	127,773
	1991	130,744
	1992	118,762
	1993	110,693
Unknown	1987	1,057
	1991	2,400
	1992	1,351
	1993	637
Total	1987	602,531
	1991	520,547
	1992	455,659
	1993	423,184

Source: Statistics Canada, *Work Injuries 1991-1993*, Cat. No. 72-208 (Ottawa, 1994), Table 2, p.13. Reproduced with permission of the Minister of Supply and Services, Canada.

While it is impossible to know exactly how many species there are at present, some scientific estimates suggest that the total number of species is in the range of 30 million. In fact, researchers have identified more than one thousand species of ant (Wilson, 1991, in Macionis, 1995). The impact of the decline in biodiversity on the health of human populations is not entirely known. However, given the enormous interdependence in this complex ecosystem, the extinction of some species may very well indirectly lead to the extinction of others and ultimately may lead to the destruction of species that serve to protect human life.

Occupational Health and Safety

Canadians face a relatively high degree of danger when they go to work. Figures for 1982 indicate that 15 million work days, or 1.5 days per worker, were lost as a result of accidents on the job (Dickinson and Stobbe, 1988.) These figures, drawn from data gathered by the Workers' Compensation Board, likely underestimate the number of injuries that occur at work because many people do not receive or request compensation. Others may be transferred to lighter jobs during the time of recuperation, so that compensation is unnecessary. Nonetheless, there has been a significant decrease in the rate of job-related accidents. In 1987, there were 602,531 work-related injuries in Canada (for which claims were accepted for time-loss by eleven workers' compensation boards). By contrast, for 1993 the total number of time-loss injuries reported was 423,184, a decrease over a seven-year period of about 30 per cent (Table 4.2). This dramatic shift certainly reflects the change in the economy as globalization and North American free trade have sent primary and secondary manufacturing jobs out of Canada and the tertiary service sector has grown.

The U.S. government has estimated that as many as 40 per cent of all cancers may be caused by the work environment. Others have estimated that up to 90 per cent of

TABLE 4.3 Lung Cancer Rates in Selected Occupational Groups, White Males 20-64 years, Los Angeles County (SMR = 100 for all occupations)*

OCCUPATION	ESTIMATED POPULATION AT RISK	SMR	NUMBER OF CASES
Asbestos Insulation Workers	400	878	5
Roofers	2,000	496	11
Heat Treaters	550	433	4
Dental Lab Technicians	900	405	5
Decorators	1,200	358	7
Taxi Drivers	3,100	344	23
Mechanics (excl. Auto)	8,050	332	46
Photo-engravers	1,350	320	8
Pressmen	5,300	276	20
Clothing Ironers	1,100	267	9
Mariners, Longshoremen	3,150	266	23
Shoe Repairs	1,350	233	7
Mine Operatives	1,250	217	8
Electricians	12,400	205	58
Bartenders	7,100	204	35
Plasterers, Dry Wall Workers	3,800	200	17

*The figure under SMR (standard mortality rate) indicates the elevated risk of lung cancer for workers in each of the industries indicated.

Source: *Occupational Health and Safety: A Training Manual* (Toronto: Copp Clark, 1982), p. 71.

cancers are related to the working environment (Epstein, 1979; Doll and Peto, 1981). Exposure to the following substances had been found to increase the risk of cancer by the amount indicated in brackets: arsenic (2 to 8 times for lung cancer); benzene (2 to 3 times for leukemia); coal, tar, pitch, and coke oven emissions (2 to 6 times for cancer of the lung, larynx, skin, and scrotum); vinyl chloride (200, 4, 1.9 respectively for cancer for the lining of the heart, brain, and lung); chromium (3 to 40 for cancers of the sinus, lung, and larynx (Tataryn, 1979: 157-58). Furthermore, high exposures and consequent risks are more prevalent among the working classes and those with lower incomes (*ibid.*: 158). Tataryn has documented the extensive occupationally induced health problems from asbestos in Thetford Mines, Quebec, radiation exposure in Elliot Lake, Ontario, and arsenic exposure in Yellowknife. Table 4.3 illustrates the likelihood of developing lung cancer associated with a number of different industries.

Occupational health and safety are also major concerns for all working women, both those who work in the paid labour force and those who do not. Less publicized than the hazards associated with blue-collar work, the places where most Canadian women work, including "offices, banks, stores, restaurants, hospitals, medical laboratories, schools, child care centres and hairdressing establishments" (CACSW, 1987: 85-86), have their own peculiar health risks.

Particular health and safety problems are associated with each of these job areas. Clerical workers may be subject to poor lighting and ventilation, excessive noise, and toxic substances such as emissions from computer terminals. Often they spend long hours sitting on uncomfortable furniture, which may lead to back pain, and working at relatively monotonous jobs, which may lead to stress. Retail and service workers may be vulnerable to health hazards from bending, lifting, and carrying; varicose veins and foot and back problems

Box 4.9 Occupational Health and Safety

Workers' compensation boards work, on the one hand, to provide benefits to certain workers who have suffered ill health as the result of work, and, on the other hand, to reinforce the notion that health is commodifiable – and has a certain monetary value. Doran (1988: 460) argues that "despite the obvious advantages which come from the rise of workers' compensation legislation, an equally important loss has been suffered: workers have to battle to preserve their health at the expense of industrial production." The commodification of health has been characterized by an increasingly narrow definition of health that largely denies the experience of the sufferer/worker as becoming part of accepted medical and legal definitions of illness and health. Illness is defined not by the sufferer but by the medical/legal authorities who label a narrow set of experiences as, first, medically relevant and, second, occupationally induced. One primary modality through which this is accomplished is the bureaucratic necessity of a workers' compensation form that includes some categories of symptoms as relevant and, by exclusion, deems other symptoms irrelevant. Moreover, accidents are prioritized over long-term chronic conditions as more likely to be work-related.

---

are often experienced. Hairdressers, who usually stand all day, suffer back and foot problems along with the dangers of exposure to toxic chemicals in hair permanents, dyes, and aerosol sprays. Respiratory difficulties and skin reactions are frequent results. Teachers and child-care workers are continually exposed to a variety of contagious and infectious diseases. Health-care workers may be exposed to radiation, toxic chemicals, and contagious diseases, and may have to cope as well with lifting, bending, and much standing. Women who work at home may be subject to dangers from all sorts of household cleaning substances such as abrasives, astringents, soaps, and detergents (*ibid.*: 85-88).

Many women work at jobs that are different from the jobs men do, in female-dominated job ghettos. Many jobs done by women are considered by some to be inferior to jobs done by men. Women tend to be poorly paid – 72¢ for every $1.00 made by men. Women are often the last hired and the first fired. To the extent that women's jobs are considered less important than men's jobs, then women have less ability to demand safe, clean, and healthy working standards. Women's occupational health and safety issues are thus both different and potentially more problematic than those of men. Even if the risks women face tend not to be as dramatic and acute – that is, they are not as likely to result in immediate or almost immediate death – the long-term chronic health problems of their work are serious. As Table 4.4. indicates, women workers are exposed to a broad range of occupationally related health hazards. In clerical work, which continues to be dominated by women, the types of problems are wide-ranging.

Women's problems at work may have serious consequences for the next generation.

Box 4.10 Occupational Diseases of Hairdressers

There are 66,855 hairdressers in Canada, representing 0.6 per cent of the work force. The most frequent occupational diseases among hairdressers are skin diseases, followed by respiratory disease, certain cancers, and other miscellaneous diseases.

Hairdressers are in daily contact with cosmetic preparations and metallic instruments for shampooing, drying, bleaching, tinting, and permanently waving; these preparations can lead to dermatitis and eczema of the fingers and hands and occasionally the forearms. "Cosmetologists perform other beauty services such as massaging the face and neck with creams and oils, colouring eyebrows and eyelashes, manicuring fingernails and toenails, and carrying out depilatory techniques" (Heacock and Rivers, 1986: 109). Other skin diseases include hair implantation granulomas as a result of hair cuttings becoming embedded in the skin, infection and inflammation from cuts, abrasions, and burns, and allergic contact dermatitis of the upper eyelids from nail polish and other manicure products.

Respiratory disorders are also common among hairdressers. A 1976 study showed female cosmetologists to have an increased prevalence of chronic respiratory symptoms, small airway obstruction, and atypical sputum cytology. These increases were found to be related to duration of occupational exposure. Exposure to large amounts of hairspray appears to be a factor in these disorders. It has also been found that asthma could be a result of occupational exposure to henna dyes.

It has been found that certain types of hair dyes are mutagenic and, in laboratory testing, one type produced liver cancer in rats. The results of exposure to hair dyes for humans is less conclusive; no appreciable risk of breast cancer among female hairdressers was found in a 1977 study. However, a 1984 study found that female hairdressers have increased risk of death from multiple myeloma and cancer of the ovary.

Miscellaneous risks associated with professional hairdressing include varicose veins from prolonged standing, nervous fatigue from working long, irregular hours in sometimes adverse conditions, and occupational cramps from moving muscles constantly the same way.

Recommendations for hairdressers' health and safety, as suggested by Heacock and Rivers, include: better room ventilation, proper lighting, clean floors, freshly laundered working clothes, towels, and aprons, sitting or reclining breaks at regular intervals, and regular hours of work with limited overtime.

Beauticians should be warned of potential hazards associated with hairdressing products, and protective gloves should be worn to prevent contact with irritating chemicals or solutions. Hands should be washed with non-irritating or neutral Ph soap, and hand moisturizers should be readily available at sinks. More research should be done to identify real health threats to hairdressers, including lifestyle factors. Chemical constituents of hairdressing products in Canada should appear on product labels.

So may those of men. Damage to the reproductive organs, to the developing fetuses, and to sperm quality and quantity, and the potential for sterility, miscarriage, or genetic problems in offspring are among the most devastating effects of occupational health and safety inadequacies. Unfortunately, concern over potential reproductive hazards has

TABLE 4.4 Office Health and Safety Hazards

HEALTH HAZARD	EFFECT
Video display terminals	Headaches, eyestrain, temporary colour blindness, possible permanent eye damage, neck and back pain.
Photocopying machines	Ozone gas, which irritates the eyes, nose, throat, and lungs. May damage chromosomes. At least one chemical used, hitropyrene, may cause cancer.
Lighting	Fluorescent lighting is uncomfortable and has been found to cause hyperactivity in children.
Ventilation	Inadequately cleaned or poorly designed ventilating systems can spread viruses and disease, and can result in lung, ear, nose, and throat problems.
Noise	Acceptable noise level has been set at 67 decibels for office machines.
Excessive wrist action	Painful muscle strain, and problems such as key puncher's wrist (tensoy novitis) and carpal tunnel syndrome may occur.
Stress	Low pay and status and lack of control over workload and decision-making all contribute to making clerical work a very stressful job. It was ranked second in respect to stress out of 130 jobs (by 22,000 people).

Source: adapted from Charles E. Reasons *et al.*, *Assault on the Worker: Occupational Health and Safety in Canada* (Toronto: Butterworths, 1981), pp. 90-91.

focused almost exclusively on women, so that women have been banned from some jobs entirely and from others during their childbearing years or during pregnancy. Such legislation both discriminates against women and simultaneously ignores the real danger to the reproductive health of men. Instances of such stereotyping and discrimination have been frequent enough that the Canadian Advisory Council on the Status of Women has recommended that the federal government amend the Canadian Human Rights Act and the Canada Labour Code to prevent discrimination in hiring, job replacement, promotion, and other conditions of employment based on factors related to reproductive physiology, such as reproductive capacity, pregnancy, or childbirth; that exclusionary policies and practices arising from such issues be prohibited by law; and that the legislation be monitored and enforced on a continuing basis (CACSW, 1987: A14.3).

Estimating the actual prevalence of

Box 4.11 Rotational Shiftwork: What Are the Adverse Effects?

Shiftwork affects about one-quarter of the population in North America. It is common in industrial work, mining, hospitals, transportation, and food service work. A great deal of evidence suggests that shiftwork can disrupt the family and personal life, and can lead to a myriad of health problems. The information upon which this report is based is limited to rotational shiftwork, which means that the working hours vary regularly from day to day, week to week, or according to some other schedule.

Among the negative consequences are the following:

1. Persistent fatigue is common.

2. Gastro-intestinal and digestive problems are frequent.

3. Shiftworkers are more vulnerable to heart disease and heart attacks. In general, shift workers have lifestyles that are associated with ill health, including smoking, obesity, little recreation or regular exercise, and poor diets.

4. Medication may affect the person on shiftwork in an unpredictable way.

5. Shiftwork has negative effects on family activities and relationships. This can lead to depression, isolation, and loneliness. The lack of day care associated with most shiftwork may mean that children are sometimes left unattended. Participation in "normal" parent-child, husband-wife, and family socializing is severely restricted because of the unpredictability of the schedule.

6. There is some evidence that there are more accidents at work among shiftworkers.

7. Working conditions can be poorer (lighting, ventilation, cafeteria services, and opportunities for socializing may be restricted).

---

occupationally related disease is problematic for a number of reasons (Dickinson and Stobbe, 1988). First, there may be a long period of latency between exposure to the damaging substance or activity and the resultant disease. Second, there is a lack of information, and, indeed, a great deal of misinformation, about which chemicals are being used and which chemicals or activities have damaging long-term effects. Even when information is available about the negative consequences of a substance, the information may be withheld. Moreover, the effects may be difficult to monitor. One additional set of problems has to do with the fact that physicians are often poorly trained in recognizing occupationally related diseases.

As Table 4.5 shows, the average Canadian in 1991 lost 6.2 days per year on average as the result of perceived exposure to workplace hazards. The most important cause of missing work was accident or injury. This was more important for male than for female workers: 18.5 days as compared to 10.7 days annually. Women were more likely than men to lose days of work as the result of excessive job demands, poor interpersonal relations, and other physical hazards, chemicals, noise, or dust or fibres in the air.

Time-Loss Work Injuries

Some idea of the extensiveness of occupational injuries is to be found in the figures

TABLE 4.5 Annual Days Lost from Work, by Perceived Exposure to Workplace Hazards, Canada, 1991

| | TIME LOST FROM WORK | | |
Perceived Workplace Hazard	Both Sexes	Male	Female
Risk of accident or injury	16.6	18.5	10.7
Poor air quality	8.6	8.6	8.3
Computer screens	6.8	7.5	6.0
Excessive job demands	8.1	7.0	9.6
Dust/fibres in air	6.8	6.8	7.0
Loud noise	6.8	6.8	7.0
Dangerous chemicals/fumes	6.5	6.2	8.3
Poor interpersonal relations	7.3	5.7	9.4
Other physical hazards	6.8	5.5	9.9
Total	6.2	5.7	7.3

Source: *Health Status of Canadians* (Ottawa: Statistics Canada, March, 1994).

given by workers' compensation boards and commissions in Canada. According to Statistics Canada fully 455,659 work-related time-loss injuries were accepted by the compensation boards in 1992 (Statistics Canada, 1993a). Keep in mind that "accepted injuries" is most probably a (much) smaller category than "actual injuries" and definitely a (much) smaller category than injuries reported or "experienced" by workers. Still, it is worth noting that almost one-half million Canadians missed work because of on-the-job injuries in 1992. This troubling statistic is of serious consequence to male workers. Also, the statistic reflects a disproportionately rapid increase in the rate of time-loss accidents for female workers, from 19.1 per cent in 1984 to 26.1 per cent in 1992.

Agricultural Work

Agriculture has long been associated with a pastoral, idyllic, and healthy style of life. In fact, after mining and construction, agricultural work is the most health-threatening (Bolaria and Bolaria, 1994: 684). Not only do agricultural workers suffer a high rate of accidents and associated fatalities, but the working conditions include handling dangerous farm machinery, the intensification of the farm labour process, poor housing and sanitation, low wages, and long hours of heavy, hard labour. As the ozone layer thins the rate of skin cancer is bound to increase among farm labourers.

Pesticides are another major threat to human health as a result of both direct ingestion via pesticide-coated fruits and vegetables and indirect ingestion because of long-term accumulation in soil, water,

Box 4.12 Westray

Amidst much celebration and optimism, the Westray coal mine in Pictou County, Nova Scotia, began production on September 11, 1991. The mine had been presented as a no-lose situation, in that it had guaranteed funding, a guaranteed market at good prices, and the people of Pictou County were guaranteed jobs.

The celebrating ended in the early morning of May 9, 1992. The 26 men working the overnight shift at the Westray mine were killed as a furious explosion ripped through the mine. One of the questions to arise out of the Westray disaster is whether or not the development of the mine was motivated strictly by political and economic concerns. Some of the events leading up to the disaster are as follows.

The coal seam in the coal mine in Pictou County was one of its most attractive features. Termed the Foord seam, it is unusually thick, varying from two to eight metres; its sulphur content is below 1 per cent; and it is a high energy producer. The Foord seam also has quite a few health and safety risks attached to it. The area in which the seam is located is widely known to produce methane gas and there are significant geological fault lines. These problems added to the increased risk that the roofs in the mine's underground rooms would collapse. Methane gas is a highly flammable carbon-based gas that is colourless and odourless, therefore making it difficult to detect. The Foord seam had a higher methane gas content than was evident in most other mines in North America. Coal dust, also a highly flammable substance, is extremely hazardous if gathered in large quantities. Combine the two and there is potential for major disaster if proper safety measures are not taken.

The time during which the mine in Pictou County was beginning to be developed coincided with an upcoming federal election. The mine was located in the provincial riding of Central Nova, the riding of Donald Cameron, who eventually became the Premier of Nova Scotia. Cameron became the mine's strongest advocate, pushing for a formal agreement on the mine in the final days before the federal election. In 1990, after intervention from Prime Minister Brian Mulroney's office, the agreement was passed.

The mine had been dormant in Pictou County for approximately two decades. There were so few trained miners in the area that Westray had to recruit from outside the immediate area. From the beginning, the mine was plagued by cave-ins, high levels of methane gas, and coal dust. These problems were certainly not new ones, but this fact tended to be overlooked. Because of repeated construction delays, due primarily to financial stalemates, completion of the mine was rushed to meet the scheduled opening date of September, 1991. Not surprisingly, this pressure to meet the deadline resulted in a number of safety problems.

- *December 12, 1990:* The Department of Labour issued its first order to Westray after finding that the mine had been undergoing underground blasting without qualified people being present.
- *May 23, 1991:* An independent engineering company was hired to investigate an incident in which 24 metres of roof collapsed – no orders were issued regarding the collapse and no charges were laid.
- *July 29, 1991:* A report to Nova Scotia's Chief Mine Inspector stated that the levels of both methane gas and fine coal dust were fairly high and therefore presented a risk. No action was taken on this issue.
- *October 20, 1991:* Three more rock falls occurred at the mine during the weeks leading up to this date.

Box 4.12 Continued

- *March 28, 1992:* Another cave-in caused ministry officials to visit the mine once again. Once there, they noticed that the air sample taken for testing contained 4 per cent methane gas. Methane gas becomes explosive when it reaches the level of 5 per cent. The law requires work to be halted and workers to be removed if the methane level reaches 2.5 per cent. When it reaches 1.25 per cent, electricity is to be shut off. On this date, production was not stopped. The amount of coal dust on the floors of the mine had been of serious concern. A method of preventing a potentially dangerous situation is to put down limestone dust to cover the coal dust. During construction, inspectors pointed out on at least nine separate occasions that there was a need for more storage bags of limestone.
- *April 29, 1992:* Ten days prior to the explosion, the Ministry of Labour issued orders for Westray to comply with statutory requirements regarding this issue. As is now understood, the

ministry's inspectors had not checked back to see whether or not these orders were being followed when the explosion occurred on May 10, 1992.

The mine eventually closed due to the abundance of inquiries regarding the events leading up to the disaster and due to a lack of funding.

In June, 1995, a Nova Scotia Supreme Court justice ordered a stay of the four-month-old trial of two mine managers charged with manslaughter and criminal negligence "because of lack of disclosure by the crown." This judicial move appeared to end the charges against the men, who allegedly had "ignored safety procedures and tampered with monitoring equipment to boost production at the colliery" (Mansour, 1995), but on December 1, 1995, the Nova Scotia Court of Appeal ordered a new trial. The two mine managers, who are no longer in Nova Scotia, have refused to participate in the inquiry ("Westray probe's scope curbed," *Toronto Star*, November 17, 1995).

---

and air. Pesticides also are known to have a tendency to break down or to combine with other compounds over time that may be even more dangerous to human health. Short-term effects of mild pesticide poisoning include nausea, vomiting, and headaches, as well as more serious permanent damage to the nervous system, miscarriage, birth defects, sterility, and reproductive disorders. Long-term pesticide exposure has been found to be associated with various cancers of the lungs, brain, and testicles (the herbicide 2,4-D, for example, has

been found to be associated with a type of lymphoma). It has been suggested that pesticide (whether insecticide, herbicide, fungicide, or rodenticide) use constitutes a potentially catastrophic experiment with human life (Raven *et al.*, 1993).

Compounding the problems resulting from agricultural work itself is the fact that much of the hired labour force is composed of migrant (temporary), immigrant, illegal, or undocumented workers (Bolaria and Bolaria, 1994: 440). The tenuous nature of their presence in Canada has meant that

Box 4.13 Arctic Pollution

It may come as a surprise, but one of the next regions of great environmental concern is the Arctic. The build-up of chemical residues in Canada's North is due to a process called the grasshopper effect, which renders the North the next "garbage dump" of the more southern and developed world. Initially, chemical and metal emissions – pollutants – are absorbed into the atmosphere, where they are next experienced as rain, snow, or atmospheric gases. Then they are absorbed by the lakes, oceans, soils, and plants. When the weather warms, they become gases. The total amount of substance that returns to the air depends on the volatility of the chemical. The more volatile the chemical the further it travels. Since 1978 there has been a decline of PCBs in Lake Superior of about 18 per cent per year. However, 26.5 tonnes of PCBs have evaporated from the Great Lakes and travelled to the Arctic. The process of initial emission, absorption, re-emission, and re-absorption threatens the environment of the North next. Meanwhile, the unwanted chemicals do not disappear but accumulate and spread throughout the world (Strauss, 1994; Rabe, 1992).

---

many have had to take whatever job was offered and to accept its working conditions without complaint or condition. Racism, a lack of language facility, and/or a lack of skills renders these people particularly vulnerable. "In summary, immigration laws, contractual obligations, lack of protection by labour legislation, lack of alternative job opportunities, poverty and unemployment in the country of emigration and the absence of union organization place many foreign workers in a vulnerable position and render them powerless vis à vis the employer" (*ibid.*: 442).

Other Accidents and Violence

Accidents and violent deaths are the major causes of potential years of life lost among Canadians from 1-75 years of age. To some extent accidents result from human error – driving under the influence of alcohol and drugs, failure to wear seat belts, and driving at excessive speed. Many traffic fatalities and deaths from accidental falls and fires result from alcohol-related impairments. Sports-related accidents comprise 23 per cent of all accidents. Most of these occur to men (65 per cent), and in 1990 such accidents resulted in almost as many out-patient hospital treatment visits (535,000) as did work-related accidents (591,000). Clearly, sports injuries are an important public health problem that needs to be more clearly defined. This problem is particularly serious because its incidence is related to age. The 15-24 age group is considerably more likely to experience sports injuries than those in any other age group. Among people 15-24, sports injuries were responsible for approximately 42 per cent of all injuries. Approximately 8.7 million activity-loss days and 1.5 million bed-disability days in 1987 resulted from sports injuries. The most dangerous sports in order were: ice hockey, cycling, and skiing and baseball (Statistics Canada, 1991).

Box 4.14 Index on Cars

• Canada's fleet of passenger cars grew from 10.2 million in 1981 to over 12.8 million in 1989. The car population grew faster than the human one and the number of kilometres travelled by car rose by 22 per cent between 1981 and 1988. Fuel efficiency has improved, but is outstripped by the growth in car traffic.

• Two-thirds of all trips to work in major Canadian cities are by car. Less than half of these work trips by car are for households with incomes lower than $30,000, while 77 per cent of workers in households with incomes $60,000 and over in 1991 commuted by car.

• Canadian motorists get a largely free ride. Public subsidies go far beyond building and maintaining roads and bridges. They include snow removal, health care for accident victims, traffic law enforcement, administration of car and driver registration, and tax losses from paved land, to name just a few items. Pollution Probe estimates the car cost to Ontario taxpayers alone at $8.3 billion annually versus $3.4 billion in revenues, or a $4.9 billion deficit.

• Cars kill several thousand Canadians each year in motor vehicle traffic accidents – 4,180 in 1987 (including 612 pedestrians and 113 cyclists). About 250,000 people are injured or disabled every year. In Ontario, more than three people a day – one every seven hours or so – die in car accidents.

• An average car driven 14,700 kilometres a year sends 4,300 kg of carbon dioxide, the most common greenhouse gas, into the atmosphere every year. One litre of gasoline produces almost 2 kg of carbon dioxide – the extra weight comes from the oxygen molecules added to the carbon in the fuel during combustion. The same car also emits 350 kg of carbon monoxide, 30 kg of nitrogen oxides, 50 kg of hydrocarbons, and 2 kg of potentially poisonous or disease-producing particulates.

• Pollutants from car exhaust have serious or fatal health impacts. They are linked to heart disease, brain damage, emphysema, bronchitis, asthma, and cancer, among other illnesses. A by-product of car exhaust, ground-level ozone, is regarded by some researchers as the third greatest cause of lung disease after smoking and passive smoke. Failure to meet air quality standards in the United States carries a price tag of $50 billion per year, according to the American Lung Association.

Sources: Angus Reid Group, Jeffery Patterson, Statistics Canada, Environment Canada, *Ottawa Citizen*, *Clouds of Change* (Vancouver), Capital Regional District (Victoria), Pollution Probe. From *Canadian Forum*, April, 1993. By permission of the publisher and the author, Neale MacMillan.

Of every 1,000 ice hockey injuries, 106 were serious, involving, for example, spinal or neurological damage. Figure 4.2 portrays the age and sex incidence of sports-related accidents.

Violence against women and children may be an important factor in ill health and death that has long been under-reported. While actual incidence is difficult to estimate or measure, it can be said that it happens too frequently (MacLeod, 1987). Evidence available today supports the view that violence against women and children is widespread and not an instance of aberrant behaviour committed by a few "sick" individuals. Violence reflects prevalent socialization patterns and culturally based value systems: it is extolled as glamorous and exciting; it is

FIGURE 4.2 Rates of Sports Accidents per 1,000 Population by Age Group and Sex, Canada, 1987

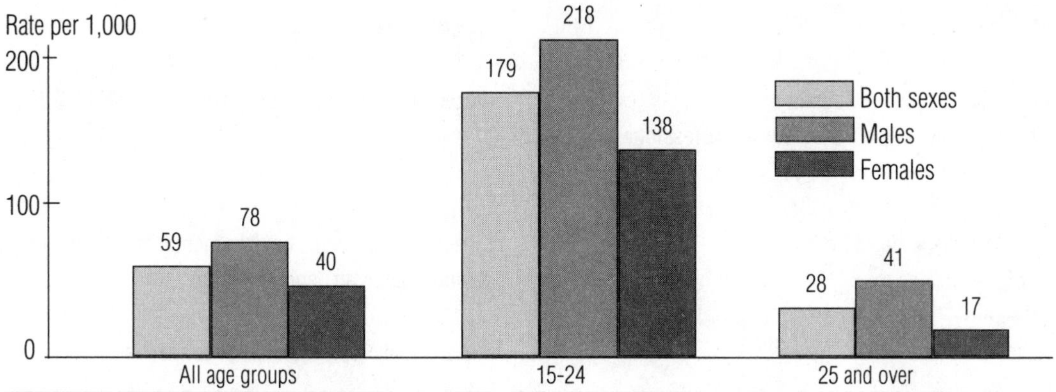

Source: Statistics Canada, General Social Survey, 1988.

featured in popular television shows; and young boys are given toy guns and military equipment to play with, while macho super-heroes are provided as role models.

Violence is sometimes used to control women and children – to keep them in their place. Long subservient, first to their parents and then to their husbands, women are logically the victims of the greater strength and power of their "keepers." In a society that gives men the dominant roles in economics, politics, and religion, it is no wonder that men frequently dominate women and children physically as well. Nor is violence restricted to the home. Women and children are vulnerable to sexual harassment, rape, and physical abuse on the streets and in the homes of friends, neighbours, and other family members, too. *No Place to Hide* (Olson, 1984) and *Even in the Best of Homes: Violence in the Family* (Scutt, 1983), the titles of books on violence against women and children, are phrases that capture the ubiquity of this violence.

Summary

(1) There are three fundamental parts of the environment – air, land, and water. All affect our health both directly and indirectly.

(2) The major environmental issues facing Canadians today include: carbon dioxide production and climate change, acid precipitation, the depletion of the ozone layer, chemicals, and the disposal of hazardous wastes.

(3) There are other significant environmental issues related to air quality, including air pollution, second-hand smoke, and medical pollution.

(4) Water pollution in the Great Lakes is discussed.

(5) With respect to land, the issue of waste disposal is raised.

(6) Occupational health and safety issues are discussed. It has been estimated that 90 per cent of all cancers are related to the working environment.

(7) The health and safety issues of

Box 4.15 A Comparison of Traditional and Popular Epidemiology

Traditional Epidemiology	Popular Epidemiology
Professional epidemiologists Health problem noted by medical professionals, scientific literature	Citizens notice a health problem
Formulate hypothesis from literature	Citizens notice an environmental problem (e.g., obvious water, air, land pollution)
Assemble data	Citizens make potential causal connection
Perform statistical analysis for proof of association	Citizens share information in community and develop communal perspective
Publish results	
Statistical significance criterion of association	Citizens meet formally, read, ask questions, talk to government, corporate officials, and scientists about putative connection Citizens organize to pursue investigation Government agencies respond and usually find no association Statistical significance and/or "reasonable person argument" criterion of association
In face of uncertainty traditional epidemiology favours "no-association" interpretation	"Yes-association" interpretation

Phil Brown, "Popular Epidemiology and Toxic Waste Contamination: Lay and Professional Ways of Knowing," *Journal of Health and Social Behavior*, 33 (September, 1992), pp. 267-81.

working women, while less dramatic (and less studied) than those of predominantly male occupations, are nevertheless myriad and consequential.

(8) The particular health and safety issues related to agricultural work are outlined.

(9) Sports accidents, a significant cause of morbidity and mortality, especially among young people, are described.

(10) The impact of accidents and violence on the morbidity and mortality rates of Canadians are considered.

CHAPTER FIVE

Social Inequity, Disease, and Death: Age and Gender

ARE WOMEN MORE often sick than men? Do women suffer from different illnesses than men? Are women discriminated against in medical treatment? Does old age bring increasing infirmity and illness? Is it true that "men die and women get sick"? Are physicians more likely to medicalize women's bodies and their bodily experiences than those of men? Are the elderly overprescribed medicines by their doctors? Do the elderly themselves select too many over-the-counter remedies from local pharmacies? These are some of the questions that this chapter addresses.

Health, illness, and death are not randomly distributed in a society. Rather, their incidence and prevalence are inextricably linked to the social organization of the society. One aspect of this social organization is the extent of inequity in the social structure. Inequity causes different life chances and experiences, as well as unequal access to fundamental social resources such as food, recreation, satisfying work, and adequate shelter. Because of that unequal access, people who differ in age, sex, income, class, occupation, ethnicity, marital status, rural or urban background, and religiosity differ in their rates of sickness and death. This chapter will examine how illness and death rates vary depending on age and sex. It will then discuss some of the possible explanations for the observed variations. As a first step to this analysis, however, it is essential to put the relationship between illness and death and the social structure into a broad, comprehensive context. Table 5.1 is a model of the major social variables and their possible connection with rates of illness and death.

On the broadest level, cultural differences between societies manifest themselves through such things as varying definitions of health and illness and varying views regarding appropriate types of medical treatment, for example, the degree to which a society is "medicalized" – that is, influenced by medical explanations and treatments for social problems, such as hyperactivity, which used to be regarded as naughtiness but is now treated with a drug. This concept will be developed in greater depth in later chapters. An historical illustration will suffice here. The handling of "problem drinking" is a good example of the process of medicalization.

117

TABLE 5.1 A Model for Analysis of Morbidity and Mortality Rates at the Societal Level

CULTURE the degree of medicalization	POLITICAL-ECONOMIC SYSTEM	ECOLOGICAL SYSTEM	SOCIAL STRUCTURE	SOCIAL PSYCHOLOGY	MICRO MEANING the definition and meaning of health, disease, and death
	capitalism socialism communism	environmental quality and condition (water, air) quality of agricultural land and foodstuffs transportation and communication systems	gender age ethnicity education religious affiliation religiosity working limitations employment and unemployment rates	stress type A behaviour* sense of coherence* perceived social support	

*These will be defined fully in Chapter Seven.

When Canadian society was less medicalized and religious concepts were more prominent, for instance, at the turn of the century, problem drinking was considered to be a sign of weakness or a sin. Moral responsibility and societal opprobrium were advocated to control drinking. Today, in a more medicalized society, alcoholism is considered not a sin but a sickness. Alcoholism is treated medically, and addictive behaviour is explained in terms of biochemistry, or psychiatrically, as the result of continuing interdependence.

Differences in mortality and morbidity are related to the political-economic system as well as to cultural differences. The level and distribution of numerous resources such as food, shelter, access to meaningful work, environmental quality, and satisfying social relations may vary within a society. If the political-economic system causes differences in people's access to resources, there will be structured inequities in rates of death and illness. Important examples are to be found in the effects of the communist revolution in China. Chinese public health measures, along with ubiquitous "barefoot doctors," conquered or controlled formerly epidemic diseases such as syphilis, leprosy, and schistosomiasis (Horn, 1969). Until 1949 chronic malnourishment, overcrowding, and primitive sanitation all helped to create massive health problems (Liang *et al.*, 1973). Nearly one-third of the population died before age five. Tuberculosis is estimated to have caused 10 to 15 per cent of all deaths. Revolution led to a redistribution of resources, including food and housing, and to public health measures such as sanitation and immunization. One consequence is that the life span of the average Chinese person has greatly increased. In Shanghai, one city for which statistics are available, life expectancy has risen from 40 years to 70 years. Furthermore, the leading causes of death today are the same as those in the developed countries – cancer, stroke, and heart diseases – and not the epidemic, contagious, and infectious diseases of developing nations.

As another example, Zimbabwe became an independent nation in 1980. From 1980 to 1989, the infant mortality rate fell from 140 per 1,000 live births to 50. This was largely the result of a massive government commitment to improving primary health care through the provision of safe water, public sanitation, and an increase in the standard of living through what the government called growth with equity. Free health and education were provided for those with incomes of less than U.S. $26 per month.

Analysis within a society focuses on social-structural positions such as age, sex, and class. Such variables are associated with different health patterns for different parts of the population. This chapter will examine this level of analysis in greater depth. In particular, it will examine age, gender, and disease and death.

Age

The first socio-economic characteristic, age, is associated with morbidity and mortality rates in both predictable and surprising ways. The most basic method used to describe the age distribution of a population is the population pyramid. The most obvious difference in the population pyramids in Figure 5.1 over time is the overall aging of the population and the corresponding decrease in the proportion of the population who are younger. It is also evident that the population is expected to continue to age significantly until at least 2036. This rapid growth of the older population is especially evident among women. As is indicated in Figure 5.1, the proportion of the population under twenty years of age is expected to decline significantly by the year 2036.

A number of explanations have been offered for what is called the "aging of the population." The most important factor in population aging is the overall decline in the birth rate. As people have fewer babies, there are correspondingly fewer young people. With fewer young people in the population there is necessarily a greater proportion of elderly. The second, and somewhat less important factor, is the increase in life expectancy that has occurred over the past century and a half in Canada.

Life expectancy increases are the result of a number of changes. As discussed in Chapter Three, the most important factor is the rapid decline in the infant mortality rate. Table 3.1 showed that the average life expectancy for Canadians over the last 160 years or so has grown from 39.0 to 77.8. As well, women's life expectancy has grown far more than that of men: an average of 39.8 years around 1831 to an average of 81.0 years in 1991 for women; for men, an average of 38.3 in 1831 to an average of 74.6 in 1991. Men have gained 36.3 years over this period of time and women have gained 41.2 years. Significantly, while men have, in all of the years recorded, lived shorter lives than women, the difference between the average life expectancies of men and women has

FIGURE 5.1 Age Pyramids of Canada, Selected Years

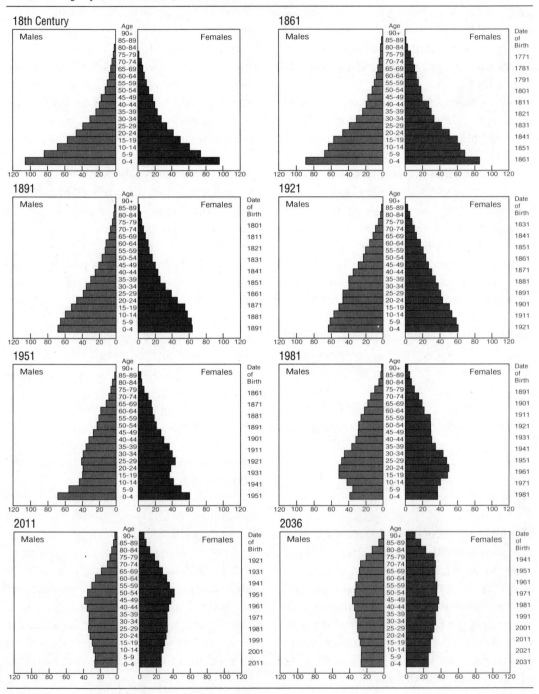

Source: Macionis *et al.*, 1994: 385.

Box 5.1 The Politics of Breast Cancer

When the president of the American Society of Clinical Oncology gave the presidential address to the annual meeting of the association, he vigorously chastised those other than medical practitioners and research scientists who dared to claim an interest in the direction and outcomes of breast cancer research and treatments. He called them "rascals who mine the road to progress with misguided causes" (in Weiger, 1995: 41). The "rascals" to whom he was referring included sociologists and virtually all social scientists, ethicists, sex therapists, policy-makers, policy implementers, and particularly breast cancer activists. He was critical because they were beginning to change the direction of breast cancer research and treatment in the U.S.

Breast cancer activists such as Pat Kelly and Sharon Batt are leading the challenge to rethink breast cancer treatment and research agenda in Canada – Pat Kelly through her passionate commitment to women suffering with the disease and their paramount need for emotional and social support and Sharon Batt through her challenges to the way things are and the way things are done in the breast cancer establishment. Batt critically analyses the effectiveness of mammography, surgery, radiation, chemotherapy, hormones, theories of heredity, and alternative types of treatment. She examines the credibility and the promise of research and prevention. With astuteness and wit she describes cancer charities, capitalism's use of cancer, and the medical treatment of the disease.

Charles Weiger, "Our Bodies, Our Science," *The Sciences*, May/June, 1995.
Sharon Batt, *Patient No More: The Politics of Breast Cancer* (Charlottetown, P.E.I.: Gynergy Books, 1994).

been greater in the past twenty-five years than at any point in history.

Life expectancy figures from the last seventy years demonstrate another feature of the changing causes of death. If we compare life expectancy at birth with life expectancy at age 60 (see Table 5.2), an important fact comes to light. Most of the gains, both for men and women, have been in the ages below 60 years. This information tends to support the inference that the rapid decline in the infant and child mortality rates has been the major cause of increased life expectancy rates. The data in Table 5.2 provide one answer to the question of the differences in life expectancy between the sexes: women have gained more years of life than men, and this is true both at birth and at age 60.

Table 5.3 explains this increase in life expectancy by portraying the decline in mortality over two age groups. The most significant decline for both sexes is clearly in the early years of life. However, while women's mortality rate continues to decline (1921-90) to almost half the earlier rate, this is not the case for men. An important part of women's increasingly greater longevity is the result of improved nutrition and other public health measures for pregnant women, as well as advances in neonatal medical procedures that are associated with the subsequent decline in infant mortality.

Life expectancy has grown significantly over the last century and a half. Some obvious questions are: How much more is it likely to grow? Will new diseases and

TABLE 5.2 Expected Years of Life Remaining, 1921-1991

| | At birth | | At age 20 | | At age 40 | | At age 60 | | At age 80 | |
	male	female	male	female	male	female	male	female	male	female
1921*	n.a.	n.a.	49.1	49.2	32.2	33.0	16.6	17.1	6.0	6.1
1931	60.0	62.1	49.1	49.8	32.0	33.0	16.3	17.2	5.6	5.9
1941	63.0	66.3	49.6	51.8	31.9	34.0	16.1	17.6	5.5	6.0
1951	66.3	70.8	50.8	54.4	32.5	35.6	16.5	18.6	5.8	6.4
1956	67.6	72.9	51.2	55.8	32.7	36.7	16.5	19.3	5.9	6.8
1961	68.4	74.2	51.5	56.7	33.0	37.5	16.7	19.9	6.1	6.9
1966	68.8	75.2	51.5	57.4	33.0	38.2	16.8	20.6	6.4	7.3
1971	69.3	76.4	51.7	58.2	33.2	39.0	17.0	21.4	6.4	7.9
1976	70.2	77.5	52.1	59.0	33.6	39.7	17.2	22.0	6.4	8.2
1981	71.9	79.0	53.4	60.1	34.7	40.7	18.0	22.9	6.9	8.8
1986	73.0	79.7	54.3	60.7	35.5	41.2	18.4	23.2	6.9	8.9
1991	74.6	81.0	55.7	61.7	36.9	42.3	19.4	24.1	7.4	9.5

*Excludes Quebec.

Source: Statistics Canada, *Report on the Demographic Situation in Canada 1994,* Cat. no. 91-209E (Ottawa: Minister of Industry, Science and Technology, 1994), Table A6, p. 106.

TABLE 5.3 Death Rates by Sex and Age, Canada, 1921-1990

| | Overall Mortality Rate | | | Infant Mortality Rate | | | 85+ Mortality Rate | |
	Male	Female		Male	Female		Male	Female
1921	10.9	10.2		98.2	77.4		228.2	224.9
1931	10.5	9.6		94.4	74.4		228.1	212.6
1941	10.8	9.1		67.0	51.9		241.9	229.3
1951	10.1	7.8		42.7	34.0		235.1	212.0
1961	9.0	6.5		30.5	23.7		208.9	192.2
1971	8.5	6.1		19.9	15.1		198.6	163.3
1981	8.0	6.0		11.0	8.5		188.5	141.6
1990	7.9	6.5		7.5	6.1		182.8	139.5

Notes:

(1) Deaths per 1,000 (male and female) population.

(2) Data not available for Newfoundland prior to 1949, Quebec (1921-25), the Yukon and Northwest Territories prior to 1950.

(3) Infant mortality: number of deaths per 1,000 live births.

Source: *Selected Mortality Statistics, Canada 1921-1990* (Ottawa: Statistics Canada, March, 1994).

Box 5.2 Burlington Breast Cancer Support Services

One of the most effective campaigns for women's health in the late eighties and early nineties in Canada has been social activism in relation to breast cancer. One of the most important examples of the potential for women's organizing can be found in the establishment of Burlington Breast Cancer Support Services (BBCSS) in Burlington, Ontario. Presently, BBCSS includes a storefront, a library, a newsletter, a board of directors, a telephone answering service, paid staff, ongoing sharing and support groups, mechanisms for fund-raising, and an international profile. It is located in a busy shopping mall within about an hour's drive of two major Canadian cities – Toronto and Hamilton. The organization has spawned at least two province-wide efforts – the Breast Cancer Support Network for Ontario and the Breast Cancer Resource Centre, Willows, presently situated in Toronto.

None of this would have happened without the tire-less and passionate commitment of Pat Kelly. When Pat Kelly was recovering from breast cancer surgery, she wanted to know what services were available to her. The diagnosis had been a surprise. She was young, in her mid-thirties, and did not have a family history of breast cancer. She was unprepared for the shock and the disruption and she had numerous ques-tions about how to manage. She says she felt that her life was threatened and she was terrified. So she did what many Canadians would likely think of first, she called the local chapter of the Canadian Cancer Society to ask what they might be able to offer her by way of support and information. The Cancer Society con-tacted a Reach to Recovery volunteer who telephoned Pat. As she says, the conversation could only be seen

as ironic. She was standing holding her 18-month-old when the telephone rang. She answered and was happy to hear the voice of the Reach to Recovery vol-unteer. She soon learned, however, that the volunteer was in her seventies and could not identify with Pat's situation. She said to Pat that the only piece of advice she could give her was not to lift anything with the arm on her affected side. Pat was carrying her baby with just this arm; she quickly realized that such advice was not what she wanted or needed.

When she had the time, she talked with friends about her situation. One friend, a public health nurse, knew a woman, Barb Sullivan, who had been diag-nosed the previous year and would, she thought, be willing to talk to Pat about her experiences. Pat called Barb and they talked on the telephone and later met informally to talk about their common experiences. After six months of meeting and talking and sharing stacks of books, popular articles, and articles culled via Medline from medical journals, including their "Bible," Rose Kushner's *Alternatives: New Advances in the War Against Breast Cancer*, they invited other women to join them. By March, 1988, Pat and Barb had con-tacted the local Burlington Family Y and began to advertise support group meetings at the Y for women diagnosed with breast cancer. In the process of reading books and articles on breast cancer, Pat and Barb came across references to Y-Me, a Chicago-based American national support group with a 1-800 telephone line and an educational service network.

By September 15, 1988, the "group" had a name, Burlington Breast Cancer Support Services, a newslet-ter, and the beginnings of other services.

Box 5.3 New Reproductive Technologies

Among the most controversial of medical interventions available today are the new reproductive technologies. Largely unheard of except in science fiction literature before the birth of the "test-tube baby" Louise Brown in 1978, these technologies fall roughly into four groups. They are (a) those concerned with fertility control (conception prevention), (b) labour and delivery "management" (high-tech deliveries in hospital by obstetricians/gynecologists), (c) pre-conception and pre-natal screening for abnormalities and sex selection (ultrasound, amniocentesis, genetic screening), and (d) reproductive technologies *per se* (conception, pregnancy, and birth management via technical, pharmaceutical, and medical intervention) (Eichler, 1988: 211). This discussion will be limited to the last category.

New reproductive technologies have separated gestational, genetic, and social parenthood for both men and women; they have eliminated the need for intercourse between a man and a woman for reproduction; they enable men and women to reproduce without an opposite-sex social parent, through surrogate mothering arrangements, on the one hand, or sperm-bank use on the other. With the new reproduction technologies, conception, gestation, and birth can be entirely separated from social parenting.

In 1985 the Canadian Medical Association decided that in-vitro fertilization was no longer experimental. This became a major factor in medicare's financial support of the procedure. The number of clinics for the treatment of the 15 per cent (or so) of Canadians who are infertile grew, along with the numbers of specialists and researchers interested in treatment of infertility, which has been defined as one year of attempting to achieve pregnancy without success (Achilles, 1990: 287). While cost estimates are difficult to assess, one study suggested that between 1985 and 1988 the

Ontario government spent $77 million in directly funding IVF clinics. This is a significant underestimate of the overall costs to society because it does not include physicians, hospitalization, drugs, medical devices, lost employment days, or other associated costs. By 1989 the Canadian federal government established a royal commission on new reproductive technologies to examine the medical and scientific developments related to IVF as well as the social, ethical, research, legal, and economic implications. The government was responding to the increasingly widespread conviction that technological developments were outpacing society's ability to understand and control them.

This is an immensely complex issue with extensive ramifications. Sociologists have long been concerned to understand the relationships between technological innovation and social change. In particular, they have sought to understand the social organization and social control of technological innovation. To simplify, in this case the fundamental questions are: In whose interests are the new reproductive technological developments? Whose interests ought they to serve? To what extent has the availability of reproductive technologies reinforced and even exacerbated a pronatalist philosophy? What are the effects of such a philosophy on men and women, particularly those who are infertile?

With respect to just one new intervention, in-vitro fertilization (IVF), there are enormous costs for the "mother," "father," and the fetus/embryo. For the mother and the fetus/embryo, these costs include significant short- and long-term effects. The treatments are very invasive and they involve administration of hormonal drugs at extraordinary levels. In the short run, the "mother's" body may experience pregnancy/non-pregnancy symptoms in turn, causing numerous

Box 5.3 continued

minor side effects such as nausea, headaches, cramping, and the like. In the long run, these drugs may lead to a greater vulnerability to cancers – particularly of the reproductive system. As no long-term studies have yet been done, these effects are now only speculative. In addition to the biological costs are the social and emotional costs of being preoccupied with bodily functioning, with motherhood, and with the experience of repeated failure to conceive. Estimates range from 0 to 8.5 per cent successful conceptions (Burstyn, 1992: 13). Women have likened the experience of being involved with IVF to a roller-coaster of emotions, including recurrent hope and despair. Even when

fertilization is successful, the offspring have higher than average neonatal and perinatal mortality rates; there is a much higher incidence of multiple births (with all the attendant risks), eleven times the risk of low birth weight (associated with numerous long-term developmental problems) (Jonas and Lumley, 1994: 661), perhaps higher rates of childhood cancer (Burstyn, 1992: 14), five times the rate of spina bifida, and six times the rate of transposition (an unusual heart defect).

Given these side effects and the ethical issues, it is no wonder that new reproductive technologies have been the subject of widespread debate.

---

new epidemics arise to threaten the health of the population as in the past? Or will the advances of the past continue on into the future? And if these advances do continue, can we expect that these additional years will be lived in a state of disability, or, as some have speculated, will people live an increasingly lengthier and healthier life, becoming ill only in the few months before death? (Fries, 1980).

Interestingly, none of these causes of the dramatic decline of mortality are explicitly sex-specific, with the exception of the relatively unimportant (numerically speaking) matter of birth control. It might be argued that one reason for the growing disparity between the mortality rates of men and women has been the decline in fertility and thus the smaller number of pregnancies for women. Clearly, pregnancy, childbirth, infant nursing, and early childcare responsibilities have a particular cost for the

health of women. Lowering the frequency of these events has had an impact on improving the health and life expectancy of women. If all else were equal, however, lowering the health-threatening impact of this one sex-specific event (pregnancy and childbirth) should result in equalizing the mortality rates of the two sexes. Instead, it has contributed to the increase in the life span of women as compared to men.

The question must then be raised, why is it that men today live shorter lives than women? First, the genetic superiority of women is an aspect of the explanation. More males are conceived and yet more male fetuses die (Waldron, 1981). Males, in a number of different species, have higher death rates than females (although this is not universal). However, even if genetics does play a significant part in the sex-mortality differential, it cannot be the only factor. For one thing, as Rutherford (1975) points out,

Box 5.4 Men, Steroids, Sports, and Health

Anabolic steroids are synthetic testosterone. They have been used by some athletes to build mass and muscle in order to improve their competitive status. Their deleterious effects are widespread. Steroid use is associated with a number of different changes. The increased amount of testosterone is, for instance, related to increased violence, sexual interest, and prowess. One researcher, in fact, has suggested that the use of steroids is a factor in the overall increase in violence – violent rapes, murders, and violence toward gays in the U.S. (Taylor, 1991: 69). Steroid use is also related to megorexia – the opposite of anorexia. Megorexia involves a distorted body image and a voracious, even insatiable appetite. The steroid user then would have an unattainable desire to acquire, not thinness as in the case of anorexia, but mass. A constant preoccupation with appearance can sustain dependency on steroid use. Steroids are also addictive and may cause serious psychological symptoms upon withdrawal. Moreover, when the user stops the body returns to the pre-steroid strength and shape. The dramatic physiological change can have devastating physical and psychological consequences. There are, too, a number of physical health conditions associated with steroid use, including liver and cardiovascular diseases, hypertension, acne, fluid retention, and sleep disturbance. Because steroid use is often not made public, its association with other health conditions may be invisible but widespread. Finally, in the age of HIV and AIDS, the needles used for injecting steroids may become infected and passed from athlete to athlete in a particular gym or on a particular team.

William N. Taylor, *Hormonal Manipulation: A New Era of Monstrous Athletes* (Jefferson, North Carolina: McFarland and Company, 1985).
William N. Taylor, *Macho Medicine: A History of the Anabolic Steroid Epidemic* (Jefferson, North Carolina: McFarland and Company, 1991).

the increase in the sex-mortality differential over the past 50 years could not be due entirely to genetics because genetic structures do not change that quickly. Furthermore, men are less likely to claim to be ill. If the cause of the mortality differential were genetic, surely it would be paralleled by morbidity rates for men and women.

Second, in order to explain this anomaly – that men are more likely to die, even while women are more likely to get sick – we must look at the causes of mortality by sex. Do men and women die for essentially the same reasons? The following tables provide some evidence in answer to these questions. The first describes the sex differences in PYLL (potential years of life lost – Table 5.4) and the second (Table 5.5) portrays the leading causes of death for men and women. The data in the tables show that men are more likely to die from diseases of the circulatory system, cancer, respiratory disease, accidents, and diseases of the digestive system. Death rates are higher among women for diseases of the endocrine and nervous systems.

The most important contributions to higher male mortality can be argued to be related to the following causes. The higher rate of cigarette smoking among men has a significant impact on the sex differential for lung cancer and also for respiratory diseases (Waldron, 1981), although the sex differential for lung cancer is narrowing.

TABLE 5.4 Potential Years of Life Lost (PYLL) by Selected Causes and Sex, 1985

Cause of Death	PYLL between 0 and 75 years				Deaths between 0 and 75 years (both sexes)	
	Males No.	Females No.	Total No.	%	No.	%
All malignant neoplasms	222,862	194,178	417,040	24.2	30,592	31.8
Diseases of the heart	220,537	75,422	295,959	17.2	26,582	27.7
Motor vehicle accidents	122,275	45,667	167,942	9.8	3,999	4.2
All other accidents	97,012	26,679	123,691	7.2	3,656	3.8
Suicide	87,827	21,316	109,143	6.3	3,113	3.2
Congenital anomalies	47,794	43,108	90,902	5.3	1,369	1.4
Causes of perinatal mortality (excluding stillbirths)	51,930	38,505	90,435	5.3	1,214	1.3
Respiratory disease	37,462	22,583	60,045	3.5	5,115	5.3
Cerebrovascular disease	26,450	22,855	49,305	2.9	4,370	4.5
All other causes	198,096	117,892	315,988	18.3	16,098	16.8

Source: *Canada Year Book*, 1988, Table 3.8, Catalogue 11-402E/1987 (Ottawa: Ministry of Supply and Services, 1989), p. 3-21.

TABLE 5.5 Leading Causes of Death, 1990

	MALE		FEMALE	
	No	Rate*	No.	Rate*
Diseases of the circulatory system	38,823	296.3	36,266	269.0
Cancer	28,865	220.3	23,560	174.8
Respiratory diseases	9,351	71.4	6,921	51.3
Accidents and adverse effects	9,064	69.2	3,993	29.6
Diseases of the digestive system	3,691	28.2	3,303	24.5
Endocrine diseases, etc.	2,533	19.3	2,939	21.8
Diseases of the nervous system	2,375	17.4	2,580	19.1
All other causes	9,358	71.2	8,434	62.7
Total	103,960	793.3	87,996	652.8

*Per 100,000 population.

Source: *Canada Year Book*, 1994.

Box 5.5 AIDS, Men, and Women

Today in Canada the death rate from AIDS (as well as the morbidity rate) is much higher for men than for women. This reflects the fact that AIDS was, initially, largely a disease of homosexual men, and, secondarily, of intravenous drug users. Today, however, the disease is growing rapidly among young heterosexual women. In the Third World, there are many places where HIV and AIDS exist in a 1:1 ratio of men to women. Female prostitutes and male migrant labourers are among those who have the highest rates. The symptoms and the progression of the disease seem to be somewhat different among males and females. Medical researchers are presently trying to match their level of understanding about HIV and AIDS in women with the increased incidence and prevalence. Social scientists are working to develop strategies for empowering women and educating men and women in sexual relationships so that they use barrier methods of birth control when sexually active or avoid sexual intercourse and the exchange of bodily fluids. A movement to promote chastity premaritally has sprung up across North America and groups promoting this choice have been speaking in high schools across the nation. HIV and AIDS prevention is among the most important health social issues facing young people today. What do you think are the best strategies for prevention?

Significantly, the rate of cigarette smoking among women is also increasing. Another important contribution to the respiratory and lung cancer differential is likely linked to the higher risks occurring at men's employment venues, as discussed in Chapter Four (see Tataryn, 1979). Men are more prevalent in the work forces of all of the industries that work with carcinogenic substances. (It is worth noting, however, that cleaning products used primarily by women in the home may be found to be carcinogenic and of significant cost to women's health.)

Risk factors for heart disease include smoking, high fat diet, and stress (particularly type A behaviour). These risk factors are more prevalent among men than women (Waldron, 1981). Higher alcohol consumption among men is implicated in several causes of mortality – but chiefly accidents (Clarke, 1990). Men drive more than women and less safely (Waldron and Johnston, 1981: 19). They consume more alcohol than women (Health and Welfare Canada and Statistics Canada, 1981). At least half of all fatal motor vehicle traffic accidents involve drunken drivers. The probability of other accidents and suicides is also increased by alcohol consumption.

Thus, while there may be gender-based genetic differences, it is clear that the male lifestyle, including cigarette smoking, industrial employment, excess alcohol consumption, and high fat diet, contributes to male mortality. In addition, the male tendency to engage in "machismo," violent, aggressive, and high-risk activities such as speeding in an automobile, cannot be overlooked in gender-specific mortality rates.

Age and Morbidity

Health problems affecting the elderly are of growing concern as the population ages. Estimates suggest that approximately 78 per cent of Canadian men and women who are over 65 have at least one chronic condition and 30 per cent have three or more (Williams and Rush, 1986). Although the elderly do not appear to go to the doctor more frequently than young people, they are more likely to be hospitalized and to use prescribed medications (Jennett *et al.*, 1991). In addition, there is some evidence that the elderly are more likely to be given prescriptions inappropriately (Ferguson, 1990; Shorr *et al.*, 1990; Brook *et al.*, 1989) and that as many as 19 per cent of hospital admissions may result from inadequate or inappropriate drug prescriptions (Grymonpre, 1988). In addition to overprescribing by doctors there is evidence that some of the problems associated with medicine use among the elderly result from metabolic changes due to aging that affect the absorption rates of drugs, side effects from multi-drug use associated with the simultaneous treatment of several problems, and drug-taking mistakes made by the elderly themselves as the result of problems with their visual, motor, or memory abilities (Tamblyn *et al.*, 1994).

The elderly have more chronic conditions, are more likely to be hospitalized and medicated, and yet they are still likely to report good health and health satisfaction (Table 5.6). This may be because they see and evaluate themselves not in absolute but relative terms in comparison with others of the same age.

TABLE 5.6 Self-Reported Health and Happiness Status of Canadians Ages 55 and Over, 1990

	Men	Women
HEALTH STATUS		
Excellent	32%	27%
Good	46	48
Fair	17	19
Poor	4	6
Not stated	1	1
HAPPINESS STATUS		
Very happy	56	49
Somewhat happy	39	43
Somewhat unhappy	2	5
Very unhappy	1	1
No opinion, not stated	3	2

Source: Statistics Canada, *Canadian Social Trends* (Summer, 1992), p. 26 (Catalogue No. 11-008).

Gender and Mortality

Throughout the nineteenth and twentieth centuries, where records are available for industrializing and industrialized societies, they show that males have often experienced greater mortality rates than females. This difference begins even before birth, because there are more stillborn male babies than stillborn female babies (Lewis and Lewis, 1977). In fact, there are significant differences between men and women in both mortality and morbidity rates. Morbidity rates are higher for women. While men are more likely, at every age during their life cycle, to die, women are more likely to be ill. However, this has not

Box 5.6 Males, Risk-Taking, and Mortality

At every age (except infancy) males are much more likely to die as the result of an accident than females. During the teenage years (15-19), young men are more than four times as likely as young women to die as the result of an accident. During the next five-year period, 20-24, the disproportion is even greater – almost five times as many young men as young women die accidentally. Even at the last stages of life, 80 and over, men are more likely to die than women.

Among the types of accidents included in these statistics are motor vehicle, falls, poison, drowning, choking, guns, and electrocution. A number of explanations for gender differences in health outcomes are described in this chapter. Which ones do you think best explain these figures? Are there other explanations that you could suggest? What social changes might have an impact on the accident rates?

---

always been the case. Before industrialization and capitalism, the mortality rates for men and women were approximately the same (Eyer, 1984). At times, the female rate was higher than the male rate. The higher rate for women was often associated with the high rate of maternal mortality because of women's limited access to important health-giving resources such as adequate nutrition, foodstuffs, and birth control. Over the past century, improvements in nutrition, the spread of more effective contraception, and the overall improvement in the status of women have had a dramatic effect on their life expectancy.

Table 5.6 portrays the different morbidity rates of men and women. A comparison of the overall life expectancy with the overall disability-free life expectancy reinforces the point made earlier: that men are more likely to die, women to get sick. The difference in life expectancy between Canadian men and women overall is 6.4 years. The rank ordering of overall causes of death is the same for both men and women (Table 5.5). The most important cause of death is circulatory

system disease. The second is cancer or all malignant neoplasms. Within this category, lung, bronchial, and tracheal cancers are the largest causes of male death, while breast cancer is the largest cause of female death. This is, however, followed closely by lung, bronchial, and tracheal cancer for women. Moreover, these two cancer rates are growing for women and as women continue to take up smoking, their lung cancer rate will continue to increase while the male rate will continue to decrease. Another interesting difference is that the death rate by accident is more than twice as high among men as among women. This figure includes automobile and other accidents, suicide, and homicide.

Gender and Morbidity

Women can expect to live longer than men, yet a higher percentage of their years alive involve the experience of some disability (see Table 5.7). Expected years of life free of severe disability represent 97 per cent of

FIGURE 5.2 Prevalence of Health Problems by Sex, Age 15+, Canada, 1991

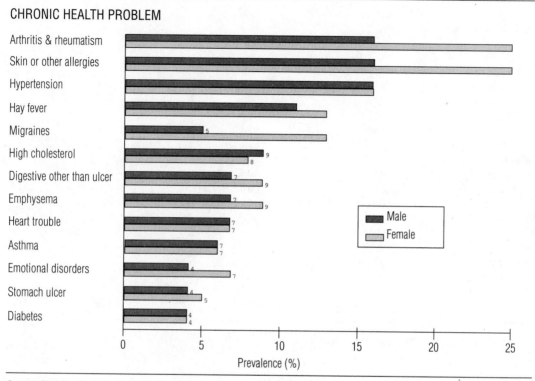

CHRONIC HEALTH PROBLEM

Source: Statistics Canada, *Health Status of Canadians*, Cat. 11-612E, No. 8, p. 27.

total male life expectancy and 94 per cent of female life expectancy (*Canada Year Book*, 1994). Women's greater life expectancy is clearly at a cost of more years of the experience of disability. Women are more likely than men to be ill, no matter how the illness rate is determined: by self-report, clinical records, physiological testing, days of disability, doctor visits, or hospitalization. In short, the incidence of illness among females is higher than it is among males. A small part of this difference is accounted for by the fact that pregnant women are expected to visit the doctor for regular

prenatal checkups and to deliver the baby in hospital. In addition, childbirth is followed by routine medical checkups for both mother and baby (medical "care" for pregnancy and childbirth is included as illness in the statistics). The sex difference for acute illnesses is between 20 and 30 per cent greater for women than men overall (Verbrugge, 1985), including childbirth-related illness. It is only slightly lower when childbirth-related illnesses are excluded.

Most non-fatal chronic illnesses are also generally more prevalent among women. Figure 5.2 shows some of these health

TABLE 5.7 Mean Disability Days, by Sex and Age Group, Age 15+, Canada, 1978-79, 1985, 1991

SEX AND AGE GROUP	1978-79	1985	1991
	MEAN DISABILITY DAYS		
BOTH SEXES			
Population 15+	0.72	0.74	0.64
15-44 years	0.49	0.59	0.57
45-64 years	0.97	0.80	0.66
65+ years	1.40	1.39	0.90
MALE			
Population 15+	0.55	0.63	0.56
15-44 years	0.33	0.52	0.53
45-64 years	0.84	0.71	0.53
65+ years	1.21	1.07	0.72
FEMALE			
Population 15+	0.88	0.86	0.71
15-44 years	0.65	0.66	0.60
45-64 years	1.08	0.90	0.79
65+ years	1.54	1.64	1.02

Source: Statistics Canada, *Health Status of Canadians*, Cat. 11-612E, No. 8, p. 52.

differences between the sexes. There is a remarkably higher incidence of reports of certain problems among women than among men: for instance, migraine, arthritis and rheumatism, and skin and other allergies. It is notable that the health problems reported more often by men are more often causes of fatality, e.g., high cholesterol associated with heart disease.

When the leading types of illness for men and women are ranked, they are largely comparable. In effect, men and women generally suffer from the same sorts of illness and disability, but men experience them more severely and are more likely to experience the more severe types of heart diseases. Furthermore, men's illnesses proceed more quickly to death. In summarizing a decade of research based on morbidity and mortality data from a variety of sources in the United States, Verbrugge (1985: 668) concludes as follows: "Women have more frequent illness and disability, but the problems are typically not serious (life-threatening) ones. In contrast, men suffer more from life-threatening diseases, and [there is] more permanent disability and earlier death from them."

The morbidity rates for men and women are potentially inaccurate for three known reasons. In the first place, male rates are likely under-reported. Household surveys

Box 5.7 *Shape* and *Men's Fitness*

Shape and *Men's Fitness* are two popular fitness maga-zines; the first is directed toward females and the second toward males. Published by the same pub-lisher, the circulation figures of *Shape* and *Men's Fitness* are 781,000 and 240,000, respectively. A comparative analysis of the two magazines over 1992 indicated a contradiction of stereotypical images of the health and fitness of males and females. Some examples of stereotypical differences follow. (1) Cover models in *Men's Fitness* tend to be action-oriented males observed biking, running, hiking, weight-lifting, walking, and skiing. In contrast, the models on the cover of *Shape* tend to be in inactive poses such as sitting by the pool or standing on the beach. (2) The attire of the male models in *Men's Fitness* appears to be appropriate to the particular sports in which they are involved. However, the female models wear costumes that would often be banned, as too revealing, in a typical fitness club. (3) Each month, the magazines present "success" stories submitted by readers. Although the ostensible purpose of the magazines is the promotion of health and fitness, the success stories for women tended to describe weight-loss achievements. By contrast, the success stories of males tended to describe weight gain. (4) An analysis of the portrayal of holidays noted that Valentine's Day was an important issue for females (*Shape*) in that two articles were dedicated to it. On the other hand, there were no articles in *Men's Fitness* on Valentine's Day. This seems to reflect a continuing focus on men as instrumental and woman as socio-emotional specialists.

This is a very brief overview of a report that com-pared two gender-specific health and fitness maga-zines in the 1990s (Johnstone and Robinson, 1995). It is not to be generalized from but to be read as a series of hypotheses you might test as you observe your own selection of mass media.

---

tend to rely on the answers of available sub-jects, and these have tended to be women. Interview subjects are known to under-report the morbidity of the absent person (Nathanson, 1977: 20). In the second place, health statistics probably minimize reports of women's illness because health statistics focus on the most publicly visible health problems rather than relatively minor, private complaints. Yet women are more likely to suffer any number of minor yet uncomfortable sensations associated with non-fatal chronic or acute conditions (Verbrugge, 1985). The American National Health Interview Survey has shown that women complain more often than men of minor health problems such as headaches, insomnia, palpitations, and tremors. Such problems are likely to lead to preventative health actions such as visits to a doctor, which then get counted in morbidity sta-tistics. Third, preventative or early stage medical care is also more frequent for women because of visits to the doctor for pregnancy, childbirth, and their children's health care. But should preventative action be considered morbidity?

The conclusion that must be drawn, then, is that women are more likely to suffer from a variety of illnesses than men and that men are more likely to die at every age than women. Which is the better measure of parity? To live (and be ill longer) or to die earlier? What will the future bring? Are women actually "dying to be equal," as evidenced by their growing rates of cigarette smoking, alcohol consumption, and increased involvement in non-traditional, high-risk, and high-stress careers? These questions will be addressed in the last section of this chapter.

Explanations for Differences in Disease and Death

Before attempting to discuss the reasons for age, gender, class, and ethnic differences in morbidity and mortality, it is important to understand the causes of illness and death. The 1977 Health and Welfare publication, *A New Perspective on the Health of Canadians*, distinguishes among four components of health: human biology, environment, lifestyle, and health-care organization. Human biology covers all aspects of the mental and physical health of the body, including genetic inheritance, the processes of maturation and aging, and the many complex internal systems in the body. Environment covers all factors external to the body that affect health: clean drinking water, clean air, adequate foodstuffs, garbage and sewage disposal, and also the social environment: gender, social class, and ethnic and cultural differences. Lifestyle refers to the aggregate of individual health habits such as exercise, diet, smoking, alcohol use, seat belt use, and promiscuity and sexual carelessness. Finally, health-care organization covers the technologies, facilities, personnel, and organizations devoted to medical care.

The question then becomes, what is the link between inequities in the social structure and the components of health? Table 5.8 simplifies and summarizes some of the explanations for health inequities for each of these components.

Age

Of all of the four causes for differences in mortality and morbidity rates examined in this chapter, age seems the least controversial explanation. It is a biological fact that people age and their bodies undergo some degeneration. Many kinds of chronic diseases seem to be largely associated with old age, such as cancer, arthritis, stroke, and heart disease. Certainly, biological factors are part of the explanation for age differences in mortality and morbidity. But there are social and economic causes as well. These are listed in Table 5.8 and described below.

One: differences in lifestyle. People of different age groups or generations have lived through different historical and political/economic circumstances. Such events as war and depression have significant and long-term consequences for the health and disability levels of those who live through such calamities.

Two: psychosocial aspects of symptoms and care. Different historical periods have different cultures and social norms regarding the recognition of symptoms and signs and the action to be taken to respond to them (e.g., whether to doctor oneself, do yoga, or go to

TABLE 5.8 Summary of Explanations of Inequalities in Morbidity and Mortality

FOUR FACTORS IN THE SOCIAL STRUCTURE THAT ARE LINKED TO HEALTH

AGE	GENDER	CLASS	ETHNICITY
Older people more likely to use medical facilities and experience chronic illness, but less likely to perceive themselves as ill	High incidence of minor illness in females High incidence of serious illness in males	The lower the social class, the greater mortality and morbidity	Native people have high incidence of morbidity and mortality from contagious and infectious diseases and from alcohol and violence

HYPOTHESIZED CAUSAL LINKS

1. Biological differences 2. Differences in lifestyle 3. Psychosocial aspects of symptoms and care-taking behaviours 4. Material inequities	1. Biological differences 2. Acquired (lifestyle) differences 3. Psychosocial aspects of acknowledgement symptoms 4. Willingness to talk about health and illness 5. Frequent use by women of medical care system because of childbirth procedures as well as illness 6. Differing responses by doctors to male and female patients	1. Biases in measuring and recording processes 2. Illness causes downward mobility 3. Class-based cultural and behavioural differences 4. Class-based material inequities	1. Racism 2. Lifestyle differences 3. Class-based material differences laeding to inequities 4. Biological differences 5. Environmental hazards

the doctor). Any and all of these things may result in different health outcomes.

Three: differential access to health-giving resources. There is no doubt that there are significant political and economic differences between different age groups. It is also clear that poverty accompanies aging, particularly for women. Income disparities lead to differences in nutrition, stress levels, density in living quarters, access to transportation, and the like. Such resources affect health status.

Gender

Verbrugge (1987) has made a major contribution to understanding gender differences in morbidity and mortality. After a thorough examination of the research literature available on the effects of differences in biology, lifestyle, preventative measures taken, and use of medical facilities and treatment, Verbrugge (1985) reached the following conclusions.

(1) Because of both biological factors and

Box 5.8 Health Research and Its Relationship to Women

In 1990 the Office of Research on Women's Health (ORWH) at the National Institute of Health was established by the American government. In 1992 the ORWH began to ask and answer a growing number of questions about differences in the health of men and women. It concluded that the health problems of women were getting worse and that scientific research to understand and prevent this worsening was not occurring. The report concluded that: (1) women constituted the majority of the population and would continue to do so, and that their health was getting worse; (2) women's health was poorer than that of men; (3) some health problems were more common in women than in men; and (4) certain health problems affected women and men differently. The report also noted how women's quality of life was inferior to that of men.

Among the explanations for the state of women's health was the argument that medical research and even the culture of science had excluded and minimized the importance of women's health. David Noble has put this point of view very forcibly: "Throughout most of evolution, the culture of science has not simply excluded women, it has been defined in defiance of women and in their absence." This has resulted in both a relative absence of female scientists and in a marginalization of research on women's bodies. Women's bodies have been assumed to be like those of men, but more difficult to research because of their reproductive changes such as menstruation, pregnancy, and menopause. They have been seen as difficult to study and therefore understudied.

But by 1994 significant policy changes were made in the NIH to advance the cause of women's health research. One impetus to these policy changes was a significant conference held by the Women in Science section of the New York Academy of Science in New York City in 1992 and the subsequent publication of the presented papers (Sechzer, Griffin, and Pfafflin, 1994). There is evidence of similar critical analyses in Canada – for instance, the National Forum on Breast Cancer held in Ottawa in 1993 and the Canadian Advisory Council on the Status of Women meetings on women's health in 1994. Thus women's health issues are becoming politicized in North America.

---

lifestyle, women have more illness of a mild, transitory type and men of a more serious type. Mild illness accumulates over time, so that women suffer more bed-days and disability days and are more likely to see themselves as ill than men are. When men get sick, however, it is more likely with a serious or fatal condition.

(2) Women are more attentive to bodily sensations and more willing to talk about them. They tend to take more care for each episode of illness. When the illness is serious (e.g., cancer) men and women are equally likely to take action.

(3) While the sexes have similar levels of ability to remember major health problems, women are better "describers" of mild problems because they are willing to talk about them. Women are more likely to include their feelings in their descriptions of their health.

(4) Women's greater attention to minor signs and symptoms, and their greater willingness to take preventative and

TABLE 5.9 Relative Rankings in the Causes for Differences in Illness Rates in Women and Men

RANK	EXPLANATION
One	Differences in lifestyle
Two	Psychosocial differences in the acknowledgement of signs and symptoms and associated behaviours
Three	Prior medical care
Four	Biological differences
Five	Willingness to report and talk about illness
Six	Differing responses of doctors to males and females

healing actions (i.e., bed rest, diet) mean that their health problems tend not to become as severe as those of men of the same age. This greater carefulness regarding their health helps women extend their lives. Table 5.9 separates out six known causes of gender differences in morbidity and ranks them (*ibid.*).

Summary

(1) Illness and death rates vary depending on social-structural conditions: cultural differences, the political-economic system, the socio-demographic structure, social psychology, and micro meaning.

(2) One socio-economic characteristic that affects morbidity and mortality rates is age. Typical of most of the industrialized world is the aging of the population, which occurs with an overall decline in fertility rate and an increase in life expectancy, among other factors.

(3) Life expectancy increases are the result of many changes, including a rapid decline in the infant mortality rate and medical advances and their impact on the later stages of the life cycle. Both men and women have undergone great increases in life expectancy in the past 150 years, yet the excess of male mortality over female mortality has been greater in the last 25 years.

(4) Potential years of life lost (PYLL) is a useful figure because it calls attention to causes of death among the younger population. From this statistic, we can see that males are more than twice as likely as females to die prematurely from causes such as motor vehicle accidents, ischaemic heart disease, other accidents, and suicide.

(5) As people age they are more likely to be ill, yet older people are more likely to perceive themselves as "healthy for my age" and to be fairly satisfied with their health.

(6) At every stage in their life cycles, men are more likely to die and women are more likely to be ill. This disparity seems to be related to industrialism and capitalism.

CHAPTER SIX

Social Inequity, Disease, and Death: Class, Race, and Ethnicity

WHY DO SOME PEOPLE seem to be sick more than others? Why are others usually well? Does health differ in different regions of the country? What does the level of education have to do with health, illness, and death? Do poor people live shorter lives? Are the poor more often sick than their wealthier counterparts? These are some of the questions that this chapter addresses. Health, illness, and death are linked to the social organization of the society. One crucial aspect of this social organization is the extent of inequity in the social structure. Inequity is associated with different life chances and unequal access to such resources as food, recreation, satisfying work, and adequate shelter. Because of that unequal access, people who differ in age, sex, income, class, occupation, ethnicity, marital status, rural or urban background, and religiosity differ in their rates of sickness and death.

This chapter will examine how illness and death rates vary depending on specific social conditions: income, race, and ethnicity. It will then discuss some of the possible explanations for the observed variations. This chapter's analysis of inequity is limited to Canadian society and a focus on social-structural positions such as class, race, and ethnicity. Such variables are associated with different health patterns for different parts of the population.

Social Class

One of the most consistent of epidemiological findings supported by a wide variety of evidence is that there is an inverse relationship between social class and mortality. Low socio-economic status has repeatedly been associated with a high incidence of infectious and parasitic diseases and a high rate of infant mortality. Antonovsky (1967) noted that the high mortality of the lower classes from all causes of disease has been recorded since the twelfth century. The data highlighted in the following section show how this general finding has particular relevance for Canada. In spite of a national health service of more than twenty-five years' duration, there continues to be a significant and consistent class gradient in health outcomes. Figures 6.1 and 6.2 portray the differences in life expectancy

Box 6.1 Income Distribution within a Society and Its Effect on Health

In developed countries, death rates are lowest when the degree of egalitarianism between classes is highest. "Income distribution is now probably the single, best predictor of longevity among developed nations" (Wilkinson, 1990: 391). Moreover, there is a tendency for causes of death for which nations have unusually high rates to be those that show the greatest socio-economic differences internally. This pertains to cirrhosis of the liver in France, respiratory disease in England, and homicide in the United States.

The absolute GNP or the average income of the people within the country is less important than the relative equality of wealth of people in a given society.

In the developing world, health is a function of the physical effects of living circumstances as well as the meaning of the social conditions. Moreover, the social tensions exacerbated by inequality impact on the poor and the wealthy alike. "It looks as if it may damage social relations among people at all levels of society, and lead, where relative deprivation is greatest, to a complete breakdown of the social fabric" (*ibid.*: 409). The following figure portrays this finding against a selection of countries. Those countries where the poorest 70 per cent of the population have the most, tend to have the highest rates of life expectancy.

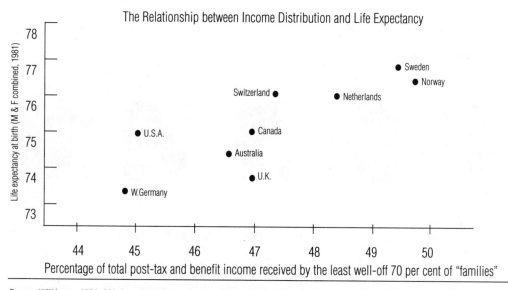

The Relationship between Income Distribution and Life Expectancy

Source: Wilkinson, 1990: 392, from: Luxembourg Income Study, Working Paper 26, World Bank, World tables.

at birth for males and females of different neighbourhood income levels in Canada (based on 1986 data). As these figures indicate, there is a direct relationship between life expectancy and average neighbourhood income. This is true for both men

FIGURE 6.1 Life Expectancy of Males by Neighbourhood Income, 1986

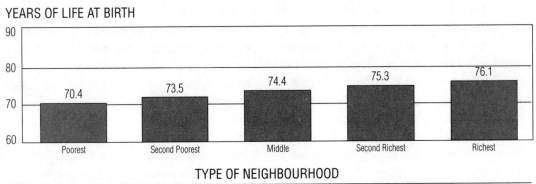

Source: *Health, Health Care and Medicine: A Report by the National Council of Welfare*. Ottawa, 1990.

FIGURE 6.2 Life Expectancy of Females by Neighbourhood Income, 1986

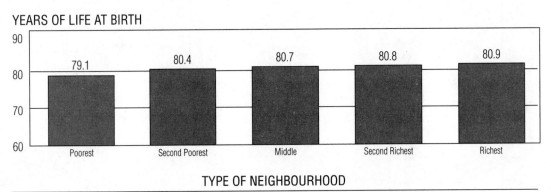

Source: *Health, Health Care and Medicine: A Report by the National Council of Welfare*. Ottawa, 1990.

and women, although the class gradient is steeper for men than women. Moreover, within every neighbourhood "type," women live longer than men: those of the highest social class can expect to live the longest. These findings are true for both men and women. The life expectancy range for men from highest to lowest neighbourhood income is 5.7 years; for women, 1.8 years. Figures 6.1 and 6.2 also indicate that class differences reinforce sex differences. The gender disparity is 8.7 years between men and women of the lowest income group and only 4.8 years in the highest

TABLE 6.1 Infant Deaths per Thousand Births in Urban Canada, by Neighbourhood Income

	1971	1986
Poorest neighbourhoods	20.0	10.5
Second poorest	16.6	8.0
Middle	15.2	7.7
Second richest	12.4	5.7
Richest neighbourhoods	10.2	5.8
All neighbourhoods	15.0	7.5

Source: *Health, Health Care and Medicine: A Report by the National Council of Welfare*. Ottawa, 1990.

TABLE 6.2 Prevalence of Three Health Status Indicators by Income Adequacy, Age 15+, Canada, 1991

	HEALTH STATUS INDICATOR		
Income Adequacy	Activity Limitation (%)	Two-week Disability (Mean no. of days)	Very Satisfied with Health (%)
Total	11	0.64	55
Lowest	25	1.34	37
Lower middle	19	0.96	47
Middle	13	0.70	54
Upper middle	9	0.53	57
Highest	7	0.48	65
Not stated	10	0.56	55

Source: Statistics Canada, *Health Status of Canadians*, 1991, Cat. 11-612E, no. 8, p. 46.

income group. Regardless of gender, statistics for life expectancy, computed on the basis of the quality of the neighbourhood, indicate that those from a poor district, as compared to those from a middle-class district, live shorter lives.

This association is reinforced by the data in Table 6.1 on infant death rates by neighbourhood income in 1971 and 1986. The first thing to notice here is that there have been large declines in the infant death rate from 1971 to 1986. The second is the clear class gradient demonstrating the highest levels of infant deaths in the poorest neighbourhoods

TABLE 6.3 Prevalence of Selected Health Problems by Sex and Income Adequacy, age 15+, Canada, 1991

Sex and income adequacy	Total population 15+(000s)	Any health problem	Hyper-tension	Heart trouble	Diabetes	Arthritis/ rheuma-tism	Asthma	Emphy-sema, ect.	Hay fever	Skin or other allergies	Stomach ulcer	Other digestive problems	Recurring migraines	High blood cholesterol	Any emotional Disorders
BOTH SEXES															
Total	20,981	100%	63%	16%	7%	4%	21%	6%	8%	12%	21%	5%	8%	9%	8%	5%
Lowest	799	100	73	22	15	4	37	8	19	14	24	9	12	16	9	17
Lower middle	1,633	100	71	22	12	7	31	8	15	9	21	7	11	12	12	9
Middle	4,766	100	63	17	8	3	23	5	9	9	19	6	10	10	8	6
Upper middle	5,743	100	60	14	5	3	16	5	6	13	22	4	7	9	8	4
Highest	2,171	100	61	14	4	3	12	6	4	16	18	2	6	7	9	2
MALE															
Total	10,266	100	59	16	7	4	16	6	7	11	16	4	7	5	9	4
Lowest	261	100	61	19	10	–	31	–	15	17	17	10	10	–	–	10
Lower middle	686	100	70	19	13	9	29	10	17	9	17	8	11	6	13	6
Middle	2,264	100	59	16	8	4	21	4	8	9	14	6	9	6	8	5
Upper middle	3,067	100	56	16	5	3	13	6	5	12	16	3	6	4	9	3
Highest	1,340	100	62	18	4	3	10	6	3	16	14	–	6	5	10	–
FEMALE															
Total	10,715	100	66	16	7	4	25	6	9	13	25	5	9	13	8	7
Lowest	538	100	79	23	17	5	40	9	20	12	27	8	13	19	11	20
Lower middle	947	100	72	24	12	6	33	6	13	9	24	7	10	15	10	11
Middle	2,503	100	66	17	8	3	25	5	9	10	23	5	12	14	8	7
Upper middle	2,676	100	64	13	5	3	20	4	6	15	28	5	8	14	7	5
Highest	831	100	60	9	3	–	16	7	4	15	25	–	4	10	7	–

Source: Statistics Canada, *Health Status of Canadians*, March, 1994, Cat. 11-612E, no. 8.

in both time periods. Moreover, as Table 6.2 indicates, income adequacy is directly related to activity limitation, disability in the two weeks prior to the survey, and satisfaction with health. Again, the poorest suffer more disability and are least satisfied with their health.

Class or income differences are also evident in the prevalence of illness. Table 6.3 illustrates the relationship between income adequacy and selected health problems for men and women. Overall, for both sexes, of those in the lowest income bracket, 73 per cent have health problems, while 61 per cent of those in the highest income bracket have health problems. When this is looked at by sex it is interest-ing to note that the overall class difference is the result of differences among women, not men (i.e., 79 per cent of women in the lowest income category report health problems while only 60 per cent of those in the highest income category report health problems). This pattern is reversed among men, where those in the highest income category report more problems than those in the lowest (62 per cent as compared with 61 per cent). However, when specific conditions are considered, income ade-quacy appears very important for both men and women. Canadians with the lowest income adequacy are more than three times as likely to experience arthritis and almost nine times as likely to experi-

TABLE 6.4 Female Population by Time Since Last Pap Smear Test, by Age and Education, Canada, 1978-1979

EDUCATION		TOTAL (000s)	LESS THAN ONE YEAR	1-2 YEARS	MORE THAN TWO YEARS	NEVER	UNKNOWN
Age 15 and over:							
Total	No.	8,907	3,701	1,559	1,305	1,826	516
	%	100.0	41.6	17.5	14.7	20.5	5.8
Secondary or less	No.	6,666	2,512	1,168	1,028	1,493	465
	%	100.0	37.7	17.5	15.4	22.4	7.0
Some post-secondary	No.	697	333	133	61	157	33
	%	100.0	47.7	16.2	8.8	22.6	4.8
Degree or diploma	No.	1,498	839	272	205	165	18
	%	100.0	56.0	18.2	13.7	11.1	1.0
Unknown	No.	47	17	–	11	–	–
	%	100.0	37.4	–	24.2	–	–

Source: *Health of Canadians.* Report of the Canada Health Survey, Catalogue 82-538, Table 101 (Ottawa: Statistics Canada, 1983), p. 185. Reproduced with permission of the Minister of Supply and Services, Canada, 1989.

ence emotional difficulties as those in the highest income categories. By comparison, hay fever tends to be slightly more common among those with higher income.

The relationship of these selected health problems to income adequacy is usually stronger for women than for men. For instance, men in the lowest income group are two and a half times as likely as men in the highest income group to experience heart problems, and women in the lowest income group are almost six times as likely as women in the highest income group to experience heart problems.

Power is often considered a component of social class or the effect of social standing. The relationship between power and illness was raised by the findings of the British-based Whitehall Study, which compared the health of more than 10,000 British civil servants over nearly two decades. None were impoverished. None were deprived. All were white-collar workers in office jobs. Nevertheless, there was a direct decrease in mortality associated with higher status and more power in the organization. A similar finding was noted in a larger Swedish study – heart disease was more prevalent among those whose work was psychologically demanding and lacking in decision-making power. In a further refinement of the British study that intended to explain the relationship between health and empowerment more clearly, researchers noted that while the level of stress was as high for those in the more powerful positions, it declined dramatically when they left work and went home. They seemed much more able to turn off the stress response (Taylor, 1993).

TABLE 6.5 Female Population by Frequency of Breast Self-Examination, by Age and Education, Canada, 1978-1979

EDUCATION		TOTAL (000s)	MONTHLY	QUARTERLY	LESS OFTEN	NEVER	DON'T KNOW HOW	UNKNOWN
Age 15 and over:								
Total	No.	8,907	1,884	1,840	1,642	2,736	584	222
	%	100.0	21.1	20.7	18.4	30.7	6.6	2.5
Secondary or less	No.	6,666	1,341	1,270	1,116	2,260	481	198
	%	100.0	20.1	19.1	16.7	33.9	7.2	3.0
Some post-secondary	No.	697	153	157	166	178	34	–
	%	100.0	22.0	22.5	23.8	25.5	4.9	–
Degree or diploma	No.	1,498	378	402	351	288	66	14
	%	100.0	25.3	26.8	23.4	19.2	4.4	9.9
Unknown	No.	47	–	12	–	22	–	–
	%	100.0	–	25.0	–	24.6	–	–

Source: *Health of Canadians*. Report of the Canada Health Survey, Catalogue 82-538, Table 102 (Ottawa: Statistics Canada, 1983), p. 186. Reproduced with permission of the Minister of Supply and Services, Canada, 1989.

Education

Education is an important resource in Canadian society. It opens doors to certain (and often safer) occupations; it increases the chances of promotion to a more satisfying occupation. It may also have a direct effect on health. Evidence suggests that the level of education is associated with preventative action on the part of women. As Tables 6.4 and 6.5 show, women with more education are more likely both to have Pap smears done and to do regular breast self-examination for breast lumps. Both of these preventative actions are known to be associated with a decreased mortality rate from cancer.

However, taking all things into consideration, lower-income Canadians do not have poorer health habits than higher-income Canadians; in fact, if anything, their health habits are slightly better (*Active Health Report*, 1987: 36). Higher-income earners are more likely to drink heavily, to drink and drive, to use marijuana, and to experience high stress levels. But those with lower incomes are less likely to have blood pressure checkups, and they get less exercise and use more tranquilizers. Lower-income Canadians also face more barriers to health in that they are more likely to be unemployed and poorly educated. In addition, they and their friends are more likely to smoke, which may indirectly influence their other health-related behaviours. Furthermore, they are much more likely to be exposed to occupational health hazards.

Health disparities in children are also associated with income differences. In the Ontario Child Health Study the marked

FIGURE 6.3 Weighted Prevalence Rates of Chronic Health Problems per 1,000 Children, by Socio-economic Status

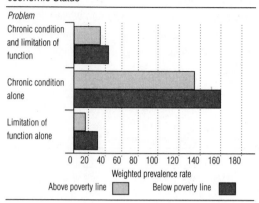

Notes:

1. Prevalence is the proportion of people in a population that have the condition under study. "Weighted prevalence" means that the prevalence in a sample is weighted to represent the prevalence in the total population.

2. Sample based on *Census of Canada*, 1981.

3. Poverty line as established by Statistics Canada, 1983.

Source: D. Cadman *et al.*, "Chronic Illness and Functional Limitation in Ontario Children: Findings of Ontario Child Health Study," *Canadian Medical Association Journal*, 135 (1986), pp. 761-67.

differences between children below and above the poverty line with respect to various chronic conditions were noted, as shown in Figure 6.3. Whether determined by a chronic condition, limitation of function, or both, children below the poverty line are in a poorer state of health.

Race, Ethnicity, and Minority Status

Race, ethnicity, and minority status are also important factors in Canadian society and around the world. They should be defined and distinguished. "Race involves a cluster of biological traits that form three broad, overarching categories: Caucasians, Mongoloids and Negroids. There are, however, no pure races. Ethnicity is based on shared cultural heritage. Minorities are categories of people who are socially distributed – including those of certain races and ethnicities – and who have a subordinate social position" (Macionis *et al.*, 1994: 345).

There is a close and persistent relationship between ethnicity and class. A significant amount of economic inequality can be attributed to differences in ethnic background (Reitz, 1980). Inequality has significant implications for social relations between people and among those of differing backgrounds. Furthermore, the way that others view an individual has important consequences for the individual's self-esteem. To the extent that ethnicity is related to occupational status and income and education, then people of different ethnic groups will differ in their morbidity and mortality rates. Table 6.6 portrays the wide variation in average income, education, labour force participation, and unemployment rate found in various ethnic groups. While data on the life chances of specific ethnic groups is not available, Bolaria and Bolaria (1994) compared death rates for the Canadian-born and foreign-born in Canada in 1985-87 and found that the death rates favour the foreign-born (Table 6.7). It is important to note that the most appropriate comparison here is between standardized death rates because they control for the age distributions within the population. Thus,

Box 6.2 Race and Life Expectancy: The International Context

The following table portrays a little-known conceptu-alization of race and health. The researchers calculated the racial composition of each of 118 countries around the world and then categorized these countries as mainly white and mainly non-white. In comparing the average life expectancy in each group they found that people in mostly white countries live longer, on average, than those in mostly non-white countries. Males live an average of ten years longer in "white" countries and females live an average of thirteen years longer.

Gender-Specific Life Expectancy Rates and Infant Mortality Rates of Countries by Racial Composition, for 118 countries

	All Countries		Countries of Mainly White Populations		Countries of Mainly Non-White Populations	
	Mean	SD*	Mean	SD*	Mean	SD*
Average male life expectancy at birth	64.0	9.0	71.6	2.8	61.1	8.8
Average female life expectancy at birth	68.6	10.2	77.9	2.6	65.0	9.7
Infant mortality rate per 1,000	48.0	40.8	12.7	9.9	61.7	40.1
No. of countries	118		33		85	

SD = standard deviation
Source: Bolaria and Bolaria, 1994.

the major reason the crude death rates for foreign-born are higher is that more foreign-born Canadians are in the age groups with the highest mortality rates.

The one group for which there are some data available is Canadian Native people. There is some debate about whether Aboriginal Canadians ought to be consid-ered an ethnic group or not. Those who favour the ethnic group designation argue that there are cultural similarities among all fifty-five or more sovereign Native peoples, and all are a colonized people, so that Native people constitute an ethnic group. Others, however, argue that because the cultural diversity of the Native peoples at colonization has been seriously compro-mised by the dominance of the state, they cannot be considered a cultural or ethnic group. Regardless of which side of this complex controversy one takes, the evi-dence remains clear that Native people

TABLE 6.6 Education, Employment, and Income among Selected Ethnic and Racial Categories, 1985-1986*

	British	French	Asian	Black	Native
Total Population	6,332,725	6,093,160	600,530	174,970	373,260
Population 15 and over	5,156,505	4,869,515	465,160	128,540	241,325
Post-secondary certification**	29.2%	29.7%	37.2%	35.8%	13.3%
Labour force participation**	62.7%	63.1%	70.8%	75.2%	50.1%
Unemployment rate	9.9%	12.7%	8.6%	12.5%	30.5%
Worked all year, full time**	32.9%	30.6%	36.5%	36.1%	15.0%
Average income (full-time all-year workers)	$27,602	$25,143	$23,798	$22,098	$20,988
Median income (all with income)	$13,759	$12,814	$13,132	$13,030	$7,591
Income $35,000 or more**	11.9%	8.8%	8.9%	6.2%	2.8%

*Latest data available. Note that the 1986 census asks for income and employment information for the previous year.
**Based on population 15 years of age or older.
Source: Macionis *et al.*, 1994, from: Statistics Canada, *Profile of Ethnic Groups – Dimensions*, 1989.

TABLE 6.7 Crude and Standardized Death Rates for the Canadian-Born and Foreign-Born Populations of Canada, 1985-1987

	Canadian-Born		Foreign-Born	
	Male	Female	Male	Female
Crude Death Rates (per 1,000)	7.24	5.59	13.24	11.63
Standardized Rates (per 1,000)*	7.90	6.06	6.87	5.45

*Directly standardized using the Canadian population in 1971 as the standard.
Source: Bolaria and Bolaria, 1994: 134.

have experienced different life chances and different life expectancy rates than non-Native Canadians have.

There are consistent and extensive differences in a number of indices between Natives and non-Natives in Canada. The average life expectancy of registered Indians has been and continues to be lower than the life expectancy of other Canadians. Table 6.8 shows the differences in birth rate, crude death rate, and infant mortality rate between all Canadians and registered Indians.

TABLE 6.8 Birth Rates, Crude Death Rates, and Infant Mortality Rates, Selected Years, 1960-1990, Registered Indian and Canadian Populations

Year	Birth Rate (live births/1,000 pop.)		Crude Death Rate (deaths/1,000 pop.)		Infant Mortality Rate (deaths of children one year or under/ 1,000 pop.)	
	Canada	Registered Indians	Canada	Registered Indians	Canada	Registered Indians
1960	26.8	46.7	7.8	8.8	27.3	82.0
1967	18.2	39.9	7.4	8.4	20.8*	48.6*
1976	15.7	29.0	7.3	7.3	13.5	32.1
1978	15.3	28.0	7.2	7.4	12.0	26.5
1979	15.5	27.3	7.1	7.1	10.9	28.3
1980	15.5	27.5	7.2	6.8	10.4	24.4
1981	15.3	27.2	7.0	6.3	9.6	21.8
1982	15.1	28.7	7.1	6.2	9.1	17.0
1983	15.0	27.1	7.0	5.6	8.5	18.2
1984	15.0	26.3	7.0	5.7	8.1	19.0**
1985	14.8	30.3†	7.3	6.0	7.9	17.9†
1986	14.7	26.1	7.2	5.6	7.9	22.2†
1987	14.4	27.6	7.2	5.4	7.3	13.7
1988	14.5	29.3†	7.3	5.4	7.2	12.8†
1989	15.0	N.A.	7.3	4.8‡	7.1	9.9‡
1990	15.3	N.A.	7.2	3.8‡	6.8	10.2‡

*Figures are for 1958.
**Data not available for Yukon.
†Data not available for Pacific Region.
‡Data do not include Northwest Territories Indians because of the transfer of health services to the government of the Northwest Territories.
Source: Bolaria and Bolaria, eds., 1994: 249.

If we break these comparative mortality rates down even further, we can see that a major contribution to the differential mortality rate is the wide differences that occur between reserve Natives and non-Natives in childhood and early adulthood (see Table 6.9). For male and female Natives on reserves, the difference in mortality rates

TABLE 6.9 Indian Reserve Areas and Canada, 1977-1982: Age-Specific Mortality Rates per 100,000

Age	MALES			FEMALES		
	Reserves	Canada	Ratio	Reserves	Canada	Ratio
1-9	118.4	49.8	2.38	91.3	37.2	2.46
10-19	218.1	92.5	2.36	85.1	37.1	2.29
20-29	438.7	158.0	2.78	184.2	52.1	3.53
30-39	517.1	163.9	3.16	276.2	84.2	3.28
40-49	716.4	391.1	1.83	424.8	215.7	1.97
50-59	1,131.4	1,054.4	1.07	854.4	535.3	1.66
60-69	2,507.3	2,544.7	0.99	1,578.8	1,245.7	1.27
70+	5,883.6	7,943.9	0.75	4,228.8	5,449.2	0.78

Source: Y. Mao *et al.*, "Mortality in Canadian Indian Reserves, 1977-1982," *Canadian Journal of Public Health*, 77 (1986), p. 264.

from that of the rest of the population is particularly high for ages 1-39, and at its peak period is over three times that of the rest of the population. Once a Native person has reached 40 years of age, the differential mortality rate begins to decline. At 65 and over, the average number of deaths is just about the same or even less than in the rest of the Canadian population.

Not only is the life expectancy lower among Canadian reserve Natives, but also the major causes of death are different (Table 6.10). In part, these are a mirror of the age differences in mortality rates just examined. Accidents, violence, and poisonings account for over one-third of all deaths among Native persons as compared with 11 per cent in Canada as a whole (*Indian Conditions*, 1980). A major killer of Canadians, one of the so-called diseases of affluence, is disease of the circulatory system; these account for 60 per cent of non-Native Canadian deaths and only 21 per cent of Native deaths (*ibid.*).

While the incidence of coronary heart disease is almost equally a major cause of death for male reserve Natives as it is for non-Native male Canadians, what is particularly noteworthy in Table 6.10 is the high rate of death for Indian males from motor vehicle accidents, suicides, and cirrhosis/alcoholism. Drownings, homicide, fires, and water transport accidents, which cause so many deaths among Native males, are not even mentioned as major causes of death among non-Native males.

The mortality rate comparison between reserve Native women and other Canadian women shows a similar pattern, except that coronary heart disease is even higher among Native women than among non-Native women. But women on Native reserves have a particularly high mortality rate from motor vehicle accidents and cirrhosis/alcoholism. Breast cancer, lung cancer, and ovarian cancer, all significant diseases for non-Native women, are not even listed as causes of death for Native

TABLE 6.10 Age-Specific Mortality Rate (ASMR) Ranking of Major Causes of Death Ages 1-69, 1977-1982, per 100,000 Population

MALES

INDIAN RESERVES		CANADA	
Cause	ASMR	Cause	ASMR
1 Coronary heart	104.5	Coronary heart	116.1
2 MVTA	68.3	Lung Cancer	35.3
3 Suicide	53.0	MVTA	29.9
4 Cirrhosis/alcoholism	32.6	Suicide	19.9
5 Drowning	26.1	Cirrhosis/alcoholism	17.3
6 Homicide	24.1	Cerebrovascular disease	16.6
7 Cerebrovascular disease	22.9	Cancer of the large intestine & rectum	10.8
8 Fires	18.5	Chronic obstructive lung disease	10.5
9 Lung cancer	17.7	Pancreatic cancer	5.5
10 Water transport accidents	14.0		

FEMALES

INDIAN RESERVES		CANADA	
1 Coronary heart disease	41.6	Coronary heart disease	33.4
2 MVTA	29.6	Breast cancer	18.6
3 Cirrhosis/alcoholism	25.0	Cerebrovascular disease	12.1
4 Cerebrovascular disease	24.0	MVTA	10.9
5 Suicide	17.0	Lung cancer	10.3
6 Diabetes	16.0	Cancer of large intestine & rectum	8.8
7 Fires	12.6	Cirrhosis/alcoholism	6.5
8 Cervical cancer	10.7	Suicide	6.4
9 Pneumonia	8.7	Ovarian cancer	5.3
10 Homicide	8.5	Chronic obstructive lung disease	3.8

Source: Mao *et al.*, "Mortality in Canadian Indian Reserves," p. 265.

women. On the other hand, fires, cervical cancer, and homicide are so insignificant among non-Native female Canadians that they are not listed.

Aside from the much greater death rate from violence for reserve Native men and women, several other causes of death deserve attention. The first is the noticeably high rate of death by cirrhosis/alcoholism for both male and female reserve Natives. It is almost double the rate of that of other Canadian males and four times the rate of other Canadian females. And for women aged 20-29 the rate of death from cirrhosis/ alcoholism is 17.6 times that of Canadian women in this age group and 4.6 times that of reserve-dwelling Native males of this age. In addition to the direct evidence of the

Box 6.3 The Native People of Grassy Narrows

The experience of the Ojibwa Indians of the Grassy Narrows Reserve provides a poignant and trenchant critique of the disastrous impact that the Canadian state and Canadian industry can have on a people and their way of life. Until 1963, the Ojibwa Indians lived a settled, traditional life, hunting and fishing on and around the English-Wabigoon River in northwestern Ontario. Then, in 1963, they were relocated by the Department of Indian Affairs, in order that the reserve be nearer to a road, and thus nearer to a number of services and amenities, such as schools, various social services, and electricity. Uprooting and moving the Indians had a tumultuous impact on the health and lifestyle of the people. Before the Ojibwa had time to adjust to this crisis, another hit. This time it was the discovery that the English-Wabigoon River system, which had been their main source of livelihood for many, many years, was poisoned by methyl mercury.

Before 1963, over 90 per cent of all deaths among the Ojibwa Indians were attributed to natural causes. By the middle of the 1970s, only 24 per cent of the deaths resulted from natural causes. By 1978, 75 per cent of the deaths were due to alcohol-induced violence. Homicide, suicide, and accidental death rates soared. Child neglect and abuse grew rapidly, and numerous children were taken into the care of the Children's Aid Society and placed in foster homes. As Shkilnyk says: "Today the bonds of the Indian family have been shattered. The deterioration in family life has taken place with extraordinary swiftness." Despite the good intentions of the Canadian government in relocating the people, their socio-economic conditions deteriorated. "All the indications of material poverty were there – substandard housing, the absence of running water and sewage connections, poor health, mass unemployment, low income, and welfare dependency" (Shkilnyk, 1985: 3).

The Indians of Grassy Narrows experienced too much change, too quickly. Their autonomy and cultural traditions were destroyed. Because of mercury pollution, they were robbed of their means of livelihood. There are many lessons to be drawn from this situation. Still, the solutions remain complex.

---

effects of alcohol on mortality, it has been estimated that over 40 per cent of all other Native deaths result from the misuse of alcoholic beverages. An even greater proportion of the violent deaths can probably be attributed to alcohol consumption (see Table 6.11).

The second noticeable cause of death is the high rate of cervical cancer. This is likely due to several characteristics of the lives of reserve-based women. They are less likely to go to physicians for cervical cytology testing (Pap smears), and Pap smears are a very successful and important preventative screening device for cervical cancer. In addition, the longer the duration of active sexual intercourse, the higher the incidence of cervical cancer. Some evidence suggests that early sexual intercourse may be more typical of young reserve-based women than other young Canadian women. For instance, one study indicated that the proportion of first births to women under 20 was over three times higher on reserves than for Canada as a whole (Mao *et al.*, 1986). Whether or not this is primarily or

TABLE 6.11 Causes of Violent Native Deaths by Alcohol Involvement as Determined by Significant Others

	PROBLEM WITH ALCOHOL	UNDER INFLUENCE AT TIME OF DEATH	DRINKING PRINCIPAL ACTIVITY PRECEDING DEATH	MEASURABLE BREATH ALCOHOL CONTENT
	%	%	%	%
Motor Vehicle Accidents	57.1	78.6	17.6	70.6
Fire	54.4	72.7	5.6	90.0
Other Accidents	29.4	55.6	50.0	63.2
Suicide	62.5	53.0	85.7	72.3
Homicide	75.0	100.0	80.0	100.0

Source: George K. Jarvis and Menno Boldt, "Death Styles Among Canada's Indians," *Social Science and Medicine*, 16 (1982), p. 1349.

entirely the result of different frequencies of sexual intercourse or of differing frequencies of the use of birth control needs to be determined. The higher rate of cigarette smoking among young reserve-based women is an additional risk factor in the cervical cancer rate because of its effect on the immune system. Overall, post- and neonatal deaths on reserves are almost four times higher than in deaths in these categories off reserves. Furthermore, Native people are more likely to die from a pattern of illnesses generated by poverty, inadequate nutrition, lack of clean drinking water, crowding and social stress, and illnesses that characterize much of the Third World. These illnesses include infective and parasitic diseases such as meningitis, streptococcal disease, and their complications (including rheumatic fever), hepatitis A, and tuberculosis. The costs of environmental problems to the culture and way of life, as well as to the very physical and mental health, of Native people cannot be overestimated (see Table 6.12).

Explanations for the Health Effects of Inequalities

The Black Report (1982), an evaluation of Britain's National Health Service and its impact on the health of the population, highlighted four different types of explanation for class differences in health. These differences have persisted for the past 40 years, since the introduction of the National Health Service, which was designed to equalize health status in Britain. These four explanations are: measurement artifact, natural or social selection, cultural/behavioural differences, and materialism. The authors of the Black Report prefer the materialist explanation, which sees health as the result of political-economic differences or differences in the way that members of different social classes are constrained to lead their lives.

One: measurement artifact. The findings of class-related health differences are merely the result of the biases involved in the measurement and recording processes.

Box 6.4 Race, Political Power, and Health

Political empowerment may become an important consideration in efforts to understand the crisis of morbidity and mortality. In a recent study, LaVeist (1992) examined all U.S. cities with populations of more than 50,000 residents and with blacks comprising at least 10 per cent of the population. The selection process resulted in 176 cities in 32 states. The two crucial variables were the absolute and relative degree of black political power and the post-neonatal mortality rate (deaths occurring between the second and twelfth months of life). Black political power was measured as the proportion of black representatives on the city council divided by the proportion of blacks in the voting age population. The sum measure of black political power is an indication of absolute black political empowerment – i.e., the percentage of city council members who are black. The results underscore the importance of *relative* political power to mortality rate. Where blacks were well represented on city councils, the black post-neonatal mortality rate was relatively low. LaVeist discusses some of the processes that could be implicated in the correlation, including the most obvious explanation that black people's needs (water, welfare, hospitals, protective services, etc.) would be more attended to.

Certainly measures both of class and of health are of imperfect validity. The argument is that the association itself is false because it is due to a measurement bias that affects the measurement of class and health simultaneously (see Blane, 1985, for a discussion of some of the finer points of this explanation).

Two: natural or social selection. An explanation based on cause and effect is questionable. It is argued here that perhaps class differences result from human biological differences rather than that the human biological differences result from the class inequities. One view is that resources unequally available to people in different social classes cause changes in human biology so that the poorer classes, lacking adequate nutrients, clean drinking water, safe working conditions, and the like are more likely to become ill. The competing view is that people suffer from ill health first and then drop down in the social class hierarchy. Illness itself, because of resultant disability, unemployment, or demotion, according to this argument, causes the decline in social class. A number of studies have suggested that this explanation has some validity and that illness certainly may cause a drop in class level for some. Overall, however, the impact of ill health on downward mobility is very slight and tends to be limited to certain sexes and age groups, namely, men in later middle age.

Three: cultural/behavioural. Class (and here minority racial and ethnic status is also relevant) does cause illness; the explanation stipulates that the mechanism through which this occurs is class differences in lifestyle preferences and behaviours, including such things as the consumption of harmful commodities (refined foods,

Box 6.5 Racial Differences in Preventable Deaths

Black Americans suffer higher rates of sickness and death than whites. This finding is particularly evident among health problems that are essentially preventable. A study based in Alameda County, California, provided a number of details that help explain this racially based inequity (Woolhandler *et al.*, 1985).

(1) Blacks are less likely than whites to have private health insurance.
(2) A greater proportion of blacks have never been to a physician for a checkup.
(3) Blacks are four times as likely as whites to rely on public-sector care.
(4) More than twice as many blacks as whites have not received immunizations.
(5) Fewer black women receive Pap smears.
(6) Fewer black than white people with hypertension are under treatment.

(7) When admitted to hospital, blacks are more likely admitted through an emergency room.
(8) Blacks are more likely to be transferred to a public emergency room.

These disparities in health care are an important part of the explanation for the morbidity and mortality differential between blacks and whites. While this study is based in California, where universal health care is absent, a similar study is not available to compare Canadians of different racial background. Nevertheless, it is possible to put forth the hypothesis that although the morbidity and mortality inequities may not be as strong in the Canadian situation, both because of racism and economic inequalities, it is likely that black Canadians have higher age-adjusted rates of morbidity and mortality than white Canadians.

---

tobacco, alcohol), leisure-time exercise, and the use of preventative health measures such as contraception, "safe sex," pre-natal monitoring, and vaccination. This explanation implies that lifestyle behaviours are the result of a number of individual, free-choice decisions. The assumption is that because of the culture of poverty, those in the poorer classes choose to live for today, to ignore preventative health guidelines, and to indulge themselves in smoking and eating fatty, rich foods, all the while lying around on the couch and neglecting to exercise. The problem with this notion is that individual decision-making must always be seen in the context of the social structure and of the constraints that impede the behaviours of

the people placed in different locations in the social structure. Furthermore, there is no evidence that the lower classes or minorities tend uniformly to fail to practise good health habits. To take just one example, class is inversely related to alcohol consumption and alcoholism.

Four: materialist. Illness is the result of lifestyle. In this case, however, the lifestyle and class differences result from conditions of work, adequate supply of money to provide for nutritious foods, amount of leisure time, availability of transportation, housing quality, air pollution, and clean drinking water. Here lifestyle differences are not thought of as based on individual choice, as they are above. Blane (1985)

TABLE 6.12 Selected Contaminants and Their Impact on First Nations Health Conditions

SOURCE AND CONTAMINANT	AREAS OF MAJOR CONCERN (NUMBER OF PROJECTS OR DEVELOPMENTS)
Impact: Destruction of wildlife; restrictions on hunting and fishing rights; contamination of food, air, and water	
Flooding of First Nations lands through dams and hydroelectric developments	Atlantic (8) Northern Quebec (11) Ontario (17) Manitoba (4) Saskatchewan (2) British Columbia (9)
Acid rain and toxic chemicals from smelters, coal-fired electricity, transportation, and industrial processes	Quebec and Ontario (43% of lakes contaminated) Ontario (Serpent River, Big Trout Lake, Weagamow, Wawa-Sudbury, 65% of headwaters in Muskoka-Haliburton area) Arctic and Northern Canada (lakes and coastal regions contaminated) Canada (40% of forest affected, a dozen rivers no longer support trout or salmon)
High water temperature from large-scale forest harvesting	British Columbia (Meares Island, Lyell Island, Moresby Islands, Stein watershed)
Aquaculture and fish farming in marine water	Bays traditionally harvested by First Nations in maritime waters
Oil and gas exploration, drilling, pipelines, refineries, and potential for spills	West coast offshore High and eastern Arctic (Beaufort Sea, Mackenzie Delta) Northern Alberta
Noise from military	Northern Canada
Impact: Water contamination and destruction of fisheries	
Mercury and other heavy metals from mining, smelters, and acid rain	Northwest Territories (lakes and rivers) Ontario English-Wabigoon River system, St. Clair River, Sarnia
Toxic chemicals, including PCBs, DDT, dioxin, and endusulfin	Great Lakes (1,000 chemicals) Ontario (Niagara River) Quebec (St. Lawrence River system) Northern Canada
Impact: Social and economic disruption	
Dislocation of whole communities, disruption of industries, depletion of resources	All regions noted above

Source: Bolaria and Bolaria, 1994: 263.

reviews a number of studies documenting the materialist argument and concludes that the materialist explanation appears the most promising and yet has been largely neglected.

Ethnicity

The final question is, why are there ethnic differences in morbidity and mortality rates? The explanations offered are similar to those offered for age, gender, and class, particularly class. The most important additional explanation is racism, which through prejudice and discrimination may have an additional impact on the health of Canadian Indians (and American blacks) and probably other visible minorities. Ethnicity is largely explained as a subset of class.

Racism leads to bias in the way that people are treated in all aspects of their lives. It limits their job, educational, religious, recreational, marital, and family choices and chances. Among the most important effects of racism is the environmental destruction of the lands of First Nations people. As Table 6.12 shows, the threats to the environment include a wide variety of such things, such as flooding as the result of the construction of dams and hydroelectric projects and acid rain.

Canadian Indians on reserves suffer from isolation, remoteness, and limited power in the control of their own housing, location, education, and occupation. Furthermore, there are numerous examples of the pollution of Native waterways and the destruction of their previously staple foodstuffs.

Economics and Health

Why are there pervasive correlations between death, disability, and disease and age, gender, class, and ethnicity? What do these social-structural characteristics have in common? Is the relationship between the social-structural variables and health variables the result of one overriding mechanism? Or are there different relationships between each pair of variables?

Marxist analysis attempts to integrate all discrete and superficially unique explanations into a unified explanation. A substantial body of scholarship documents the relationship between the economy and health status. Beginning with Frederick Engels's, *The Condition of the Working Class in England*, analysis has focused on the state and the economy as they affect health outcomes. Engels noted the contradictions between the workers' need, in the earliest industrialist economies, to sell their labour power, and the capitalist factory owners' need for profit. Because industrial profit was necessary to maintain factory-based production, the wages of the workers were kept low and their working conditions poor. Shelter for the poor was frequently inadequate, crowded, unheated, and unsanitary. Poor hygiene and inadequate nutrition resulted in widespread contagious diseases such as tuberculosis, typhoid, and infectious diseases.

The first principle of capitalism is that business requires profit. Profit, as the colloquial saying puts it, is the bottom line. According to Marxist analysis, all capitalist states are arranged in such a way as to

Box 6.6 Karoshi

Overwork itself is an insured cause of mortality in Japan. Karoshi is the Japanese term that means death from overwork. The Japanese Labour Ministry has resisted the organizing and lobbying efforts made by spouses and family members of victims of Karoshi through the organization called the Association of Families of Karoshi Victims. The government has, until now, refused to keep statistics. But the United Nations Human Rights Commission has recognized Karoshi sufficiently to organize a hearing and to accept testimony. This action may lead to formal human rights investigations. The Japanese Ministry of Labour has been forced to respond. It now offers compensation to the families of overwork victims only if certain conditions demonstrating excessive overtime are met. These include the following: the victim must have worked for 24 hours preceding death or seven days straight with more than four hours' overtime a day.

---

maximize profits. Furthermore, competition forces capitalists to maximize outputs while minimizing the costs of production. The growth of profits results in the accumulation of capital. Excess capital is invested in new or more efficient production units. New and greater profits are then realized. Increased production invokes a need to find and create new markets. When domestic markets are saturated, new foreign markets must be opened up. Profits from these markets are reinvested, and the cycle of profit, competition, accumulation of capital, investment, and expanded markets repeats itself.

When the state is based on capitalism, state decisions are guided by the necessity of supporting the most successful profit-making initiatives. Usually when a choice between profitability and human need is confronted, profitability must, in the final analysis, come first. The argument is that the good of the whole, not the good of individuals, is advanced when economic productivity and profitability prevail. Consequently, when the need for profit is the determining principle, the health needs of the population assume secondary importance. The worth of individuals in a capitalist economy is measured by their relationship to the means of production. Because of upturns and downturns in the economy, a reserve labour force is required. The reserve labour force is made up of people who are peripheral to the paid labour force and can be brought into it or dismissed from it depending on market demands.

The more marginal the worker is in the capitalist economy, the more easily he or she can be replaced. Occupational health and safety precautions are expenses to the capitalist owner. To the extent that the cost of such precautions threatens the level of profit, and in the absence of state legislative requirements, health and safety standards will be minimal. Accidents, job-related sickness, and alienation will be most prevalent in the workers who are most marginal and most replaceable. One overriding reason for income, gender, age, and ethnic differences in health is the variation in access to safe,

Box 6.7 The Great Dying

The numbers have been debated, but the most reliable recent estimate is that the New World peoples likely numbered between 90 and 112 million before Europeans arrived. Not only were the Amerindians numerous, they also enjoyed good health. In fact, they were, before contact, apparently in better health than Europeans. They did not have measles, smallpox, leprosy, influenza, malaria, or yellow fever. For most, staying healthy and living well were essential to their religion. In an age before Europeans' hygiene included regular bathing, the peoples of the New World kept clean and the Europeans could not help noticing the "good smells" emanating from the healthy, robust people. Apparently, they especially admired their even, white teeth and clear complexions, "something most pockmarked Spaniards, Portuguese and French had

lost at an early age" (Nikiforuk, 1991: 80). At this time, the life expectancy of Europeans was about twenty years less than the Amerindians, who on average lived well into their fifties. It is suggested that the good health was the continuing result of the origins of these people, who had had to cross the frozen Bering Strait to the New World. Because of the harsh cold, the diseased immigrants and their germs were killed.

Into this context of good health the new disease smallpox entered – an unknown yet voracious predator that spread, as an epidemic, until it and other strong killers such as the plague and tuberculosis killed probably 100 million people. This, the great dying, occurred over about one century. Some historians rank this as the greatest demographic disaster in the history of the world.

satisfying, adequately remunerated work. Because job-related illness and death are significant in Canadian society, this Marxist analysis provides an important explanation for inequities in health outcomes.

To conclude: often, health is the outcome of access to fundamental resources. Poverty, poor nutrition, inadequate housing and transportation, and the lack of effective birth control all contribute in known ways to ill health. In a capitalist economy there are significant differences in access to these life-giving resources in respect to differences in age, gender, income, and ethnic group.

Commodification

One other contribution of Marxist analysis to understanding inequities in death,

disease, and disability is through the concept of commodity or commodification. Commodities are objects or activities having an "exchange value." They can be bought and sold in the marketplace and can be used to acquire other things. Their value is determined by market factors such as supply and demand. Health can be seen as a commodity. It is no longer simply the individual experience of well-being but, in a capitalist economy, is subject to supply and demand. The buying and selling of organs for transplantation into unhealthy bodies is perhaps the most blatant example of commodification. Health is purchasable for those with the money. There is also a way in which health itself is an object that reflects value and worth back on the individual. In this case, health is used to indicate the productive value of the person to the society. Healthy people are thought to be good people. Health is believed to embody a certain level of conspicuous consumption and a degree of much valued self-control (Crawford, 1984).

Whenever a government or a corporation decides to allow the continuance of a practice that is destructive to the health of a population in the interests of financial benefits such as taxes or profits, health is being commodified. Cigarette smoking is a case in point. There is absolutely no doubt that cigarette smoking is a significant cause of disability and death in Canadian society. Yet neither the cigarette industry nor the government has been willing to pull cigarettes from the market or to limit their production to low-tar cigarettes. Cigarettes are a fundamental and crucial aspect of contemporary economies. According to Doyal (1979), the state made a deliberate decision to allow the tobacco industry to continue. The British Department of Health showed that a reduction in cigarette smoking would be costly to the state. Not only would the tax on cigarettes be lost, but there would be the additional financial costs of caring for people longer into their old age, which would be the inevitable result of increased life expectancy.

Summary

(1) As social class increases, life expectancy increases. Class is also related to the length of life that is lived with a disability. Low-income earners do not have poorer health habits overall yet they are less likely to engage in some preventative health measures. They are also more likely to be exposed to occupational health hazards.

(2) Race and ethnicity affect mortality and morbidity rates. For example, the life expectancy rate is lower and the PYLL higher among Canadian Native people than among non-Native people. The causes of death also differ; Natives tend to die from illnesses related to poverty, inadequate nutrition, lack of potable water, crowding, and social stress; non-Natives tend to die from "diseases of affluence," such as diseases of the circulatory system.

(3) Biological risks are part of the explanation for age inequities in mortality and morbidity. Historical differences in lifestyle, differences in culture and cultural norms, and differential access to health-giving resources are also factors.

(4) Other factors that contribute to different rates of morbidity and mortality are class, ethnicity, lifestyle, and environmental contaminants.

CHAPTER SEVEN

Getting Sick and Going to the Doctor

W HAT IS THE ROLE of interpersonal relationships in illness? Is it true that people can die of a broken heart? Does the repression of feelings, particularly anger, cause cancer? Can a person choose to live or die? What is stress? Is stress good or bad? Why do people go to the doctor? When do people choose to visit a physician rather than "carry on" or go to bed with an aspirin? The symptoms of a head cold, or backache, or digestive difficulties, or influenza will take one person to bed, another to the doctor, and others to the drug store, acupuncturist, nurse practitioner, or neighbour.

On any day, for every person at the doctor's office, there are at least three times as many people not at the doctor's office who are suffering from the same symptoms. The enormous expenditures on over-the-counter medications bear testimony to the frequency of self-help care. So does the prevalence of folk remedies such as "feed a cold and starve a fever." Yet doctors' offices are often filled with people with conditions that cure themselves or conditions for which the doctor cannot provide medical assistance.

Moreover, many people continue to visit doctors for psychosocial rather than strictly medical reasons. Some studies, for instance, show that 50 per cent of patient visits to primary care providers include psychosocial complaints (Ashworth *et al.*, 1984). Most patients who go to physicians do not have serious physical disorders (Weiss and Lonnquist, 1992). However, evidence suggests that many people go to doctors for non-biomedical reasons, including life stress and emotional distress, diagnosable psychiatric disorders, social isolation, and information needs (Barsky, 1981). In addition, most people today (70 per cent) report that they believe it is appropriate to seek help from primary care physicians for psychosocial problems (Good *et al.*, 1987).

This chapter is divided into two parts. The first part will deal with just a few of the many socio-psychological factors found to be associated with illness, including stress, coronary-prone behaviour, social support, and sense of coherence. The second part will discuss the processes through which people come to define themselves as ill and as needing care, and the actions they take in regard to their health.

Stress

Stress occurs when an organism must deal with demands that are either much greater than or much less than the usual level of activity. As such, stress is ubiquitous. All of us are stressed to some degree or we would not be alive. The presence of at least some stress in life is beneficial. Stressful experiences can be healthy and can fit us for positive and flexible adaptations to stress later on. Or stress can be so overwhelming that it leads to serious illness or death. Too much change, in too short a time, can overtax the resources of the body.

Two major writers were involved in the early articulation and measurement of stress: Cannon (1932) and Selyé (1956). Cannon (1932) suggested that health is ultimately defined, not by the absence of disease, but rather as the ability of the human being to function satisfactorily in the particular environment in which he or she is operating. Human beings must constantly adapt to changes and assaults – to changes in weather, conflicts at work, failure in school, great success on the hockey team, promotion, flu germs, and so on. The body adapts to such changes by maintaining a relatively constant condition, e.g., when the body becomes overheated, it will evaporate moisture to help keep it cool; when affronted with bacteria, it will produce antibodies. The process of maintaining a desirable bodily state (the constant condition) is called homeostasis (Cannon, 1932). The body is thus prepared to meet threats by adapting in ways that will return the body to the desired state.

Cannon described the typical bodily reaction to stress as "fright or flight," and detailed the accompanying physiological changes. Somewhat later, Selyé defined stress as a state that included all the specific changes induced within the biological system of the organism. It is a general reaction that occurs in response to any number of different stimuli. Both "positive" and "negative" events can cause stress. It does not matter whether the event is the happy decision to become engaged to be married or the disappointing failure of a grade in university – each requires adaptation.

Building on several decades of research on the pituitary-cortical axis, Selyé proposed the General Adaptation Syndrome (GAS) as the body's reaction to all stressful events. The "syndrome" has three stages: (1) an alarm reaction, (2) resistance or adaptation, and (3) exhaustion. During the first stage the body recognizes the stressor and the pituitary-adrenal cortical system responds by producing the arousal hormones necessary for either flight or fright. Increased activity by the heart and lungs, elevated blood sugar levels, increased perspiration, dilated pupils, and a slow-down in the rate of digestion are among the physiological responses to this initial stage of the syndrome. During the adaptive stage the body begins to repair the damage caused by arousal and most of the initial stress symptoms diminish or vanish. But if the stress continues, adaptation to it is lost as the body tries to maintain its defences. Eventually the body runs out of energy with which to respond to the stress, and exhaustion sets in. During this final stage, bodily functions are slowed down abnormally or stopped altogether.

Continued exposure to stress during the exhaustion stage can lead to what Selyé calls the "diseases of adaptation." These include various emotional disturbances, schizophrenia, migraine headaches, certain types of asthma, cardiovascular and renal diseases, ulcers, hypertension, lowered resistance to viruses, and immune deficiency. Table 7.1 provides a stress self-assessment checklist. People are not always aware that they are living under stressful circumstances. Sometimes bodily and emotional symptoms are the first sign that something is amiss. Table 7.1 lists many of the warning signs or symptoms of stress that may result in subsequent health problems. A typical person will have a score of between 42 and 75 in any given month. The higher the score, the greater the likelihood of illness.

As well as the symptoms listed, stress can also lead to death. To study such effects of stress Engel collected 170 reports of sudden death. He discovered that the deaths usually occurred within an hour of hearing emotionally intense information, which could be either positive or negative. Of the sudden deaths, 21 per cent occurred on the collapse or death of a close friend, 20 per cent during a period of intense grief, 9 per cent at the threat of the loss of a close person, 3 per cent at the mourning or anniversary of the death of a close person, 6 per cent following loss of status or self-esteem, 27 per cent when in personal danger or threat of injury (whether real or symbolic), 7 per cent after the danger was over, and, finally, 6 per cent at a reunion, triumph, or happy ending (Engel, 1971).

Stress can be short-term, medium-term, or long-term. Briefly, the short-term stressors arise from small "inconveniences," for example, traffic jams, lost keys, or waiting for the doctor. Such things usually result in a temporary sense of anxiety. These have been called daily hassles (Lazarus and Delongis, 1983). Medium-term stressors develop from such things as long winters, a temporary layoff from work, or an acute sickness in the family. Long-term stressors result from such incidents as the loss of a spouse or the loss of a job. Holmes and Rahe (1967) have systematized the stress value attached to a list of life events in a scale called the Social Readjustment Rating Scale (SRRS). Based on extensive interviews with 394 people of varying ages and socio-economic status, Holmes and Rahe were able to develop average scale values representing the relative risks of a number of specific life events. Marriage is assigned an arbitrary value of 50. The adjustment value of other items is estimated in comparison to marriage. The resulting scale itemizes 43 changes and quantifies their hypothetical impact (see Table 7.2). In general, the higher the score, the greater the likelihood of illness. It is important to notice that, while some of the events are considered negative or undesirable and others might be considered positive or desirable, they all require adjustment.

As the scale and other related research indicate, stressors can be the result of changes in any area of life and at a variety of levels, such as: (1) the individual level

TABLE 7.1 Stress Self-Assessment Checklist

Use the following scale for each symptom and circle the number that best applies to you.
  1. Never
  2. Occasionally
  3. Frequently
  4. Constantly
In the last month I have experienced the following:

1. Tension headaches	1	2	3	4
2. Difficulty in falling or staying asleep	1	2	3	4
3. Fatigue	1	2	3	4
4. Overeating	1	2	3	4
5. Constipation	1	2	3	4
6. Lower back pain	1	2	3	4
7. Allergy problems	1	2	3	4
8. Feelings of nervousness	1	2	3	4
9. Nightmares	1	2	3	4
10. High blood pressure	1	2	3	4
11. Hives	1	2	3	4
12. Alcohol/non-prescription drug consumption	1	2	3	4
13. Minor infections	1	2	3	4
14. Stomach indigestion	1	2	3	4
15. Hyperventilation or rapid breathing	1	2	3	4
16. Worrisome thoughts	1	2	3	4
17. Skin rashes	1	2	3	4
18. Menstrual distress	1	2	3	4
19. Nausea or vomiting	1	2	3	4
20. Irritability with others	1	2	3	4
21. Migraine headaches	1	2	3	4
22. Early morning awakening	1	2	3	4
23. Loss of appetite	1	2	3	4
24. Diarrhea	1	2	3	4
25. Aching neck and shoulder muscles	1	2	3	4
26. Asthma attack	1	2	3	4
27. Colitis attack	1	2	3	4
28. Periods of depression	1	2	3	4
29. Arthritis	1	2	3	4
30. Common flu or cold	1	2	3	4
31. Minor accidents	1	2	3	4
32. Prescription drug use	1	2	3	4
33. Peptic ulcer	1	2	3	4
34. Cold hands or feet	1	2	3	4
35. Heart palpitations	1	2	3	4
36. Sexual problems	1	2	3	4
37. Angry feelings	1	2	3	4
38. Difficulty communicating with others	1	2	3	4
39. Inability to concentrate	1	2	3	4
40. Difficulty making decisions	1	2	3	4
41. Feelings of low self-worth	1	2	3	4
42. Feelings of depression	1	2	3	4

TOTAL SCORE

Source: J. Neidhardt, M.S. Weinstein, and Robert R. Coury, *Managing Stress* (Vancouver: Self-Counsel Press, 1985).

TABLE 7.2 The Stress of Adjusting to Change

EVENTS	SCALE OF IMPACT
Death of spouse	100
Divorce	73
Marital separation	65
Jail term	63
Death of close family member	63
Personal injury or illness	53
Marriage	50
Fired at work	47
Marital reconciliation	45
Retirement	45
Change in health of family member	44
Pregnancy	40
Sex difficulties	39
Gain of new family member	39
Business readjustment	39
Change in financial state	38
Death of close friend	37
Change to different line of work	36
Change in number of arguments with spouse	35
Mortgage over $10,000	31
Foreclosure of mortgage or loan	30
Change in responsibilities at work	29
Son or daughter leaving home	29
Trouble with in-laws	29
Outstanding personal achievement	28
Wife begins or stops work	26
Begin or end school	26
Change in living conditions	25
Revision of personal habits	24
Trouble with boss	23
Change in work hours or conditions	20
Change in residence	20
Change in schools	20
Change in recreation	19
Change in church activities	19
Change in social activities	18
Mortgage or loan less than $10,000	17
Change in sleeping habits	16
Change in number of family get-togethers	15
Change in eating habits	15
Vacation	13
Christmas	12
Minor violations of the law	11

Source: T.H. Holmes and R.H. Rahe, "The Social Readjustment Rating Scale," *Journal of Psychosomatic Research*, 11 (1967), p. 214.

(e.g., bacterial infections); (2) interpersonal situations (e.g., loss of a spouse); (3) social-structural positions (e.g., unemployment, promotion at work); (4) the cultural system (e.g., immigration); (5) the ecological system (e.g., earthquake); or (6) the political/state system (e.g., wars). The greater the number of stressors, and we might hypothesize the greater the number of levels of stressors, the more vulnerable a person is to the possibility of disease and emotional and bodily dysfunction.

It should be emphasized that stressors are not to be thought of as objective things that affect an unthinking organism in a monolithic way. A person's evaluation of the stressful situation, the strategies available for coping, the degree of control felt, and the amount of social support experienced all mediate the impact that the stressor ultimately has on that person. As Victor Frankel (1965) (see also Bieliauskas, 1982) has pointed out, some people, even in the most atrocious of circumstances such as a concentration camp, have been able to use their experiences in a manner that was meaningful to them, and thus these people ultimately became stronger and healthier as a result of this most extreme of stress-filled situations.

The SRRS scale has been used with college students and with blacks and Mexican Americans. It has also been used in a number of cross-cultural studies that have included Swiss, Belgian, and Dutch peoples (Bieliauskas, 1982). Overall, researchers have found that there are significant similarities among different cultural groups in their evaluation of the impact of various events. Such studies have shown the SRRS to be a remarkably stable or reliable instrument. It has also been used to "predict" illness and symptoms of distress. Holmes and Masuda's 1974 study concluded "that life-change events . . . lower bodily resistance and enhance the probability of disease occurrence. . . ." Several researchers have correlated high SRRS scores with symptoms and with illness, including major illnesses such as heart disease (Theorell and Rahe, 1971; Rahe and Paasikivi, 1971). High SRRS scores have also been found to be associated with psychological distress in a number of studies (Bieliauskas, 1982). Several researchers have used SRRS successfully to predict the onset of illness. One interesting example of this research is the study that examined the SRRS of 2,600 navy personnel prior to their departure on voyages of six-eight months; the researchers found significant correlations between the levels of stress before the start of the voyage and the subsequent levels of disease during the voyages (Rahe, Mahan, and Arthur, 1970).

The SRRS has also met its share of criticism for the following reasons.

(1) It ignores differences in the meaning people place on the various events. There is evidence that the impact of the death of a spouse, for instance, varies depending on whether or not the death was sudden or occurred after a protracted period of illness.

(2) Some of the events listed may be signs of illness or the results of illness, such as changes in eating habits, sleeping habits, or personal habits or sex difficulties. Thus the scale is, in part, tautological and its ability to predict illness is questionable because some items on the scale are related to illness.

(3) Some research has found that distinguishing between desirable and undesirable events enhances the predictive value of the scale. Marriage is generally taken as a reason for celebration, death as an occasion for mourning. It could be argued that the stress-related effects of death are therefore more pronounced than those of marriage.

(4) The ability to control events has been shown, in a number of studies, to be an important factor in determining the degree of stress experienced.

(5) Whether stress affects the incidence of disease or merely behaviour during illness has been questioned. It may be that life events affect the likelihood of people reporting illness rather than affecting the disease process.

(6) The SRRS asks about events that have occurred during a specified period of time. Some research has shown that the association between stress and subsequent illness or disease cannot be studied separately from the previous stress level and the history of past illnesses. Thus experiences of the years before the time period referred to in the SRRS may also have a powerful effect on the level of stress experienced. A car accident followed by the loss of a driver's licence and a household move in the years before the designated SRRS time period might exacerbate whatever level of stress is experienced during the year of the SRRS measurements.

There have been a series of critiques of and improvements on the SRRS in the 1980s

TABLE 7.3 Indications of Excess Stress

PHYSICAL	PSYCHOLOGICAL	BEHAVIOURAL
Rapid pulse	Inability to concentrate	Smoking
Increased perspiration	Difficulty making decisions	Medication use
Pounding heart	Loss of self-confidence	Nervous tics or mannerisms
Tightened stomach	Cravings	Absent-mindedness
Tense arm, leg muscles	Worry or anxiety	Accident-proneness
Shortness of breath	Irrational fear or panic	Hair-pulling, nail-biting, foot-tapping
Tensed teeth and jaw		Sleep disturbance
Inability to sit still		Increased use of alcohol or other recreational drugs
		Careless driving
		Uncalled-for aggression

Source: Adapted from J. Neidhardt, M.S. Weinstein, and R. Coury, *Managing Stress* (Vancouver: Self-Counsel Press, 1985), pp. 5-6.

and 1990s. Turner and Avison (1992) refer to 16 different reviews and critiques that have altered the way that life events are measured. Zimmerman (1983) describes 16 alternative inventories designed to measure life stress through life events. These critiques and improvements address the problems and issues of the SRRS. Two such refinements will be briefly described here. Turner and Avison (1992) note that crisis situations (life events) can pose opportunities as well as problems. When life events are managed well and resolved, the researchers hypothesize, they may not contribute to stress for the individual. Using large samples of both physically impaired and community-based comparison groups, Turner and Avison (1992) found support for the idea that resolution of a life event generally eliminates or minimizes its stress impact. Thoits (1994) continued and replicated the findings of the elaboration of the relationship between resolved and unresolved life events and health outcomes. She emphasizes that men and women must be seen as actively involved in managing stress more and less successfully. Moreover, she notes that people with the personality characteristics of mastery and self-esteem manage stress more successfully.

Conger *et al.* (1993) compared the reports of exposure and vulnerability to specific types of life events made by men and women. Their findings were consistent both with the social-structural positions of men and women (primarily with respect to labour force participation and other daily role responsibilities) (Aneshensel and Pearlin, 1987) and with the identity perspectives (primarily with respect to the self-concepts and identities of men and women (Thoits, 1983, 1991). Men were more likely than women to report exposure to and be distressed by work and financial events. Women, by contrast, were more likely to be upset by exposure to negative events within their families. Further, men and women responded differently to negative events. Men were more likely to become hostile whereas women's somatic complaints increased.

Stress research has continued to dominate the field of the sociology of health and illness, especially in the United States. In fact, in an analysis of the relative ranking of the "top" American journals in the field of sociology the *Journal of Health and Social Behavior* was ranked as the second most important journal to sociologists on the basis of the "objective" measure of citation impact. Johnson and Wolinsky (1990) compared *Social Forces*, the *American Journal of Sociology*, and the *American Sociological Review* over an 11-year period beginning in 1977 and found that the *JHSB* has always outranked *SF*, has outranked *AJS* 7 of 11 years, and has outranked *ASR* 4 out of 11 years. They argue that this ranking is largely attributable to the widely cited work on stress, coping, mental health, and social support and to the reputations of the editors over this period of time as superb stress researchers. While the claim made by Johnson and Wolinsky is passionately debated in a series of articles in the same volume of the journal, the point that stress research has been among the most central in the field remains plausible.

Social Support

Social support is something that almost everyone thinks she or he understands, on an intuitive if not on a cognitive level. If you were to ask your friends whether they knew the meaning of social support, you would probably find unanimous agreement. If you were then to ask each to define its meaning, you would likely hear as many definitions as there are people giving them. For some, social support would be defined as "a feeling that you have or don't have." For others, it would be defined as friends with whom to party. Another would define it as someone with whom to do things. Still another might see social support as having someone on side in case of an argument. Material and practical aid might be necessary components of social support for others.

When sociologists attempt to measure social support, the varieties and idiosyncrasies of meaning become apparent. An early and often used definition of social support comes from the work of Cobb (1976), who thought of social support as information that would lead a person to believe (1) that he or she is cared for and loved, (2) that he or she is esteemed and valued, and (3) that he or she belongs to a network of communication and mutual obligation. A group of researchers used this definition as the basis for a scale that was later revised and subjected to a number of uses (see Turner *et al.*, 1983). The scale asks people to describe themselves in comparison to others with respect to the amount of social support they feel they have. It is, in essence, a subjective measure based on each person assessing his or her felt degree of social support. It includes both social support in the sense of a person's feeling of being loved and esteemed by others, and social support as an experience of being part of a network of people.

This definition of social support emphasizes the subjective perception of the respondent. Others have considered social support as something that can be objectively measured – social support exists to the extent that a person can count on others to offer specific services such as cooking, cleaning, snow shovelling, or transportation when the need arises (Thoits, 1982). In this case, the degree of social support is reflected in the reliability and extensiveness of the actual aid supplied rather than in the subjective feeling of the individual. In a further refinement, others have pointed out that different kinds of "support" are desired and expected from different "kinds" of others. That is, people may expect different things from their friends than from their kin. Whether or not the network is made up of people who know each other or not may also affect the experience of social support.

A number of studies have examined the impact of social support on health outcomes. From 1978 to 1995 the *Journal of Health and Social Behavior* published numerous articles concerning social support in a variety of situations. There were articles on all of the following: the health consequences of being unemployed, self-assessments of the elderly, occupational stress, mental health, psychological well-being,

primary deviance among mental patients, life stress, psychotropic drug use, psychological distress and self-rejection in young adults, adolescent cigarette smoking, health in widowhood in later life, teenage pregnancy, social support among men with AIDS, and depressed patients. This list gives some idea of the range of situations in which social support can act.

Some researchers have emphasized that social support has a direct relationship to health so that the person who has support is less likely to become ill; others have noted that adequate social support can minimize the harmful effects of stress on a person's mental or physical health. These two notions are compatible with one another: social support may have both direct and indirect effects on health (see Turner *et al.*, 1983).

In 1973, Gove analysed causes of mortality and noted that married people tended to live longer. Analyses of the dates of death of famous people (Phillips and Feldman, 1973) showed that death rates declined just before a significant occasion (such as a birthday, wedding, or Christmas celebration) – an occasion on which these people would have the opportunity to reaffirm social ties with the group of significant others. A very large study (Beckman, 1977) charted the lives of 7,000 people over a period of some nine years; during that period 682 of the 7,000 people died. After controlling for a variety of sociodemographic and risk factors, the data revealed that those who died tended to lack social ties (family, church, informal and formal group associations). Beckman concluded that isolation and the lack of social and community networks may increase vulnerability to disease in general.

Cassel (1974), in a series of studies, found that people with a variety of illnesses shared a similar characteristic with people who attempt suicide, drink excessively, or become involved in numerous accidents – they are deprived of meaningful social contact. Cassel explained that the missing link among those with inadequate social support was informational. People who did not have sufficiently supportive relationships suffered because they lacked cognitive evidence that their actions were leading to desired outcomes. Weisman and Worden (1975) found that survival rates of cancer patients were affected by the level of social support the patients received. Meddison and Walker (1967) documented that the health of bereaved people varied depending on how much social support they were offered. Unfortunately, as most of these studies are correlational, causal connections cannot be assumed.

One study that made an effort to move beyond the correlational connections between social support and health outcomes is based on interviews with teenage mothers during pregnancy and then following childbirth to investigate the impact of social support (Turner, Grindstaff, and Phillips, 1990). These researchers examined the impact of social support on what has been argued to be a very important indicator of the health of a population – infant birth weight. Consistent with their hypothesis, the researchers noted that pregnant teenagers who received more support from family, friends, and partners had higher birth-weight babies. Not only was birth weight positively affected by the level of social support but the psychological health of the new mother was as well. This research also found that socio-economic background influenced the relationship between social support and the health outcomes of mothers and their infants. Social support was especially helpful to young women with lower-class backgrounds, although it was not unimportant for those from higher socio-economic backgrounds.

The relationship between social support and specific diseases such as cancer and heart disease has also been examined. As early as 1931, a relationship between social factors and cancer was found: widowed and divorced women were more likely to have cancer than single or married women. A number of studies since that time have reiterated this relationship between cancer and the loss of a marriage partner (LeShan, 1978). Relationships have also been discovered between the incidence of cancer and other indicators of the lack of social connections. For example, the greater the religious cohesion, e.g., among Mormons, the lower the incidence of cancers. While the studies neither establish a causal connection nor explain all of the variance, there seem to be sufficient grounds to pursue further research on the potential association between social connections and cancer.

Professor David Spiegel, a psychiatrist and researcher at Stanford University Medical School, set out to refute the notion that the mind could be used to affect the outcome of disease. He observed 86 women with breast cancer for 10 years. To his surprise, he found that women who took part in group therapy and had been taught self-hypnosis lived twice as long as those who had not (Spiegel *et al.*, 1989).

What are the mechanisms through which support operates? How does social support affect health? Does support influence the interpretation of stressful life events? Are people who feel supported likely to believe they can manage to cope with the sudden death of a spouse because they know that others will listen and continue to care? The mechanisms through which social support affects health need to be further studied and clarified.

Coronary-Prone or Type A Behaviour and Heart Disease

Scientific interest in the socio-psychological causes of heart disease can be traced to the beginning of this century. Sir William Osler, a renowned physician, made the following observation about people who had had heart disease in his famous lecture on angina pectoris at Oxford in 1910: "In a group of 20 men, every one of whom I know personally, the outstanding feature was the incessant treadmill of practice" (Osler, 1910: 698). Osler thus commented on his own casual observation of the particular "speeded-up" character of those with angina.

Two researchers who have contributed most to understanding the relationship between personality and heart disease, Friedman and Rosenman (1974), were stimulated to examine the question after several heuristic experiences. The first was a comment made by the woman who was president of the San Francisco Junior League in the mid-1950s. Friedman and Rosenman were testing the hypothesis that women were less likely to suffer from heart disease because their diets were lower in fats and cholesterol than those of men. Women of the Junior League had agreed to co-operate by keeping diaries of what they and their husbands ate during a two-week period. Before the diary-keeping started, the president of the organization told the researchers that she thought they would find that husbands and wives ate exactly the same diets. The diaries confirmed her hunch.

The next clue occurred when the women and their husbands were to meet with Friedman and Rosenman to hear the results of the diary-keeping. Apparently many of the women were late for the meeting; they seemed unconcerned. Many of the women did not wear watches; many of the watches worn by other women were not working. In contrast, the men appeared precisely at the arranged time. They all wore watches; after twenty minutes of waiting for the results to be announced, they began to look at the researchers impatiently. Most of the men knew where the clock was when they were asked about it.

The researchers then surveyed three companies and 90 physicians about their ideas regarding the cause of heart disease. About 75 per cent of both groups replied: stress. Still another clue resulted from an innocuous remark made by an upholsterer who had been hired to reupholster the researchers' waiting-room furniture. He queried what sort of practice the doctors had. Apparently the chairs were worn in an unusual place – near the front of the seat. This indicated that those waiting were probably more restless and hurried than most patients waiting for physicians (Gasner, 1982). All these clues led the doctors to include psychological variables in their next series of studies.

The intriguing role of psychological factors was then subject to intensive investigation. With a group of colleagues, the Western Collaborative Group Study, Friedman and Rosenman undertook a number of studies of the physiological, psychological, and personality factors that might be implicated in heart disease. Having examined such well-known factors as diet, weight, exercise, and serum cholesterol levels, the researchers were surprised to discover that personality was the strongest indicator of which patients were likely to get sick and die. What they found corroborated the description given by Osler in 1910 – that those most likely to develop coronary arteriosclerosis and heart attacks, and be liable to sudden death, were those whose behaviour was aggressive, restless, and impulsive. They called these traits the type A behaviour pattern. Even the language of the type A personality was found to be distinctive. These people tended to speak rapidly and loudly, to interrupt others, and to use bodily gestures to emphasize the importance of what they had to say.

Later researchers pinpointed one of the psychosomatic mechanisms that links personality traits to physiological outcomes. In a series of experimental and clinical studies, Lynch (1977) concluded that type A speech patterns tended to cause unusually large and sudden bursts of blood to move through the arteries. Such increases in blood pressure were the immediate response to rapid speech. Regardless of the emotional content of the talk, its speed alone resulted in elevated blood pressure levels. Some research then documented that changing type A behavioural patterns can decrease the risk of heart attacks by up to one-half in people who have already had one myocardial infarction (Lynch, 1977).

Friedman and his colleagues at Mount Zion Medical Center of the University of California at San Francisco and Stanford University studied 1,035 patients who survived their heart attacks to determine if they could alter their type A behaviour and whether this intervention would increase their survival time. They found that 98 per cent of the patients demonstrated type A behaviour in that they displayed a continuously harassing sense of time urgency and were readily hostile and suspicious. The patients were divided into three groups. One group received traditional post-heart attack information on diet, weight, and exercise. Another group received this advice plus counselling, biofeedback, and individual and group education on how to change their type A behaviour. Those in the third group were simply examined and interviewed annually by their physicians. The study continued over five years, at which time the effectiveness of the intervention was evaluated. It was noted that the group given assistance with modifying their type A behaviour had the lowest rates of heart attacks (Friedman, 1982).

Sense of Coherence

Still on the socio-psychological level, Antonovsky (1979) reversed the usual questions about what sorts of things cause illness. Instead he asked: How do people stay healthy? Rather than look at the deleterious effects of such things as type A behaviour, the lack of social support, and the consequences of stress, Antonovsky focused on the positive and beneficial. Citing evidence from many studies, he argued that a "sense of coherence" or a belief that things are under control and will work out in the long run is a crucial component of the state of mind that leads to health. People who have good or excellent health are, all things being equal, likely to have a strong sense of coherence, which is defined as follows.

> . . . A global orientation that expresses the extent to which one has a pervasive, enduring though dynamic feeling of confidence that (1) the stimuli deriving from one's internal and external environments in the course of living are structured, predictable, and explicable; (2) the resources are available to one to meet the demands posed by these stimuli; and (3) these demands are challenges, worthy of investment and engagement. (Antonovsky, 1979: 19)

The "sense of coherence" concept draws attention to the fact that the extent to which a person feels that he or she can manage whatever life has in store is an important factor in health.

There are three components to this concept. The first is comprehensibility – the basic belief that the world is fundamentally understandable and predictable. Such a belief in the comprehensibility of the self and of human relationships is absolutely essential for coping. People with a high sense of coherence feel that the information they receive from the internal and external environment is orderly, consistent, and clear. People with a lower sense of coherence tend to feel that the world is chaotic, random, and inexplicable. A person with a high sense of coherence feels that his or her actions in the past have had the expected consequences, and they will continue to do so in the future. The student who knows how hard to study or how many drafts of a paper to do to get the desired A or B (or, heaven forbid, a C) grade is someone with a high sense of comprehensibility.

Susan provides an example of someone with a strong belief in comprehensibility. Susan was interviewed during a study of women who had received a diagnosis of cancer; she said:

> Oh sure, I'm very sick, I've lost both breasts and I'm scheduled for eighteen months of chemotherapy, but it'll be okay. My mother and sister had breast cancer. They were caught early enough. I have been having regular checkups for the past ten years, and so when the

doctor first noticed that something wasn't right he immediately booked me for a mammography and then exploratory surgery, and finally, almost as a preventative thing, he took off both breasts, but he believes that I will be fine now, and I guess I do too. (Clarke, 1995)

During the same study Rose was also interviewed; she could be characterized as having a low sense of comprehensibility. Rose said:

I don't know why everything always happens to me. I go to church regularly. I am a good wife and mother. I keep the house clean, and as a grade three teacher, I've certainly done my share for the neighbourhood, the PTA, and the church. It's not fair that I am the one to get this colon cancer. I've even watched our diets – fibre, protein, low cholesterol, and so on. It just doesn't make sense. (*Ibid.*)

The second component of the sense of coherence is being able to cope. The person with a sense of coherence not only understands what the world expects, but also feels that he or she has the ability and the resources necessary to meet whatever is demanded. The student knows what is necessary to get an A, B, or C grade and also feels able to obtain the desired grade.

Data from people with multiple sclerosis provide some sense of the meaning of this component. At 20, Josh played baseball and hockey in college. He did reasonably well in school and had a part-time job and an active social life. At 21 he was lethargic, felt weak, and was often discouraged about his future. After numerous tests and false diagnoses, Josh had recently been told that he had multiple sclerosis. He didn't think he could cope with his life. Before, he had been absolutely tied up in activity and doing, doing, doing. That was all he knew and all he wanted from life. The multiple sclerosis limited his strength, mobility, and energy. He felt that there were few alternatives to his previous lifestyle, and besides, he didn't want any of them.

Jan, at twenty-three, had recently been diagnosed as having multiple sclerosis. She, too, had been very active socially. A cheerleader and an avid tennis player, she had done reasonably well in high school, had graduated as a nurse, and was working at her local hospital when she was diagnosed. She quickly realized that she would probably have a difficult time working as a nurse because of the physical strength and energy required. She began taking courses to prepare herself to work in nursing administration. A desk job on a nine-to-five basis would be much more manageable, she felt. She cut back on her sports activity but stayed active by walking to work. By making sure she had at least ten hours of sleep at night, Jan felt that she could manage whatever the future had in store for her.

The third component of a sense of coherence, meaningfulness, refers to the motivation to achieve a desired outcome. This component depends on the extent to which life makes sense, has a purpose, and is worth the effort. The student values learning, and the grade achieved is important because grades and education have a purpose to play in life's satisfactions and goals.

Laura was the mother of Alex, a severely mentally challenged eight-year-old boy who also had cerebral palsy. Apparently he hadn't received enough oxygen at birth. Laura was naturally shocked and even devastated when she was first told about his disabilities. Over time, though, she organized her life around Alex's disabilities and used them to give her life particular meaning. She started a local self-help group for parents of disabled children, was active on the local board of Extend-a-Family, and worked with the school board and the teachers in developing special programs to integrate children with special needs in the schools.

Joan, on the other hand, provides an example of a parent who could not find meaning in the severe "slowness" of her daughter. Again, the disability was the result of an accident at birth. From the time of Amanda's birth Joan had been furious. She was furious with the doctors, the nurses, the hospital, and, it seemed, with most everyone. Her fury turned inside and she became depressed. She isolated herself and continued to bewail her fate and that of her daughter. She tried one new kind of treatment after another – diet, vitamins, acupuncture, hypnosis, patterning, and prescription drugs. She tried them all, but nothing worked and nothing made sense to her. She was without hope or meaning.

The student with a high sense of coherence, when confronted with an unexpected (and seemingly undeserved) failure, does not give up in bitterness and disgust but rises to the challenge: he or she questions the mark, takes the course again, or takes a different course. In other words, the student with a high sense of coherence would be able to explain the failure to himself or herself and be willing and able to take action to improve the grade, believing that such action would have the desired outcome and that the whole process of learning and being examined is meaningful.

All of these aforementioned socio-psychological issues – stress, social support, type A behaviour, and sense of coherence appear to operate on the body through the immune system. This burgeoning field of study, focusing on mind-body events, is called psychoneuroimmunology. For an extensive overview of the research in this field that is accessible to social scientific understanding, see Kaplan (1991).

The Illness Iceberg

The distribution of illness in a population has been described using the simile of an iceberg (Last, 1963; Verbrugge, 1986). The simile implies that most symptoms of disease go largely unnoticed by the people who have the symptoms, by health-care practitioners, and by epidemiologists interested in measuring the incidence and prevalence of disease. There are a number of reasons why a great deal of illness goes undetected. In the first place, people often explain away or rationalize physical changes in their bodies in ways that seem to make sense and therefore do not require a medical explanation. Sudden or extensive weight loss and a long-standing cough that doesn't seem to get better can both be signs of very serious illness. They are both, however, easily explainable in lay terms: "I've been too busy to eat," or "If it would only stop raining my cough would go away." In the second place, some signs of latent illness develop slowly over a long period, so that the patient is not alerted to them. High blood pressure and cholesterol buildup in the arteries are two examples. Some of these diseases can only be detected by clinical tests such as x-rays, CAT scans, and blood and urine tests.

Practitioners, too, are limited in what illness they can detect. Some diseases are not observable through any specific clinical measures and can only be diagnosed by a myriad of complex tests, plus symptoms described by the patient and some element of luck or art. Also, practitioners are often limited in their ability to detect illness because their patients do not provide sufficient information. Epidemiologists face all of the obstacles described above. In addition, epidemiologists frequently rely on a wide range of data collection strategies (as discussed in Chapter Two), all of which are subject to various limitations of validity, reliability, recall, response rate, truth-telling, and the like.

In the face of such ambiguity and variability in the recognition and acknowledgement of signs and symptoms of illness, what are the processes that lead people to decide that symptoms are not a problem or to do something about them?

Why People Seek Help

What makes you decide to go to the doctor? Do you go when pain becomes severe? Is it when your symptoms interfere with your responsibilities at home, school, or work? Do you go for reassurance that a symptom is not a reflection of anything serious? Do you try to avoid going to the doctor? When you feel symptoms of a cold or a flu, do you generally just go to bed early or take a day or two off work? Do you tend to begin a course of vitamins? Do you go to the local drugstore to buy something from the shelf? Or do you go to a naturopath, chiropractor, nutritionist, allopathic physician, acupuncturist, or other therapist? The processes by which people come to notice signs they think may be symptomatic of illness, the kinds of attention they pay to these signs, and the action they decide to take are all a part of the study of illness behaviour. Whether someone seeks help and what kind of help is sought are the result of complex social and psychological determinants.

The first stage of illness is the acknowledgement or notice of symptoms or signs. Sometimes symptoms are noticeable as little more than a minor behavioural change such as tiredness or lack of appetite. Some common mild symptoms, such as a cough, cause only a mild discomfort; others may cause a searing pain. Sometimes symptoms are noticeable as measurable physical anomalies such as a heightened temperature or an excessive blood sugar reading. At times, illness is not experienced until it is diagnosed after a routine medical examination.

People with similar symptoms may respond very differently to them. One may go straight to the doctor; another may "let nature take its course" at first, even with severe symptoms. Although early responses may be quite variable, they will make sense within the context of the social, cultural, economic, and psychological conditions of each person.

An early study, "Pathways to the Doctor" (Zola, 1973), was based on interviews with more than 200 people at three clinics. For each person it was the first visit for that particular problem. According to Zola's analysis, the decision to seek treatment was based on a great deal more than the mere presence of symptoms. Rather, the decision to seek medical care was associated with one or more of the following: (1) occurrence of an interpersonal crisis; (2) perceived interference with social or personal relations; (3) sanctioning by others; (4) perceived interference with vocational or physical activity; and (5) a kind of temporalizing of symptomatology.

Each of the five motivations to seek medical care will be illustrated by a description of a case. John's case exemplifies action on the basis of an interpersonal crisis. John had been feeling tired for the last six months; he had explained it to himself as overwork. As he said:

I knew that as soon as I finished the presentation, I'd feel better. This was the big one. We'd been working on preparing this series of ads for a major soft drink company for just about two years. My promotion, my future was tied up in it. I

didn't want to let down. Then in the final week, Mac (his co-worker) was really upset one day and quit. He said it was all my fault, that I had been impossible to work with, that I didn't co-operate and that I was too busy. That night I decided I had to go to the doctor to get this thing diagnosed and fixed.

Perceived interference with social or personal relations can be illustrated by the case of Mary Beth, who had a cough that seemed to be hanging on amidst increasing tiredness. Walking the stairs to her second-floor flat seemed to be more and more of an ordeal. She managed to get to work regularly and stay through what seemed like very long days. When a skiing vacation with a group of friends came up and she realized that she wouldn't be able to go, she decided that she had had enough and made an appointment with the doctor. In both these cases, the symptoms had continued for a long period. The point of decision was not marked by new symptoms or symptoms that suddenly become more severe, but rather by changes in the social environment.

Sanctioning occurs when someone insists or ensures that the person with symptoms goes to the doctor. Men visit physicians less often than women, and it appears that a large percentage of their visits result from the urging of women (wives, sisters, mothers, or female friends). Charles's situation is illustrative. Charles had been experiencing chest pains for some weeks. He had been telling his wife Carol about them, but insisted that since they seemed to occur only after meals, they must be caused by heartburn and nothing more serious. Carol, however, thought otherwise, and made an appointment with their doctor. Faced with an appointment and a worried wife, Charles went to the doctor.

The fourth impetus to seeking medical aid is rooted in the work ethic. Sometimes the only changes that merit medical attention are those that interfere with work. Zola (1973) gives the example of a man with multiple sclerosis who, despite losing his balance and falling in a number of different locations, did nothing until he fell at work. Then he decided to seek medical advice.

The final impetus noted by Zola (1973) was temporalizing. Sometimes symptoms only become problematic when they seem to have continued "too long" or to have developed "suddenly" and "unexpectedly." For example, at least three times over the winter Larry had had a cold with a very dry cough that lasted for a few weeks. He bore with it. When the first weekend of warm, sunny weather arrived at Easter, he noticed that he had the cough again. This time he decided that he had had it too long, that spring was here, and that the cough needed "looking after." Susan's situation shows that sometimes it is the suddenness rather than the duration of the symptom that causes a person to seek help. Susan described herself as "healthy as a horse." When she woke up one day with a very bad headache, she decided quickly that "something was wrong" and went to the doctor.

Mechanic (1978) proposed that seeking help depends on ten determinants: (1) visibility and recognition of the symptoms; (2) the extent to which symptoms are perceived as dangerous; (3) the extent to which symptoms disrupt family, work, and other social activities; (4) the frequency and persistence of symptoms; (5) amount of tolerance for the symptoms; (6) available information, knowledge, and cultural assumptions; (7) basic needs that lead to denial; (8) other needs competing with the symptoms; (9) competing interpretations that can be given to the symptoms once they are recognized; and (10) availability of treatment resources, physical proximity, and psychological and financial costs of taking action.

Mechanic (1978) and Zola (1973) do not contradict one another. Each takes a somewhat different point of view and focuses on some aspects of the reasons for seeking help while ignoring others. In particular, Mechanic acknowledges the relevance of the symptoms themselves – their severity, perceived seriousness, visibility, frequency, and persistence – and the knowledge, information, and associated cultural assumptions in determining whether or not the symptoms will be acted upon. In Zola's model, symptoms *per se* are not discussed except with regard to how long they continue or how suddenly they emerge. Zola's point of departure is – given certain symptoms – what action will likely be taken. Aside from this major difference, both models accentuate the importance of the disruption in family, work, and recreation and other competing goals in determining when an individual will to go the doctor. Zola stresses the necessary role that others sometimes play in determining health actions. Mechanic notes the real constraints that may exist with regard to the costs of medical resources.

Summary

(1) Stress is a process that occurs in response to demands that are either much greater than or much less than the usual levels of activity. Historically, the bodily reaction to stress has been termed "fright or flight" by Cannon. Later, Selyé suggested that the General Adaptation Syndrome (GAS) is the body's reaction to all stressful events. It has three stages: an alarm reaction, resistance, and exhaustion. Exposure to stress in the third stage can lead to "diseases of adaptation."

(2) Some researchers have developed scales to measure degrees of stress.

(3) Various factors affect the amount of stress experienced. One important one is social support, which has been seen to be a buffer against the degree of stress experienced.

(4) Coronary-prone behaviour has been defined as that which is compulsive, dominating, and aggressive. People with these characteristics have type A personalities. Type A behaviour can be measured using three sub-scales: speed and impatience, job involvement, and hard-driving conscientiousness. The type A person is more likely to suffer premature cardiac disease.

(5) People are more likely to stay healthy when they have a feeling of comprehensibility, manageability, and meaningfulness in their lives and are able to develop a sense of coherence and a belief that things are under control.

(6) The first stage of illness is the acknowledgement or notice of symptoms or signs. What people do about the illness, how illness is experienced and handled or treated, is called illness behaviour. Illness behaviour varies according to the individual's social conditions. Factors that lead an individual to seek treatment are the occurrence of an interpersonal crisis, the perceived interference with social or personal relations, sanctioning by others, the perceived interference with vocational or physical activity, and/or temporalizing: symptoms that have continued "too long" or develop "suddenly" and "unexpectedly."

(7) A number of diseases involve a great deal of uncertainty. Diagnosis can be problematic as early symptoms may arise and then remit. Diseases also tend to affect individuals differently.

The Experience of Being Ill

WHAT IS IT LIKE TO acknowledge for the first time symptoms of a potentially serious illness? For instance, how do women feel and how do they talk to themselves and to others upon first noticing a lump in a breast? How do men or women manage to cope with a diagnosis of a myocardial infarction? How do people who feel awful but cannot get a diagnosis, such as some people with chronic fatigue syndrome, manage? What is it like to be told that your child has epilepsy, Tay-Sachs disease, or Down's syndrome? What is it like for the doctors, the nurses, the siblings, and significant others? How do people talk to themselves when they come to realize that they have cancer, diabetes, AIDS? How do people manage the uncertainty surrounding the diagnosis and the possible or probable future prognosis of illness? How do people tell their significant others once they have received a diagnosis from the doctor? And how do family members and significant others manage the news? How are such mild and self-limiting diseases as the flu or a cold understood in the whole context of the lives of people? These are the sorts of questions that might be asked about the experience of illness.

The purpose of this chapter is to describe and explain something of the experience of being ill in Canadian society. Most published sociology to date takes it that the object of sociological science is the observation of the institutions and structures of society that constrain people's thoughts, feelings, beliefs, and actions. This view prevails in most of the articles published in all the major North American journals of sociology.

Chapters Four through Seven are written in this positivist tradition, using quantitative and objective data and social phenomena to provide causal explanations. These four chapters do not examine the processes whereby these "external objective" forces come to be integrated into the social actions (thoughts, feelings, beliefs, and behaviour) of human beings. Nor do these chapters offer an explanation of the meanings and interpretations that people give to these factors.

Chapter Eight is written in the symbolic interactionist tradition: it draws attention to

Box 8.1 Just One AIDS Story

A report in the *Globe and Mail* (Appleby, 1989) exemplifies the kinds of experiences that had at that time become typical among those suffering from AIDS. The story concerned the judgement made in a court case that a male prisoner who had AIDS had received "cruel and unusual" punishment at Metro Toronto Detention Centre. The prisoner had complained that "he had been kept in isolation, fed food that made him vomit – including porridge containing cigarette stubs – and forced to sleep without a mattress." He also alleged that the guards told him: "You'd better watch what you eat, we're going to get you."

Apparently the prisoner, Kyle Downey, was at "level 3" of the illness, which means that the symptoms were not yet manifest. He had, however, lost twenty pounds and could only ingest liquids. He had been transferred to the detention centre from the Don Jail a few hours after a guard claimed to have been bitten by Downey. Within hours a number of prison guards who were members of the Ontario Public Service Employees Union threatened to strike over the issue of having to guard prisoners with AIDS. Some union members even suggested that Kyle Downey be charged with attempted murder. Later the bite charges were shown to be false and the guard who initiated the charges was found to have assaulted Downey. The strike was called off. But the issues surrounding the treatment of AIDS patients remain.

the meanings, interpretations, and world views of human beings in relation to illness, sickness, disease, and death. In this tradition the chapter examines the subjective reality, the consciousness of people making and finding meaning in interactive and social context. The analysis in this chapter is at the individual level. However, it must be emphasized that a person's views are affected by a particular society and by a particular place at a unique point in time in that society. Meanings are constructed out of social interactions in specific social, political, economic, and historical contexts. Meanings reflect a person's position in the social structure and that person's relationships and experience. Cultural attitudes to illness vary. The meaning of illness to people varies.

Illness, Sickness, and Disease

Sociologists generally distinguish among disease, illness, and sickness. Disease is said to refer to an "objective" phenomenon, one that is diagnosed by a physician; it is usually believed to be located in specific organs or systems in the body and curable through specific biomedical treatments. Illness, by contrast, is the personal experience of the person who acknowledges that he or she does not feel well. Sickness is the social actions taken by a person as a result of illness or disease, such as taking medication, visiting the doctor, resting in bed, or staying away from work. Patients feel illness; physicians diagnose and treat disease.

Sickness, disease, and illness may, in some ways, be independent of one another. The fact that people can feel ill and act sick

FIGURE 8.1 A Model of the Relationships among Illness, Sickness, and Disease

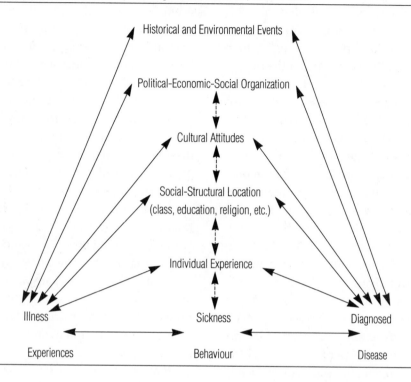

and yet not visit a physician confirms this, as in the case of a mild condition such as a cold. People can act sick without either feeling ill or having a medical diagnosis, as when a student complains of having flu to get an extension for an essay or an exemption from an examination. People may be told by a physician that they are not ill and ought not to be acting sick. This may happen when a person is experiencing sleeplessness from the stress of final exams, and thus feels ill, but does not have a medical condition. Finally, a person may have an illness diagnosed as a disease by a physician. Such a diagnosis legitimizes sick role behaviour. In this case sickness, illness, and disease may

occur together. Figure 8.1 is meant to clarify the relationships among illness, sickness, and disease. Illness, sickness, and disease are integrally related because they are all socially constructed experiences. People do not experience or talk about their illnesses in a social or cultural vacuum. Everything that people feel, say, think, and do about their illness is culturally and socially mediated. For example, a sore back conjures up one set of meanings, ideas, and actions when it happens to a person with bone cancer. A sore back conjures up a different set of meanings, ideas, and actions when it happens to a world-class skier just before the Olympics.

In fact, noticing that a part of the body,

such as the back, is sore is only possible within social-cultural context and its resultant language categories. Cultural constructions include "lay epidemiology," which can have a significant impact on the plausibility for the public of modern health promotion messages. Davison *et al.* (1991), for instance, have shown how the British public have highly developed ideas about what "type" of person is a coronary candidate and that these ideas are widely shared. For example, one respondent in their interview-based study said, "Yeah, big, fat, wheezy blokes huffing and puffing" (1991: 9) were likely candidates. Another suggested, "Fit, skinny, young. The last person you'd expect to have a coronary" (*ibid.*: 89). These lay categories are widely shared.

Sickness behaviour is also socially mediated. Social-structural and cultural factors influence whether a person visits a chiropractor, naturopath, masseuse, or the health food store. Whether a woman goes to the hospital at the first sign of labour or does not go until she is ready to deliver depends in part on cultural influences. Old order Mennonite women, according to the nurses in the obstetrics department of a local hospital, often have their babies on the way to the delivery room because they tend to wait through several "stages" of labour before travelling to the hospital. According to the same source, women from some other ethnic backgrounds usually enter the hospital at the first sign of contractions, and once there they are likely to have long, protracted, and painful labour.

Disease, too, is a socially mediated event. Doctors make their decisions from a complex mix of sociocultural, historical, economic, medical, scientific, and clinical data, and in the context of their own age, specialty, class, gender, and ethnicity. In industrialized societies a number of newly prominent diseases have come to the attention of doctors. Anorexia nervosa and bulimia are two types of eating disorders that have attained prominence as newly discovered diseases, and apparently are growing in incidence and severity in North American society (see Currie, 1988).

Outrageous Acts and Everyday Rebellions is the title of a witty satire by Gloria Steinem (1983) illustrating how illness, sickness, and disease may be sociological determinations. Steinem suggests, for example, that beliefs about menstruation are socially determined: if men menstruated rather than women, menstruation would have very different meanings and consequences. First of all, if men menstruated, it would be a laudable, dramatic event, symbolic of strength and manhood. Men would compare with one another how large their "flow" was, how long it lasted, and whose blood was the brightest. Menstruation would be celebrated as a recurring, ritual reminder of the power of masculinity. To prevent problems such as cramps and premenstrual syndrome (PMS), billions of dollars in research money would be spent. Doctors specializing in menstruation would be thought to hold the most prestigious specialty, the most highly paid and desired medical option. Sanitary napkins and tampons would be free. Epidemiological research would demonstrate how men performed better and won more Olympic medals during their periods. Menstruation would be considered proof that only men could serve God and country in war, in politics, or in religion (*ibid:* 338).

Variations in the Experience of Being Ill

In all societies people experience illness, pain, disability, and disfigurement. Sorcery, the breaking of a taboo, the intrusion of a disease-causing spirit into the body, the intrusion of a disease-causing object into the body, and the loss of the soul are all seen as possible causes of disease (Clements, 1932). All these explanations, except for the intrusion into the body of a disease-causing object, involve the supernatural or magic in an attempt to understand illness. Westerners tend to see illness as empirically caused and mechanically or chemically treatable. To a large extent, we have separated the mind, the body, and the spirit. But in most of the non-Western world, non-empirical explanations and cures for disease seem to dominate: illness is seen as a combination of spiritual, mental, and physical phenomena.

The experience of pain varies from one ethnic group and culture to another. For instance, Zborowski (1952) found that patients in New York City from Jewish and Italian backgrounds responded emotionally to suffering and tended to dwell on their painful experience. He also noted a difference in the attitudes of each of these groups to discomfort. Italians saw pain as something to be rid of and were happy

once a way to relieve their pain was found. Jews were mainly concerned with the meaning of their pain and the consequences of their pain for their future health. "Old Americans" and the Irish, by contrast, reacted stoically.

What is viewed as disease or health varies from society to society. Disfigurement or illness may or may not be seen as normal. "Afflictions common enough in a group to be endemic, though they be clinical deformities, may often be accepted as part of man's natural condition" (Hughes, 1967: 88).

Disagreement is widespread over which states of physical being are desirable and which are not. Among some people, the obese woman is an object of desire; others define obesity as repugnant or even as a symptom of emotional or physical illness. People with epilepsy are sometimes the object of ridicule and fear, but some people think they possess supernatural powers. There are many examples that illustrate how health and illness evaluations depend on value judgements (Clements, 1932; Hughes, 1967; Fitzpatrick *et al.*, 1984; Turner, 1987).

It is clear that for many people health and religion, the natural and the supernatural, are often closely related. This is true of people's conceptions of the causes of disease and accident, and of their cures. As Freidson has said, "human and therefore social evaluation of what is normal, proper, or desirable is as inherent in the notion of illness as it is in notions of morality" (Freidson, 1970: 211).

Popular Conceptions of Health, Illness, and Disease

Medical anthropologists and sociologists who have examined beliefs about illness in modern Western communities have found that beliefs vary depending on the cultural background. Non-medical people hold immensely strong beliefs about illness, its causes, and its appropriate treatments (Cornwell, 1984). Cornwell distinguished between medicalization-from-above and medicalization-from-below. The first refers to definitions of the reality of the mind and body that are the result of developments in scientific medicine. The second refers to the acceptance/rejection of such definitions on the part of lay people. It is particularly important to note that medical and lay ideas are often quite different and even, at times, contradictory. The following sections describe a number of popular conceptions of health and illness.

Illness as Choice

Illness is a choice. We choose when we want to be ill, what type of illness we'll have, how severe it will be, and how long the illness will last. Because the body and mind are connected, illness is a result of thinking and feeling as well as of bodily processes. We give ourselves illnesses to take a break from the busy obligations of everyday life. We choose to be ill to allow for "time-out" from our routines: to rest, to take stock, to escape, and to withdraw. A host of research in fields such as psychoneuroimmunology and psychosomatic medicine attests to the prevalence of this belief on both theoretical and empirical levels. A great deal of work has been done on the mind-body interaction in the case of cancer, in particular. Several physician writers have addressed mind-body relationships in various ways. For an overview of some of the most popular, consider the work of Chopra (1987, 1989) and Dossey (1982, 1991).

Oncologist Carl O. Simonton and his colleagues were the earliest contemporary popularizers of the role of the mind in both sickening and healing the body. They believe that cancer arises from a sort of emotional despair. Simonton's book, *Getting Well Again*, is full of examples of people who come to recognize, with the help of the psychotherapeutic skills of Simonton's staff, that their cancer has been, in a sense, a personal choice. Once a patient comes to understand the reason for the choice of cancer as a way of coping with a personal problem, he or she can confront the problem on a conscious level and seek to find solutions (Simonton *et al.*, 1978).

Simonton's method includes a number of techniques. One of the best known is the systematic use of imaging – conjuring up mental pictures of cells. People are directed to imagine their cancer cells as black ants and their healthy cells as knights in shining armour, for example. There is some evidence that the more powerful the imaging of the anti-cancer forces, the more likely it is that the imaging will be effective. Effective imaging usually results in one of the following outcomes: increased longevity; moderation of symptoms, including pain; increased sense of well-being, autonomy, and control; improved quality of life; and even, at times, remission of the disease.

Imaging is the most dramatic of the techniques used to combat cancer "psychologically"; several other methods are being used. Alastair Cunningham, a psychologist and immunologist at Princess Margaret Hospital in Toronto, has for a number of years been running groups for people with cancer. Included in Cunningham's program for coping with cancer and minimizing its impact are stress-management techniques, writing a journal or diary, and imaging. Dr. Cunningham has written a workbook called *Helping Yourself* that is accompanied by two audio tapes, one for deep muscle relaxation and the other on imagery for healing. These were sponsored by and are widely available through the Canadian Cancer Society. The effect of all of these on quality of life, longevity, and the disease itself is under investigation. The point to be made here is that within this model of health/illness an important part of healing (along with medical interventions) and of improving the quality of life is taking responsibility for one's state of health and choosing to try to do something about it.

Illness as Despair

Closely related to the idea that illness reflects choice is the idea that illness reflects and results from a sort of despair. Many social commentators, psychologists, and medical researchers have thought of illness as a sort of emotional despair. In this century, Lawrence LeShan (1978), a psychotherapist who counselled and then studied cancer patients, came to the conclusion that there was a common type of emotional experience among the majority of cancer patients that predated the development of the disease. Examples of this emotional experience include long-standing, unresolved grief over the loss of someone close, such as the death of a mother or father in childhood, or childhood loneliness or isolation and the loss of someone close in adulthood, or the loss, for example, of a job. For LeShan's patients, understanding their grief and expressing feelings of sadness and anger became pathways toward being healed.

Norman Cousins's work, which formed the basis for his book, *Anatomy of an Illness as Perceived by the Patient*, is the foundation of another very similar and newly popular approach to illness, health, and healing. *Anatomy of an Illness*, a long-time best-seller translated into at least thirteen different languages and "must" reading in a number of medical schools throughout the world, tells the story of Norman Cousins's reaction to and healing from a potentially fatal and severe case of ankylosing spondylitis, a degenerative disease of the connective tissues in the spine.

Cousins's story of his "miraculous" recovery focused on three aspects of his "systematic pursuit of salutary emotions" (Cousins, 1979: 35). The first was his partnership with his doctor. Cousins asked his doctor to engage in a collaborative partnership. The physician, on his side, had the traditional armaments to offer. Cousins, on his side, had done some reading about how healing could be affected by the power of positive thinking and by massive doses of vitamin C. Because he found the hospital forbidding, noisy, and thoroughly unhealthy, Cousins asked his doctor to treat him in a nearby motel. The second important component of his recovery was that his doctor agreed to his request for massive intravenous doses of vitamin C, which Cousins, having done considerable research on his own, believed in. The third aspect of his rehabilitation strategy, and the most important for our purposes, was laughter. Cousins had books of jokes and cartoons, tapes, and old movies brought to his room for his enjoyment. Laughing deeply relieved his pain and generally made him feel better. As he says, "It worked. I made the joyous discovery that ten minutes of genuine belly laughter had an anesthetic effect and would give me at least two hours of pain-free sleep. When the pain-killing effect of the laughter wore off, we would switch on the motion-picture projector again, and, not infrequently, it would lead to another pain-free sleep interval" (ibid.: 39). As he was interested in demonstrating the credibility of his experience, Cousins instituted some small experiments. In one, he found that each episode of laughter decreased the level of inflammation in his connective tissues.

Five years after recovering from this disease, and having achieved some renown in the area of holistic health (he was appointed senior lecturer at the School of Medicine, University of California at Los Angeles, and consulting editor of Man and Medicine), Cousins suffered a serious heart attack. Rehabilitation resulted in another book, The Healing Heart (1983), in which he further developed his arguments regarding the importance of positive emotions in maintaining and achieving health. Cousins's work is based on the idea that the mind and body are a single entity; positive change in one, for example in the body by the use of pharmaceuticals such as vitamin C or in the mind by mood elevation through, for instance, laughter, can lead to healing.

Illness as Secondary Gain

Sometimes illness has definite benefits. Remember the last time you were conveniently sick on the day of a big test for which you had not had time to prepare. Little kids quickly learn to complain of sore tummies when they do not really feel like going to church, to school, or to grandma's house. Illness may provide an opportunity or permit someone to behave in ways he or she would like to but otherwise would feel constrained not to.

Or illness may allow someone to meet needs that would otherwise go unmet. The benefits of illness are as varied as people's illnesses. A case example will illustrate. Brett was a high school guidance counsellor and history teacher; he had also taken on the job of basketball coach and was supervising set building for the school play. Brett was married and had two pre-school children who were in day care while he and his wife Lucy worked. Lucy was required to do a great deal of travelling in her job as a buyer for a large department store. Brett was often responsible for driving the kids to and from day care, as well as feeding and bathing them on his own. He hadn't expected so much responsibility: he had been raised to assume that his wife would handle the children, the house, the cooking, and most of their domestic life.

He felt trapped. On the one hand, he supported and applauded Lucy's success and realized that if she was really going to "make it" in her business of choice she would have to continue at her present pace for three or four years. On the other hand, he frequently felt overwhelmed by his responsibilities. The day his doctor diagnosed mononucleosis, Brett was secretly relieved. At least now he had a way out.

The question of secondary gains has been explained in a more systematic way in research on mothers on welfare (Cole and Lejeune, 1972). In this case the focus was on whether and in what situations mothers on welfare would be likely to claim that they were ill. The researchers hypothesized that illness could be seen as a legitimation for failure. Being on welfare is thought by a substantial number of welfare recipients to be the result of and also a continual reminder of a personal failure. This being the case, then women on welfare would be likely to look for a rationalization and legitimation for their situation. Indeed, the researchers found that women on welfare who felt that they had little hope of changing their status were more likely to use illness as their explanation for being "on welfare" than women who hoped or expected to move off welfare. The point here is that sometimes claiming illness seems to be a better choice and to provide a greater level of reward than the other available options. In such a case illness can be thought of as providing secondary gains.

Illness as a Message of the Body

Within the naturopathic perspective, illness is an expression of a unique person at a particular point in time and engaged in a special set of circumstances. Illness is also the expression of a whole person – body, mind, and soul. Illness and health are not polar opposites: both exist in a person continuously in a state of dynamic equilibrium. From this perspective symptoms are not signs of illness but indications that the person is responding to a challenge and is engaging in what is called a healing crisis. The healing crisis (or the symptom) is a manifestation of what has been happening for a time but has been repressed and therefore unnoticed. Fever, rash, inflammation, coughing, crying, sleeplessness, and tension are revelations of a deep disturbance of the whole person. Symptoms are valuable and useful because they allow the person to acknowledge the crisis and to seek help. The role of the healer is to support the person emotionally and to enhance the natural recuperative powers of the physical body through the administration of minuscule amounts of natural medications that are known to create the symptoms the individual "patient" is presenting (see Chapter Fourteen for a more complete discussion of naturopathic medicine).

Illness as Communication

People communicate through words and by signs or body language. People also send messages through the way their bodies are functioning. Illness sends a message that one part of the body is alienated from the "self." The body expresses the soul. One may even say that illness expresses the soul more impressively than health in the same way that a good caricature expresses essential aspects of a personality more clearly than photographs taken in uncharacteristic situations (Siirla, 1981: 3).

Through particular sets of symptoms or particular kinds of illnesses, people convey messages about themselves. A woman with breast cancer may be saying that her need and desire to nurture has been frustrated. Since the breast is important in mothering, feeding, and soothing, it becomes the most appropriate symbol for a communication about frustrated nurturing. Thus a person's needs may be expressed through illness. "If these [needs] cannot be expressed in a realistic, 'healthy' form, symbolic organ language takes over" (*ibid*.: 8).

Different diseases, it is claimed, express different frustrations. Rheumatism, a disability that affects the muscles and the joints, may express the frustration of a person who once liked to be very active and who wanted to limit his or her activities. Cold symptoms such as runny nose and sneezing may result from a frustrated desire to cry. All illnesses can be examined as attempts by the body to express feelings that otherwise cannot be expressed or are frustrated.

Illness as Metaphor

Related to the suggestion that illness is communication is the idea advanced by Susan Sontag of illness as metaphor (1978). Illness communicates by conveying a particular message for a person at a special point in his or her life. Illness as metaphor suggests that cultures bestow meanings on various illnesses. In Sontag's view, however, metaphors are most likely to be related to illnesses that have no clear or obvious cause or treatment. The two examples that she examines in some depth in *Illness as Metaphor* (1978) are cancer in this century and tuberculosis in the last century. In 1989 she examined the metaphoric meanings of AIDS. Her basic argument is that the metaphors attached to diseases are often destructive and harmful. They frequently have punitive effects on the patient because they exaggerate, simplify, and stereotype the patient's experience. Metaphors may function as stigma. They may serve to isolate the person with the disease from the community. Metaphors often imply adverse moral and psychological judgements about the ill person. They have perpetrated the view, for example, that cancer is a form of self-judgement or self-betrayal.

Media Images of Cancer, Heart Disease, and AIDS

One study of the way cancer, heart disease, and AIDS have been portrayed in the media illustrates that diseases come to have

TABLE 8.1 Images of Cancer, Heart Disease, and AIDS

CANCER	HEART DISEASE	AIDS
Cancer is described as an evil, immoral predator.	Heart disease is described as a strong, active, painful attack.	Little is said about the nature of the disease other than it debilitates the immune system.
Euphemisms such as the Big C are used rather than the word "cancer."	Heart disease, stroke, coronary/arterial occlusion, and all the various circulatory system diseases are usually called the Heart Attack.	Much is said about the moral worth of the victims of the disease.
Cancer is viewed as an enemy. Military imagery and tactics are associated with it. The whole self, particularly the emotional attitude of the person and the disease, is subject to discussion. Because the disease spreads and because the spread is often unnoticed through symptoms or medical checks, the body itself becomes potentially suspect.	The heart attack is described as a mechanical failure, treatable with available new technology and preventable with diet and other lifestyle changes. It occurs in a particular organ that is indeed interchangeable with other organs.	Acquired immune deficiency syndrome is called AIDS. The opportunistic diseases that attack the weakened immune system are often not mentioned. AIDS is viewed as an overpowering enemy, as epidemic and scourge. It is described as affecting the immune system and resulting from mostly immoral behaviour – connotes 'shameful sexual' acts and drug abuse.
Cancer is associated with hopelessness, fear, and death.	There is a degree of optimism about the preventability and treatability of the disease.	It is associated with fear, panic, and hysteria because it is contagious through body fluids, primarily blood and semen.
Prevention through early medical testing is advised.	The heart attack is described as very preventable. Suggestions for lifestyles that will prevent it are frequently publicized.	Prevention through monogamous sexual behaviour or abstinence and avoidance of unsterilized needles and drug abuse.
There are innumerable potential causes listed. They range from sperm to foodstuffs to the sun.	There is a specific and limited list of putative causes offered again and again.	Initially the causes for AIDS were very general: being homosexual, a drug user, or a Haitian.
There is little consideration of the socio-political or environmental causes, e.g., legislation that limits smoking.	There is little mention of socio-political causes.	There is little mention of socio-political causes.
There is uncertainty about cause.	There is certainty about cause.	There is uncertainty about cause.

Source: Clarke, 1992a.

unique meanings and metaphors associated with them. Disease is seen as much more than a mechanical failure, a physiological pathology (Clarke, 1992a). Table 8.1 portrays the findings.

The moral worth of the person with cancer is attacked by the invasion of an evil predator so fearsome it is not even to be named, but fought as a powerful alien intruder that spreads secretly through the body. The person with cancer is not offered much hope of recovery. By and large, the media portray cancer as associated with disgusting symptoms, mutilations, excruciating suffering, and finally death. To some extent the person with cancer is held to be blameworthy because the cancer could have been detected through early medical checkups, which are described as often successful and always wise. There is a great deal of uncertainty about the cause of cancer. There are numerous putative causes. They are usually described as the result of individual lifestyle decisions. Again, then, it is the individual who is ultimately culpable.

The media description is radically different when the disease is a heart attack. The heart attack is presented as an objective, morally neutral event that happens at a specific time and place and causes a great deal of pain. Heart disease is portrayed in optimistic terms. Not only are there very clear and precise steps to be taken to prevent it, but when it occurs it can be treated in a variety of mechanical ways, including using technology to replace a malfunctioning heart. Heart disease does not affect the whole person or the moral being of the person. It "attacks" one part only. Heart disease is an outsider that can be repelled through quick, decisive action and the use of medical marvels. The person with heart disease may experience acute fear and pain, but the period of recovery is likely to be dominated by optimism about a cure and a resolve to change the lifestyle habits that led to the disease in the first place.

The person with AIDS is portrayed as a diseased person and as somewhat morally repugnant. He (usually a "he") is described as hopelessly doomed and isolated from potentially significant sources of emotional support such as lovers and family members. The disease itself is described in mechanical and biomedical terms. The media do not dwell on the painful or debilitating symptoms of the disease. They do not focus on the inevitable terminal stages of the disease, on death itself, or on the mortality rate. Rather, they focus on the fear

of contagion and the uncertainty about the causes of contagion. The person afflicted with AIDS is stigmatized because of the connection with a deviant lifestyle and is isolated because people are afraid of the contagion that might result from being close. This fear of being close is said to affect not only the close friends and partner of the person with AIDS, but also more distant others such as medical personnel.

It must be emphasized that these are images of cancer, heart disease, and AIDS drawn from selected magazines in Canada and the United States, including *Newsweek*, *Time*, *Maclean's*, *Good Housekeeping*, *Ladies Home Journal*, and *Reader's Digest* over a twenty-year period. These magazines have the widest circulation in North America. They are not the only magazines with a mass circulation available in North America. They may or may not be the most widely read. Little research has been done on the impact of such media depictions on people in society.

The incidence of breast cancer is increasing; so is its politicization. Following the successes of the "AIDS Movement," women who have had breast cancer have met together and established national support and advocacy groups. Working together they are empowering those who have long been powerless in determining the research and treatment agenda – patients, their friends, and their families. In consequence, in 1993 the National Forum on Breast Cancer brought together "survivors," their friends and families, doctors, researchers, and policy-makers to establish new health and research parameters and priorities with respect to breast cancer. The extent to which these inclusionary and even revolutionary goals will be met in the near or in the more distant future remains to be seen. Clearly, there are well-entrenched interests, including the pharmaceutical and medical device industries as well as the particularly concerned medical specialists (oncologists and radiologists).

Deborah Lupton (1994) documents the dominance of powerful, conservative, pro-medical ideologies. In particular, having reviewed articles on breast cancer in the Australian press, Lupton notes the ways that the description of breast cancer reinforces the power of medical institutions through the valorization of medical technology, medical practitioners, public health rhetoric, and medical research at the exclusion of the needs, wants, and feelings in the general population. Individual women received press attention only when they were portrayed as triumphant, brave fighters against breast cancer, all the while extolling the benefits of medical treatments and early detection. Such emphases neglected a detailed or balanced portrayal of the risks associated with such early detection as mammography screening or the potential environmental causes of breast cancer.

Illness as Statistical Infrequency

From this point of view illness is essentially deviant mental or physical functioning. When a condition, no matter how uncomfortable, is prevalent among a group of people, it is usually not considered pathological. Illnesses such as the common cold and flu are so prevalent as to be considered unimportant. (It is clear that they are not taken seriously because few research dollars are allocated to investigating them.)

Illness as Sexual Politics

Many feminist thinkers have recently analysed the ways in which disease can be seen as sexual politics. Kohler-Reissman (1989) provides a theoretical/empirical review of the medicalization of women's bodies and lives with a focus on childbirth, reproduction and contraception, premenstrual syndrome, "beauty," and mental health and illness. Currie and Raoul (1992) discuss "women's struggle for the body" in the face of culturally and historically entrenched bodily constraints and restrictions. From this perspective, the constraints and limitations of the gender roles to which males and females are assigned are associated with the conceptualization and the subsequent diagnosis of various diseases. For example, Ehrenreich and English (1978) document the ways in which the "mysterious epidemic" experienced by nineteenth-century middle- and upper-class women was the outgrowth of the conditions that characterized their lives. The journals and diaries of women of the time give hundreds of examples of women who wasted away into lives of invalidism. The symptoms included headache, muscular aches, weakness, depression, menstrual difficulties, indigestion, among others. The diagnoses were many: "neurasthenia," "nervous prostration," "cardiac inadequacy," "dyspepsia," "rheumatism," and "hysteria." The diseases were not fatal. They were, however, usually chronic, and lasted throughout a woman's life until her death.

The feminist analysis explains that these diseases provided an appropriate role for women in the middle and upper classes. These women were utterly dependent on their husbands, and their sole purpose in life was the provision of heirs. Household tasks were left to domestic servants. Children were cared for by hired help. The wife's job was to do precisely nothing – and thus to stand as a symbol to the world of her husband's great financial success. Medical ideology buttressed this view of the appropriate role of middle- and upper-class women with the theory of the conservation of energy. This was the belief that human beings had only a certain amount of energy. Since the primary functions of women of these classes were procreation and decoration, it behooved them to save all their vitality for childbearing and not to waste any in studying or in doing good works. Such activities would drain energy away from the uterus, where it was needed, into the brain and limbs, where it served no good purpose.

Contemporary feminists have examined how the eating disorders bulimia and anorexia nervosa are logical expressions of women's role today (Currie, 1988). Eating disorders can best be understood, in this perspective, as the internalization of conflicts that result from the prevailing contradictory images of women. Anorexia is primarily a disease of young women, particularly those from middle- and upper

middle-class backgrounds, and occurs chiefly in modern Western capitalist societies. Some have applied traditional Freudian psychoanalysis to these disorders and have attributed their causes to a fear of sexuality and an attempt to avoid femininity. Others have explained them, from a family systems perspective, as a result of mothers who overprotect and overidentify with their daughters. Daughters differentiate from mothers and become separate individuals, in this perspective, by rebelling and refusing to eat.

Several feminist authors have argued that eating disorders constitute a "hunger strike" by women in protest against the contradictory images of women as independent, competent wage earners, on the one hand, and, on the other hand, as sexual objects who earn only a portion of male salaries in work that is devalued because it is done by women (*ibid.*; Wolfe, 1990).

There are two periods in the twentieth century when anorexia was a notable problem – the 1920s and the latter part of the twentieth century. Both of these periods were times when equality of opportunity for men and women was being stressed. These were times of expanded educational and employment options for women. They were also times of contradictions: while alternatives for women were opening up in theory, in practice they still earned only a fraction of men's incomes. Furthermore, women were and continue to be notoriously under-represented at all levels of the political process – municipally, provincially, and federally. Nor have their domestic responsibilities declined. Inadequate child care and excessive domestic responsibilities mean that women who work outside the home also have a full-time job inside the home. Yet a woman's identity is still tied up with her appearance. In these circumstances, eating disorders can be seen as women's hunger strike (Orbach, 1986).

The next section of this chapter moves from a consideration of models of illness in contemporary society to some empirical analyses of the experience of illness from the perspective of the subject.

The Insider's View: How Illness Is Experienced

Most sociological analyses of illness have treated it as an objective phenomenon measured by questionnaires and by biophysical or clinical tests. From this perspective the varieties of personal experiences are irrelevant. What matters is the explanation of the incidence and prevalence of various types of disease. However, in keeping with the arguments and traditions noted in the work of people such as Weber, Husserl, Schutz, Blumer, Mead, Cooley, Simmel, Garfinkel, and Goffman, some researchers have turned their attention to the symbolic meanings and the social constructions of illness in the context of people's everyday experiences. These researchers focus on analysis at the individual level. It must be emphasized, however, that individual attitudes must always be set in a social context at a particular place and at a particular point in time.

Analysing illness at the individual level

has a long and noteworthy tradition in the sociology of medicine and the sociology of health and illness. One of the first published studies, mentioned earlier, was that of Mark Zborowski (1952, 1969); he compared the understandings and meanings of pain among Jewish, Italian, and "old American" ethnic groups in New York City. Goffman's *Asylums* (1961) described the experience of hospitalization from the perspectives of the patients or inmates and from that of the staff. Later Goffman (1963) examined stigmas associated with illnesses and other "unusual or abnormal" conditions.

Roth (1963) reported on his own experience of hospitalization for tuberculosis. Myra Bluebond-Langer (1978) described the world of children who were dying of leukemia and the experiences of their parents, their nurses, and their doctors. Her work is particularly instructive in illustrating the control and spread of information. Bluebond-Langer documented the children's extensive detailed knowledge of their condition, the next stages in their diseases, the probable side effects of various medications, which medications they were likely to be given, and even when parents and hospital staff would attempt to shield them from this information. Strauss (1975) published a book based on a series of studies of the experiences of coping with medical crises, stigma, isolation, changed self-concept, and changes in family and work relationships.

Speedling (1982) expanded the description of the experience of illness from the point of view of the person with the disease to that of the family when one of its members suffers a heart attack. His research looks at the impact of a person's hospitalization on the family, on how the family defines heart attack, on the person who has had the heart attack, and on his or her later ability to cope with the changes brought about by the disease. Stewart and Sullivan (1982) have examined the processes through which multiple sclerosis comes to be diagnosed, the various phases and stages and uncertainties associated with its history, its emotional impact, and the changes it brings in interpersonal relationships.

In her study of a community, Cornwell (1984) examines common-sense ideas about health, illness, and health services held by families and households in East London. Their ideas are described within the context of the life of the respondents. Thomas's (1982) work focuses on the experiences of people with impairment, disability, and handicaps. Thomas distinguishes among impairment, which is physical or psychological pathology; disability, which is the limitations on everyday activities such as eating, dressing, and walking; and handicap, which is a socially derived concept that labels a person pejoratively. As Thomas says, these three are not necessarily inextricably linked to one another.

Thomas pays particular attention to the self-identities of the various people involved, the disabled person, societal attitudes toward disability, and the attitudes of parents and professional caregivers. As are most books in this genre, his book is filled with long, detailed quotations drawn from accounts given by the people involved.

Schneider and Conrad's (1983) study of epilepsy examines living with epilepsy – its diagnosis, living with uncertainty, managing the symptoms, concealing the disorder, strategies of relating to others, the views of parents and family members, coping with the stigma, and handling medical regimens.

Strauss (1975) studied the impact of a number of different chronic illnesses and noted a variety of common concerns that people with such illnesses and their family members had to face. Strauss distinguished all of the following issues: (1) preventing and managing medical crises; (2) managing medical regimens (taking medications, administering needles, physiotherapeutic exercises); (3) controlling symptoms and preventing symptom eruption; (4) organizing and scheduling time (including necessary rests and treatments) efficiently; (5) preventing or coping with social isolation; (6) dealing with uncertainty and adjusting to changes during the course of the disease; (7) normalizing social and interpersonal relationships; (8) managing stigma; and (9) managing knowledge and information. Each of these will be discussed in turn.

Crisis management requires constant vigilance on the part of the ill person and those who are taking care of him or her. A diabetic, for instance, must continually monitor blood sugar levels and weigh and measure food intake and insulin levels. Such monitoring cannot be completely accurate, nor are the results entirely predictable. A reaction to an imbalance may be infrequent but is a constant possibility. People with colitis or other bowel disorders have to be prepared all the time for an unexpected and potentially embarrassing and even humiliating bowel evacuation. People who are caring for those with Alzheimer's in their own homes are well aware that the patient is constantly at risk. Newspapers often carry stories of Alzheimer's patients who have left home and become lost, often without adequate clothing for protection from the elements.

Managing medical regimens is often a complex matter. Physicians tell the patient to take a certain number of pills, or to have injections at certain intervals. The patient often adapts and adjusts the quantity or frequency to a level at which he or she feels comfortable. Most medications have side effects. The patient must learn to balance the need for the medication with the need to be free of side effects. Cancer patients may have to cope with chemotherapy, radiation, and surgery. Each of these treatments has negative psychological and physical side effects. People with cancer who are scheduled for chemotherapy on a regular basis by doctors may decide to take a week or more off because they cannot bear the side effects, need or want a holiday, have a party or other special event to attend, or have any of a number of other social and personal priorities. For the person with diabetes, managing the prescribed regimen is not as simple as just obeying the doctor's orders. Diabetics must make complicated calculations about the quantity of insulin they require. The amount of insulin necessary will vary with the amount of rest or stress the diabetic is experiencing. The patient has to learn to manage the medica-

tion in order to be comfortable and yet to avert a coma. Diabetics may also experience restrictions on their social lives – they may not be able to go drinking with friends. People with ulcerative colitis and with a colostomy or ileostomy must devote considerable time to managing their diet, monitoring their liquid intake, and being prepared to cope with a loss of control (see Reif, 1975).

The control of symptoms is a related issue. This involves managing the medical regimen daily, hourly, or even more frequently so that a crisis does not erupt. It also involves taking enough medication to prevent a crisis but not enough to cause uncomfortable side effects. Minor symptoms may lead to changing some habits; major symptoms may require redesigning the patient's house and lifestyle. Someone with colitis may be severely restricted in mobility by the necessity always to be near a toilet. A person with arthritis might have to move to a one-storey house without steps up to the door.

Organizing and scheduling time can be important in many chronic diseases because available time is almost always limited. Time is needed to manage the required medications, visit doctors and hospital, change clothes or apparatus, cope with restrictions, or be able to move about, and the like. Often the symptoms, too, such as headaches and backaches, require that the ill person drop the daily routine in order to go to bed to cope with the pain. Tiredness and fatigue are almost inevitable accompaniments to most chronic illnesses.

Preventing or coping with isolation and

trying to maintain social relationships when the person can no longer do what "normal" people do is another never-ending struggle. Sometimes people restrict their friendships to others who have the same problem and therefore understand. The ubiquity of self-help groups for most every chronic disease attests to the importance of social relationships with similarly disabled people, and perhaps, also, to the isolation felt by the disabled. People with cancer and their families have repeatedly stated in interviews that they are almost glad they have had the disease because through it they have met so many wonderful people. People have also said that they can talk to others who have a similar disability in a way that they have never been able to talk to anyone else before the diagnosis.

Another underlying experience that seems to be common among people with a variety of chronic illnesses is the experience of uncertainty (see Conrad, 1987; Strauss, 1975). Uncertainty affects every stage of illness, beginning with the diagnosis and ending with cure, confirmation that an illness is chronic, or death. Most chronic illnesses have variable and unpredictable prognoses. Cancer is perhaps the archetypal disease of uncertainty. Its very cure is measured by the diminishing probability of recurrence one, two, three, or more years after the initial diagnosis and treatment.

Diseases such as epilepsy, multiple sclerosis, and Alzheimer's are characterized by uncertainty. In all three the uncertainty is endemic to the whole diagnostic process. They are each very difficult to diagnose and require the results of a number of different tests before a reasonably confident diagnosis can be reached. But uncertainty also prevails during the stages of treatment, of spread of the disease, and of resurgence and remission. It is difficult to make plans for the future, to be considered and consider oneself a reliable friend, companion, spouse, parent, or worker when faced with a disease with an unknown prognosis.

A man who has once experienced a heart attack may distrust his body for a long time. He may avoid work and exercise and radically alter his lifestyle. He may feel unable to count on being alive in six months or a year. Because a heart attack frequently occurs without warning, a man who has experienced a heart attack may lose the ability to take his body for granted. Yet this ability is often the prerequisite for a satisfying lifestyle.

The sociological use of the term "stigma" originates with the work of Goffman (1963), who examined varieties of social interactions that were impaired by people's identity problems. Goffman designated a stigma as an attribution that discredits the value of a person. He distinguished between the effects of stigma that are known only to the person involved and those that are known to others, and analysed various strategies designed to handle the stigma in each case. The discredited person, whose deviant stigma is known to others, is challenged to "manage impressions" when interacting with others in order to maintain acceptable relationships. The discreditable person whose stigma is not known has another problem to contend with: he or she must manage information to hide the stigma and prevent others from discovering it. Many chronic illnesses carry stigma that can radically affect both the sense of self-identity and interpersonal relationships. For as Cooley has said, the way that we see ourselves is related to (1) our imagined idea of how we appear to others, (2) our imagined understanding of how others view us, and (3) our resultant feelings of pride or mortification.

Cancer has been experienced as one of the most stigmatized of diseases. Among the denigrating beliefs are the following: (1) cancer is fatal; (2) cancer means mutilation; (3) cancer implies a wretched death; (4) it is traitorous; (5) it is unclean; (6) it is contagious; and (7) it is caused by emotional repression (Peters-Golden, 1982). Susan Sontag claims that the stigma associated with cancer is often more painful than the disease itself. As she says, "Since getting cancer can be a scandal that jeopardizes one's love life, one's chances of promotion, even one's job, patients who know what they have tend to be extremely prudish, if not down-right secretive about their diseases" (Sontag, 1978: 7).

Dunkel-Schetter and Wortman (1982), researchers who have reviewed substantial amounts of literature on the subject and worked as clinicians with people with cancer, confirm that one of the most painful aspects of the disease is its stigma. An incisive illustration of this point is found in Orville Kelly's book on the subject of coping with cancer. He describes a situation in which a woman approached him after he had given a public lecture and told him that the doctors had once thought that her husband had cancer. The following conversation then ensued:

> "Thank God it wasn't cancer!" she exclaimed.
> "What was it?" I asked curiously.
> "Heart disease," she replied.
> "How is he doing?" I asked.
> "Oh, he died later of a heart attack," she said. (Kelly, 1979: 5)

Recent research (Bloom and Kessler, 1994), however, has found that women with breast cancer no longer report feeling stigmatization. Indeed, women with breast cancer perceive themselves to have more, not less, social support following the diagnosis. Moreover, women with breast cancer report experiencing more social support following surgery than those with gall bladder disease or benign breast disease.

Schneider and Conrad (1983) emphasize the importance of the stigma of another disease – epilepsy. Epilepsy may imply disgrace and shame. Borrowing a term originally used by the homosexual subculture, people with epilepsy are living "in the closet." This means that, like homosexuals, those with epilepsy have frequently tried to manage their perceptions of being different by isolation and concealment. "Coming out of the closet," in the homosexual subculture, is a political move designed to end the shame and isolation and to empower and instil pride in people who formerly had felt stigmatized. People with epilepsy and their families have evolved a number of techniques designed to manage information about epilepsy and thus to come out of the closet. Rather than being entirely closet-bound, people with epilepsy go in and out through a "revolving door" (Schneider and Conrad, 1983: 115). Some people can be told, others not. Some can be told at one time but not at another. People with epilepsy have been denied driver's licences, jobs, and even housing. Some people with the disease have therefore learned to hide their diagnosis when filling out application forms or applying for jobs, or when first meeting new people. Managing the anxiety that surrounds the concern about when and whom to tell is an ongoing problem.

People with epilepsy have also developed ways of telling – as therapy and as prevention. Sometimes telling others from whom the epilepsy has been kept secret has therapeutic consequences. It can be cathartic, for example, when a person has kept up a close friendship over a period of time, all the while keeping the truth of epilepsy secret. Telling can also serve preventative purposes. Sometimes this occurs when people believe it is likely that others will witness seizures. It may be hoped that the knowledge that the seizure is due to a defined medical problem and that there are clear ways to deal with it will prevent other people from being frightened.

Epileptics have frequently said that these negative social attributions were often more difficult to manage than the disease itself. A superlative account of the processes through which mutual denial of a stigmatized disease is maintained on a verbal level is found in the work of Bluebond-Langer (1978) on children with terminal leukemia. She devotes considerable detail to documenting the ways that doctors, nurses, and parents practised mutual "pretence" to provide the "morale" for the continuation of hope through often painful and debilitating treatments. Although they all knew the children were dying, they all agreed not to acknowledge this fact. Meanwhile, the children knew in astonishing detail how long they were likely to live.

There are times when the lack of evidence that a disease exists elicits negative attributions. People with psychosomatic illness may be viewed as malingerers. Because the symptoms of multiple sclerosis can grow and then regress, people with this disease may not seem ill, and friends and acquaintances may complain that they are poor sports or "just psychosomatically" ill: "Knowing that I did have the disease was a great relief, but despite this I could not really believe it for some years to come. Other people could see nothing wrong with me and I feared that they regarded me as a malingerer" (Burnfield, 1977: 435).

Managing Knowledge and Information

Knowledge about the disease and its probable course and effects are crucial aspects of a successful adaptation to chronic illness. Full knowledge can benefit both the person with the disease and the key others who are involved with the person. Knowledge aids not only in the treatment of the disease as a physical entity, but also as an emotional and social challenge to the person with the disease. In *Having Epilepsy*, Schneider and Conrad document the ways in which knowledge is a scarce and valuable resource in coping with epilepsy. Frequently, children with a diagnosis of epilepsy are kept in the dark about the name of the disease, about its duration, and about how to manage it. Often parents are "shocked, embarrassed, fearful, and ashamed" that their children have epilepsy, or they are afraid that their child will be ostracized when the diagnosis is made known. Hiding it from the child, according to Schneider and Conrad's findings, is often a source of resentment in the child and leads to greater disability and dependence in the future. As one woman said, talking of her parents, "I mean, they, the fact that they had never told me and couldn't cope with me, I felt was a total rejection of a child by that parent" (Schneider and Conrad, 1983: 86-88).

Two newer trends in the work on the experience of chronic illness need to be highlighted. The first is the work of Corbin and Strauss (1987) on the BBC chain and the second is the work of Charmaz (1982, 1987) on the struggle for the self. Each brings to the literature a fresh focus on understanding the experiences of chronic illness. Corbin and Strauss consider that the essential social components of this experience involve what they call the BBC chain – biography, body, and self-conception. By this, they point out that all who suffer from chronic illness are affronted, through their bodies (and their illness-related constraints, pains, freedoms, change, and so on), with challenges to both self-concepts and the biography (or story that humans tell themselves about their lives). The point here is that as the body changes, so, too, are integral parts of the person – the self and the self story in historical and future context.

Charmaz focuses on just one aspect of the chain: the self. She argues that when individuals live with the ambiguities of chronic illness they develop preferred identities that "symbolize assumptions, hopes, desires and plans for the future now unrealized" (1987: 284). Moreover, coping with or managing chronic illness requires a balancing of identities and abilities through what Charmaz thinks of as a hierarchy of preference. In her studies of people with various types of chronic illness she observed a tendency for people to have a hierarchy of preferences (or levels) for identity. In particular, she suggested a hierarchy beginning with, as a first choice, a supernormal identity and ending with, as last choice, a salvaged identity. Table 8.2 portrays the hierarchically arranged preferred identities of the chronically ill and describes their meaning.

This new focus on the self has also been described by Arthur Frank (1993), based on his analysis of published, book-length illness narratives. His research uncovered three different self-change narratives that were typical: (1) the rediscovery of the self who has always been, (2) the radical new self who is in the process of becoming, and (3) the no-new self assertion. Frank, a sociologist who has survived two very serious illnesses in his late thirties, a heart attack and then cancer, has written a wonderfully sensitive book about his illness experiences (1991). He has done so as a sociologist and the book is a wealth of sociological theorizing based on bodily and health-related experience.

TABLE 8.2 Identity Preferences

Supernormal Identity	Persons seeking this identity try to do everything even better than those who are "normal."
Restored Self	Persons seeking this identity try to be like they were before the illness.
Contingent Personal Identity	Persons in this category try to achieve the above two identities at times but also recognize ongoing rules.
Salvaged Self	Persons seeking this identity try to attain some parts of their previously healthy selves.

Source: Charmaz, 1987.

Case Study: Women and Cancer

The symbolic interaction approach focuses on people's actual experiences and the meaning made of them. To provide an understanding of such research, one case study will be given in detail. It was described in the book *It's Cancer: The Personal Experiences of Women Who Have Received a Cancer Diagnosis* (Clarke, 1985). The research was based on a series of long, unstructured interviews with women who had received a cancer diagnosis at some time in their lives. For some women, the diagnosis was relatively recent; for others it had been made in the distant past. Subjects ranged in age from 17 to 85; some lived in small towns; others came from large metropolitan areas; some worked outside their homes for pay; others worked inside their homes without financial remuneration. These women also varied in religious affiliation and in their degree of religiosity. While most were married, some were single, and several were divorced or widowed at the time of the interviews. They had, however, a number of things in common. One was that once they were asked to talk about the impact of cancer on their lives they all had a great deal to say. This was in spite of the fact that many of the women had said to the researcher beforehand that they would not know what to talk about and would need a series of precise questions asked of them. Subjects were selected because of the commonality of the cancer experience.

However, the research focuses not on the disease or on its treatment. Rather, it is

Box 8.2 Becoming Ill

One day my body broke down, forcing me to ask, in fear and frustration, what's happening to me? Becoming ill is asking that question. The problem is that as soon as the body forces the question upon the mind, the medical profession answers by naming a disease. This answer is useful enough for practicing medicine, but medicine has its limits.

Medicine has done well with my body, and I am grateful. But doing with the body is only part of what needs to be done for the person. What happens when my body breaks down happens not just to that body but also to my life, which is lived in that body. When the body breaks down, so does the life. Even when medicine can fix the body, that doesn't always put the life back together again. Medicine can diagnose and treat the breakdown, but sometimes so much fear and frustration have been aroused in the ill person that fixing the breakdown does not quiet them. At those times the experience of illness goes beyond the limits of medicine. (Frank, 1991: 8)

---

concerned with the social world, the family relationships, the concept of the self, and other issues relating to the whole context of the lives of the particular women studied. Being a patient represents a crucial but nevertheless small part of the daily round during a chronic or terminal disease. For most of the women interviewed, the greater part of their lives had no connection with the medical community. Among the questions that were asked of the women in the study are the following: What is it like when symptoms that are known to be associated with cancer are first noticed – a lump, a persistent cough, or any of the "warning signs" given by the Cancer Society? How is the time at which the diagnosis was received remembered, thought of, and talked of in the whole context of life? In our cancerphobic society, how does the diagnosis affect the self-concept of the person who is ill? Is there a sense of stigma associated with cancer, as some social commentators have suggested? How do women with cancer feel they are treated by friends, family, acquaintances? How do they talk and think about their treatment by medical care personnel?

A number of experiential themes can be drawn from the analysis. Many are useful in attempting to place the experience of cancer in the context of the experience of chronic illness in modern society. The analysis highlights the major issues faced by women as they come to assimilate the diagnosis. The first stage is frequently shock: the person questions realities that were taken for granted, the continuity of the body, the image of the self as a healthy and functioning person. At this stage, women often asked questions such as, why me? And why here in this organ?

The sense of time changes after a shock. The assumptions that time is sequential and that activities occur in a managed way are radically altered. Time was experienced by some as external and chaotic. Women said such things as, I just didn't know what to do, how to handle it. Should I tell John? Should I call the doctor? Should I eat lunch? Should I go to the grocery store and finish shopping? I just didn't know what to do (Clarke, 1985: 19). While carrying out routine actions, women were preoccupied with themselves and were isolated in the midst of continuing activity. They were confused about what to do, when and how to do it, and who, when, and what to tell.

Most of us, most of the time, believe we exercise a certain degree of control over our internal and external worlds. Women who had just received a diagnosis of cancer felt that they lost their belief that the world made sense and that most things were, after all, meaningful and controllable. For some, "magical thinking" seemed to dominate. Women reported thinking that perhaps if they tried to forget about it, it would go away, or perhaps if they didn't tell anyone, if no one else knew, then it would lose its reality. The body, long taken for granted as the house of the self and the source of action, was all at once untrustworthy. Family heritage, too, sometimes became the subject of new scrutiny.

Once the diagnosis was assimilated, women talked of changed images of themselves – changed self-identities. A number of women talked of how cancer was considered a taboo subject. It was something that was not talked about except in euphemisms such as the Big C. This sense of having been touched by the untouchable led a number of women at this early stage of the disease to see themselves as outcasts and marginal to society. The self-image of a healthy person with nothing to hide quite

abruptly changed to a sense of shame and secrecy. Women talked of how at this time they felt themselves to be undesirable as women, sometimes because of the loss of a breast, at other times because of the loss of other body organs and the necessity of attaching appliances to the body, to control elimination, for example. But even if the diagnosis did not mean surgery and physical alteration of the body, the brush with mortality that cancer seemed to imply for many people was the source of a radical revision in self-concept.

A diagnosis of cancer may lead a woman to re-examine her whole life and its meaning. Existential questioning such as "why me?" is an example. Women asked themselves how they could make sense of their cancer diagnosis. By such questions they were expressing an interest in much more than the medical understanding of the causes of their disease. Rather, they asked themselves how it could have been predicted from their past lives and how it could be said to make sense and to be meaningful in the whole context of life. This process is similar to what has been called the legitimation of biography (Marshall, 1980). Marshall describes this as an inevitable process in aging people who, as their lives are drawing to a close, want their story to be a good one, not necessarily a story of success, happiness, fame, and the like, but a story that "makes sense," that is meaningful (ibid.: 114).

Such a review of life involves criticizing and editing past experiences and decisions so that they make good sense in the whole context of life. For some, cancer was a message from God to return to the fold. For others, it was a message from the fates to eliminate stress or to change life's priorities. Cancer was considered a threat to the continuity of life. Women attempted to understand it as a message of symbolic importance and to gain from it a sense of new direction for the future.

This process can also be seen as an aspect of the search for control and mastery that all human beings require. Thus, whether or not they were cognizant of the biomedical analyses of the causes of their disease, women seemed to explain it in their own ways and within the context of their own lives. Women who saw themselves as busy and productive contributors at work and in their family might, for example, explain the anomaly of cancer as a result of the stress they had been feeling. They could then try to reduce that stress. Women for whom religion provided an important sense of identity and social connectedness tended to see their illness in the context of a loss of faithfulness to God or to the church. They could then return to involvement with religion and the church. Those whose primary focus had been the needs of others often decided to put themselves first. Those who were medically more sophisticated, or had medical doctors or scientists in the immediate family, might tend to attribute the disease to external factors such as pollution or lifestyle factors such as diet. These things, too, they could attempt to alter. Not only did women understand the meaning or cause of their illness within the context of their whole life story, but they also attempted to change the direction of the life story on the basis of what they had learned from the disease.

A diagnosis of cancer generates profound self-doubt and renewed self-analysis. It is also accompanied by changed attitudes in friends, family members, acquaintances, and even medical personnel. People may relate to the person with cancer as if she or he has been changed not only physically but morally as well. People often said that other people didn't seem to know how to talk to them or what to talk about. It is as if the whole person has changed. This disease is not seen as just a malfunctioning in one part of the body, but often as a symbol of an inadequate and questionable half-person. The person in many ways becomes, as a result of cancer, marginal to the mainstream of social affairs.

Unlike normal social interaction, interaction with a person with cancer can become confusing and problematic. Cancer is such a feared disease, in fact, the most feared disease according to a survey by James and Lieberman (1975), that many people seem to respond strongly to others who have the disease. One woman talked of meeting a new person and going out to lunch to become better acquainted. She was startled when her "new friend-to-be" said that she was sorry, that she would really like to become her friend, but was not willing to invest in someone who wouldn't be around much longer. Another woman spoke of walking down her street just after she arrived home from the hospital, and seeing a neighbour hide behind a curtain rather than wave or walk outside for a chat as would have been normal

Box 8.3 My Experience with Cancer

Christmas of 1978 my mother and I lightly joked about how large the lump on her chest just above her left breast was becoming. She kiddingly said that it was probably cancer. We laughed. My younger brother interrupted our laughter with stern words about not joking about something as serious as cancer. He said that it might mean that it would come true. We laughed again.

Three months later, my mother had her left breast and the adjacent lymph nodes removed because she had been diagnosed as having breast cancer. It began in April when she noticed the lump. It was small, but since my grandmother, her mother, had had breast cancer, she quickly made a doctor's appointment to have it checked. The doctor assured her that it was not malignant.

Christmas was drawing near, and by now the lump had grown larger. It was unattractive as it was visible above the neckline of her clothing. She would remind the doctor of her lump in February when she had an appointment to see him about her hernia that was going to be surgically repaired. In February, he again assured her that even though the lump had grown larger, he was sure that it was nothing to worry about. To soothe my mother's increasing anxiety, he sent her to have x-rays on her left breast and the surrounding area. The x-rays did not show any danger signs, and my mother was somewhat relieved. In March, she went for her hernia operation and before the operation, she asked her doctor (who was a surgeon) if he could remove the lump because by now it was extremely visible and unattractive. He agreed to please my mother and remove it after he had repaired the hernia.

My mother went to surgery at 10:00 a.m. and did not come back to her room until 3:00 p.m. She awoke in her room shortly thereafter and went to reach for the area in which the hernia was to be removed, and

found the hernia intact. She immediately began to cry and called for the nurse. The nurse called for the surgeon, who told my mother that she had cancer of the breast. He would have to perform an operation designed to eliminate all traces of malignancy, thus presuming he would effect a cure.

She signed the necessary medical forms, and once again went up for surgery. The doctor performed a modified radical, which entails removing the breast and some adjacent lymph nodes under the left arm.

After surgery, she woke to see my father at her bedside crying. Three weeks later she came home and was committed to bed. After a short rest at home, she went to a hospital in London, Ontario, to undergo radiation treatment – a practical method of attacking and destroying any remaining cancer cells. My mother had adjuvant chemotherapy, which is used when cancer cells are found in the adjoining lymph nodes. This is done in the hope of eliminating undetectable microscopic foci of cancer cells in the body, and thus postponing or hopefully preventing a recurrence of the disease.

My mother came home each weekend but was emotionally and physically drained. Her stay at the cancer hospital lasted six weeks. When she returned home, she no longer received any formal counselling. She was restricted to bed rest for one month. Since then she has had to visit a specialist in London twice and sometimes three times a year. The medical professionals tell her that she is progressing well, but they cannot predict or promise anything for the future.

This is a very short description of my mother's infliction of breast cancer. The results of this still affect her and those around her today. Immediately after my father told me that Mom had breast cancer I was angry and afraid. Will she live? This fear hit me so hard that it continues to be with me today. I was having great

Box 8.3 continued

visions of great suffering, and probable death. After hours of crying and confused emotions, a great feeling of helplessness came over me. I felt that there was absolutely nothing I could do or for that matter, that anyone could do! . . .

While she was in the hospital I visited her each day and tried desperately to appear strong. A little piece of me died each day as I was seeing what this experience was doing to her. She cried constantly, also. She feared death, was angry, and felt there was to be no hope for her recovery. She felt hopeless in that there was nothing anyone could do for her now. She had cancer, and her perfect world was crumbling down. Fortunately, she and my father had a very strong marriage built upon 20 years of honesty and love. My mother was a very exceptional person. She was a very happy woman who smiled constantly and enjoyed the company of friends and family. People would approach my mother with their joys and fears, comfortable in her presence and thankful for a sympathetic ear. Her tower of strength was admired by many. She energetically maintained a demanding job and lovingly took care of three children along with the help of my father. I think that it was because of the aforementioned characteristics of my mother that I felt so much anger at this disease for attacking her, and at her surgeon for not detecting the cancer sooner. Why did such a fantastic person have to suffer so much? Why her?

During her three-week stay in the hospital following surgery, fear and anger were my dominant emotions. I feared she would not live. I feared that she would never see my children, her grandchildren, or ever experience growing old with my father and her family. If she did not live, everything that she had so vivaciously created and enjoyed would fall apart. This is why I described her hopelessness as a feeling of her world crumbling down. I felt anger each time I visited

her and saw her doctor enter the room to examine her. Why did he not diagnose her lump correctly? Isn't that his job? Isn't that what he is paid for? She (as most of us) was basically entrusting him with her life when she went to him concerning the initial sighting of her lump. I was angry because other people had had lumps that were removed, and were non-malignant. Not that I wasn't happy for those people, but why couldn't my mother have been so lucky? . . .

In summation, concerning my mother's postoperative stay in the hospital – it was a very difficult time for me. I saw my mother as a totally different person than she was a mere four weeks before. She had a different attitude, disposition, and especially a different appearance. It seemed like the strength that she had built up throughout her life was drained from her. My vivacious, loving, wonderful mother had disappeared with the appearance of the deadly malignant growth. . . .

It was a very momentous day when Mom come home. We were all so very happy to have her near us again, but we did not realize how difficult this would be for us as a family unit. This occasion forced me to come to terms with the reality of it all. My mother has had breast cancer. She may recover completely, but then again, she may not. My mother is not the same physically or emotionally as she was one month ago. Things have changed and I must learn to accept the changes and learn to deal with them the best that I can. This slow realization came to me shortly after she came home. I still took care of the family with a lot of help from my relatives. I began to cry less with each passing day. My school work had piled up, and I began to worry about my marks and academic achievement again. I realized that life must go on!

My mother was experiencing many new feelings, and of course, I did the best I could to help her cope

Box 8.3 continued

with and understand them. She felt a tremendous loss of femininity and self-image. She was sometimes bitter, angry, self-pitying, and thankful for life, all at the same time. . . .

Her chemotherapy treatments in the city took her away from us again, and this time for six weeks. I missed her very much and felt great pity for her at this time. I began to change in ways that were also obvious to my family and friends. My viewpoint on many things was changing. Living with the uncertainties of cancer made my life more meaningful. It intensified my appreciations and eliminated much of my hypocrisies. I began to trivialize the trivia in life. I started not doing things that I really didn't want to do, and not looking for what would happen tomorrow. I began living day by day, compressing time, and setting certain goals in order to live them. I, unfortunately, also became more impatient toward people who were ignorant of cancer. I also became very much more aware of my friends, and how they (like I once was) were so hung up on triviality. There was a tremendous personal growth and maturation for me during the six weeks of my mother's stay in London. . . .

When Mom came home to stay, our lives changed once again. The adjustment that was needed as a family unit was phenomenal. We lived each day to its fullest, still very uncertain of the future. I still carried fears of recurrence of the cancer, and I still feared she would not live to see her grandchildren. I worried that her self-image was so low, but fortunately she made wonderful progress in accepting and coping with the removal of her breast.

As time went on I began to become accustomed to my new life. Mom was doing amazingly well as it was her belief that a good attitude could carry her through the ordeal. As soon as it was physically possible she became active again. She trivialized the trivia and set goals that she wanted to achieve. She was by far not the self-confident woman she was two years ago, but she was working on it.

About six or seven months after her surgery, my family and I began to have many questions about the long-term effects of cancer. Mother had certain body pains and many bouts of depression, and she needed more than her family could offer her emotionally. I began to wonder why she was not recommended for postoperative counselling. Questions concerning her vomiting and physical body pains could only be answered by the doctor, and he had only planned to see her once every four months. Why didn't he send her to a counsellor? Well, he didn't send her to one because at this time our home town did not have one. . . .

I believe my experience with cancer has helped me to better appreciate the beauty of everyday life. I feel my love for life is directly related to the fact that one of the people I love most in my life feared death. I value the family unit and very close friends as the most important thing in my life. Most of my other values, beliefs, and goals stem from this.

I still feel anger toward my mother's doctor and the medical profession in general when I hear of cancer patients dying, or when my mother is not feeling well, but I also realize that no one is God and therefore, to err is human. I still fear that my mother will have a recurrence of cancer, especially since she is still in her five-year waiting period. I may feel more relaxed and worry about her less a few more years from now.

I sometimes worry for myself. Will I get cancer, too, like my mother? If I do will I be fortunate enough to lose only a breast, or will I lose my life? How will it affect my goals, and how will it influence my life? (Anonymous, 1984) before the diagnosis. Fear of future loss or unpredictable disruption in the relationship is one of the threats faced by people in contact with a person who has received a diagnosis of cancer.

Summary

(1) Disease, diagnosed by a doctor, is an abnormality in the structure and function of body organs and systems. Illness is the personal experience of one who has been diagnosed by a doctor or does not feel well; it involves changes in states of being and in social function. Sickness is the social actions or roles that are taken up by individuals who experience illness or are diagnosed with a disease.

(2) Illness, sickness, and disease are all socially and culturally mediated experiences.

(3) In the Western world, illness is thought to be empirically caused and mechanically or chemically treatable. In many non-Western cultures, illness and cure have a non-empirical basis. The experience of pain differs from culture to culture and, at times, so does what is viewed as illness and as health.

(4) Western industrialized society has become increasingly medicalized. The medical profession understands illness through a biomedical model.

(5) Life experiences of individuals as they encounter illness, such as cancer, are of concern to sociologists. People with cancer tend to experience shock and to find social interaction confusing and problematic. This could be a result of the stigma placed on cancer, which, at times, can be more painful for the patient than the disease itself. This is common to all stigmatized diseases, for example, epilepsy and AIDS.

(6) Uncertainty is another common characteristic felt by those with chronic illnesses. It affects the individual's self-image and also his/her interpersonal relationships. Uncertainty prevails at all stages of illness including diagnosis, treatment, spread and resurgence, and remission. It is difficult to make plans for the future.

(7) How to manage medical regimens is another issue for those who are chronically ill. Lifestyle changes and learning how to manage medication and treatment all serve to add to the problems of one who faces chronic illness.

(8) Knowledge and information about the disease and its probable course and effects help the individual and significant others to better adapt to chronic illness. Information that the public has about the disease is also important to the individual when dealing with normal social interaction. A significant source of information in modern society is the mass media.

(9) The media can associate certain meanings with diseases to the public. From these meanings we get from the media, we create images of people who have certain diseases.

PART THREE

SOCIOLOGY OF MEDICINE

The Social Construction of Scientific and Medical Knowledge and Medical Practice

The Sociology of Medical Knowledge

Is scientific knowledge universally true? Is scientific knowledge objective? Is medical knowledge based on science? Is medical practice based on science? Given a choice, how can a person decide whether to take chemotherapy and/or radiation or to do visualization, immunotherapy, or something else after a cancer diagnosis? Is chronic fatigue syndrome really just a "yuppy flu" experienced by spoiled middle-class and upper middle-class women or is it a "real" disease? Do allopathic doctors have a "better" theory of medicine than chiropractic or naturopathic doctors? Does the introduction of a new technology, for example the CAT scanner, occur as the final stage of a process of rational decision-making, including cost-benefit analysis and an evaluation of its efficacy and efficiency? Does medical science reflect cultural "values" such as racism, sexism, homophobia, or is it entirely a neutral and objective endeavour?

This chapter will investigate the sociology of medical science and medical practice. To say that there is such a "thing" as a sociology of medical science is to say that it is reasonable to examine the ways in which medical and scientific knowledge are determined, created, and constructed by, or at least influenced by, social conditions, on the one hand, and the way in which medical science affects or constructs social conditions, on the other hand. Moreover, in the tradition of conflict theory it is possible to ask whose interests do the present forms of medical knowledge, organization, and practice serve. The meanings and constructions of medical science and practice are also topics for discussion. This approach is consistent with another substantive field within sociology – the sociology of knowledge, described as follows:

the objectivity of the institutional world, however massive it may appear to the individual, is a humanly produced, constructed objectivity. The process by which the externalized products of human activity attain the character of objectivity is objectification. The institutional world is objectified human activity, and so is every single institution. In other words, despite the objectivity that marks the social world in human experience, it does not thereby acquire an ontological status apart from the human activity that produced it. (Berger and Luckmann, 1966: 60-61)

Science has been humanly produced. It has become embedded in an institutional structure and is maintained through processes of negotiation by some actors who live in a particular time (history) and place (culture, society, strata). Medical practice is based on this socially constructed science and is also influenced by other social forces.

Medical and Scientific Knowledge: Historical and Cross-Cultural Context

Positivism, the model of science upon which medicine is based, is described by attributes such as objectivity, precision, certainty (within a specific degree of error), generalizability, quantification, replication, and causality. Its search is ultimately for a series of law-like propositions. These, its formal characteristics, portray science as if it stands over and above the ways of the world. Science, in this view, is outside of culture and social structure. The subjects of study, the methods of studying such subjects, the findings and their interpretation, and the publication and dissemination of scientific knowledge would follow the same course everywhere and at every time in history. However, there are many reasons to doubt this typification.

A number of social theorists and researchers have demonstrated that assumptions of scientific objectivity are problematic. Kuhn (1962) has described the historical development of science and described how the methods, the assumptions, even the very subject matter of medical science are infused with cultural categories. Freund and McGuire (1991) have specified the value assumptions of contemporary medicine as: *mind-body dualism, physical reductionism, specific etiology, machine metaphor,* and *regimen and control.*

Mind-body dualism is said to have begun with Descartes, the philosopher who effectively argued the case for the separation of the mind from the body. Descartes's writing and thinking, however, arose out of a context of increasing secularization, which allowed for the separation of the body from the soul/mind. It wasn't until the Christian doctrine determined that the soul could be sent heavenwards after death, without the body, that a notion of a secular body, which was available for scientific investigation, became possible. Foucault (1975) has described the changes that occurred in the eighteenth and nineteenth centuries to allow the physician to view the patient's body directly through the "clinical gaze," and not merely indirectly through the patient's verbal description. Specific technical inventions such as the stethoscope gave physicians direct access to bodily functioning. With a stethoscope the doctor could observe, categorize, and explain the patient's body (or a part of the body) without the conscious awareness or the involvement of the patient. Dissection of cadavers opened up a new world of patienthood and medicine for

Box 9.1 Prozac

One of the most interesting public debates in the mid-nineties is the debate over "personality-changing" drugs such as Prozac (Kramer, 1993). The miraculous nature of this drug's effects on a wide variety of symptoms, coupled with its lack of side effects, has led to various debates about the ethics of prescribing, on the one hand, or refusing to prescribe, on the other, a drug that is reputed to make people feel "better than well." The availability of Prozac begs the question of whether people ought to take drugs that change their very selves (personalities). Prozac's existence also raises the question of whether doctors should have the right to prescribe or withhold this powerful "feel-good" drug.

description and explanation. Such endeavours further entrenched the distinction between the soul/mind and the body.

Physical reductionism emphasizes the physically observable at the expense of other aspects of the individual such as the sensual and the emotional. It also leads to a disregard of social, political, and economic causes of ill health.

René Dubos (1959) was the first to note that the *doctrine of specific etiology* is another characteristic of modern medical science. It developed from the discoveries of nineteenth-century researchers such as Pasteur and Koch, who noted the specific effects of particular micro-organisms on the body, and it has led to an exaggerated emphasis on the discovery of the "magic bullet" to cure one specific disease after another. Dubos noted that this emphasis is overly restrictive because it ignores the fact that a number of people may be affronted with the very same micro-organisms but only a certain proportion of these people respond by becoming ill. Such a doctrine is also problematic because it has tended to ignore the way that a treatment for one disease may lead to side effects that may cause other diseases.

The *machine metaphor* for the body emphasizes discrete parts, such as individual organs, and their relationship to other discrete parts. Resulting from this are specialization and the removal and replacement of parts of the body, such as heart, kidney, liver, blood, and bone marrow.

Finally, *regimen and control* is an outgrowth of the machine metaphor. It involves the underlying assumption that the body is to be dealt with, fixed, improved. Not only strictly medical but even health promotion policies imply that the body is perfectible and under the control of the individual through such actions as exercise and diet, and by maintaining healthy habits, such as not smoking and, at best, only moderate alcohol consumption. An emphasis on the correct number and spacing of checkups reinforces this notion of the perfectible body. Stein (1990) has noted how Western, particularly American, medicine has adapted to American cultural values. As he says,

disease conceptualization and treatment are embedded in the value system of self-reliance, rugged individualism, independence, pragmatism, empiricism,

Box 9.2 Même Breast Implants: Public Policy, Scientific Knowledge, and Medical Practice

Medical devices range from contact lenses to CAT scanners, MRIs, and hip replacement parts. One of the most controversial of medical devices in recent years is the widely used and accepted silicone breast implant. Information about its side effects and other problems associated with its insertion followed its widespread adoption. Même silicone breast implants are a case in point. Most breast implants have been available on the Canadian market since before 1982 and were thus exempt from federal regulation requiring that manufacturers of new medical devices implanted into the body for thirty days or more submit data on the safety and efficiency of proposed new products.

By 1988 a Public Citizens Health Research Group in Washington released data indicating that 23 per cent of rats injected with silicone developed highly malignant cancers. Questions were then raised in the House of Commons. An article was published in the Canadian scientific journal, *Transplantation / Implantation Today*, presenting the evidence that polyurethanes (of which the Même was constructed) were known to deteriorate in the human body. There was still no direct evidence about the long- or short-term effect of breast implants. A registry of all Canadian women who received breast implants, with a health follow-up, was considered, but it was not implemented. Nicholas Regush, a *Montreal Gazette* investigative health reporter, wrote an article questioning the health effects of the Même breast implant and also its legal status. Although it had been used since 1982, not only was there no record of its sale in Canada but there were no safety data on file. A number of women who had received implants responded quickly to Regush's article, and an organic chemist from the University of Florida called to inform him that he had found 2-4 toluene diamine, a potential carcinogen, in the Même's polyurethane cover. Research had

indicated that 2-4 toluene diamine had caused liver damage, central nervous system problems, blindness, and skin blistering.

Despite evidence from several sources that there could be dangers associated with the Même breast implant, active lobbying efforts in both the United States and Canada, and more and more horror stories from individual women and the active opposition of the (later fired) Health Protection Branch's expert on breast implants, Pierre Blais, the federal government did nothing. In one memo Blais tried to warn his bosses of the hazards of the Même:

In late January 1989, I reported to you that prosthesis coating is made from a common class of commercial polyurethane foam available from various vendors for assorted consumer product applications. Industrial (non-specific) applications of such foams allow broad compositional variations, elevated impurity levels and the incorporation of reactive intermediates of unknown biocompatibility. The safety, efficiency and quality assurance levels of a medical implant based on these foams therefore cannot be demonstrated without testing each implant. (Regush, 1993: 91)

In response to growing concern raised in the House, the Health Minister, Perrin Beatty, appointed a plastic surgeon to conduct an "independent" review. Her review concluded, on the basis of a number of poorly controlled studies, that the Même could be declared neither safe nor unsafe. In was not until a few years later that the FDA commissioner in the U.S., on January 6, 1992, declared a moratorium on breast implants. Two days later Benoît Bouchard, at the time Canada's Health Minister, followed suit with a Canadian moratorium. This was almost twelve years

Box 9.2 continued

after Regush had first raised questions about the safety of breast implants. The response of the plastic surgeons is instructive. As a group, they retaliated with a $3.88 million campaign lamenting the loss of choice for women. Later, a number of individual and class action suits were filed by women against various manufacturers of breast implants. The women won. Just recently, a new story in the long saga with breast implants broke. Dow Corning (U.S.) had filed for bankruptcy, claiming that it couldn't pay the awards to the women who had sued and remain profitable.

---

atomism, privatism, emotional minimalism and a mechanistic metaphor of the body. (Stein, 1990: 21)

In an expanded analysis of the foundations of the medical model, Manning and Fabrega (1973) articulate the elements of what they call the biologistic view of the body. This view includes the following tenets: first, organs and organ systems, and their specific functions, are identifiable; second, the normal functioning of the body goes on pretty much the same for everybody unless disturbed by injury or illness; third, people's sense experiences are universal; fourth, disease and experience of disease do not vary from one culture to another; fifth, boundaries between self and body and between self and others are obvious and shared; sixth, death is the body's ceasing to function; and seventh, bodies should be seen objectively.

Sociological research allows us to critique all of these assumptions. First, the observability of organs and organ symptoms depends directly on the tools available for measurement and indirectly on the theories of the body and the level of technology in a given culture. For example, the psycho-neuroimmunological system has just been "discovered" and methods are now being developed to study it as a new system. Awareness of its possibilities are, in part, the result of the re-establishment of the link between the mind/body that has come from the Eastern medical tradition, including meditation in particular. Second, it has become clear that much of the research on the "normal person" has been on the male person. Thus, less is known about the functioning of the female body (except the reproductive functioning) with respect to a whole range of disease categories such as heart disease. Third, cross-cultural, anthropological, and linguistic studies have shown how people's experiences arise out of their available language. Fourth, cross-cultural research has shown that what is considered disease in one culture may be considered normal in another (e.g., the Japanese have recently discovered a new disease category – Karoshi). Fifth, contagious diseases such as AIDS raise serious questions about the sometimes vulnerable boundaries between people's bodies and groups of people. Sixth, the definition of death is now very problematic in part because of the possibility that, for instance, "machines" can keep people alive even when they are "brain dead." Seventh, perhaps bodies should be seen objectively, but that is an impossible value to achieve. The values implicit in the medical model and in the biologistic view of the body reflect particular cultural histories, biases, and predispositions. Thus, medical science and practice are not objective facts that stand above everyday social practices.

Medical Science and Medical Practice: A Gap in Values

There is often a significant gap between published biomedical research and the actual practice of medicine (see Montini and Slobin, 1991). To try to minimize the distance between researcher and practitioner, the National Institutes of Health in the United States, through the Office of Medical Applications of Research, began in 1977 to convene Consensus Development Conferences (CDCs). The Canadian government and medical associations have also initiated this process at times. The goal of these conferences has been to bring together practitioners and researchers, to inform practitioners of the latest scientific findings, to inform scientists of the practical issues facing practitioners, and to work toward the development of timely, national standards of practice.

Unfortunately, the research to date has indicated that the CDCs have no effect on physicians' practice. Montini and Slobin show how differences in the work cultures of clinicians and researchers have served to limit the effectiveness of CDCs for medical practice. These are distinct value differences, including (1) certainty versus uncertainty, (2) evolutionary time versus clinical timeliness, (3) aggregate measures versus individual prescriptions, (4) scientific objectivity versus clinical experience, and (5) constant change versus standards of treatment.

(1) Doctors' work involves patients who want and need an immediate and definitive response. Scientific work has no time limit except the limits set by the funding bodies. In any case, time-related concerns are considerably different. The practitioner needs certainty or at least enough certainty to make decisions about caring for a particular patient at a specific point in time. By contrast, the scientist works within a world of probabilities – thus, uncertainty.

(2) Science does not progress by proof but rather by failing to disprove. There is always caution in drawing conclusions. Thus, uncertainty results. Scientific truth arises in incremental stages as more and more hypotheses are disconfirmed. However, the clinician must make timely decisions in response to the expressed and observed need of individual patients.

(3) While the scientist, in dealing with probabilities, deals in aggregates, the practitioner must deal with the suffering individual. Again, because of the immediacy of the sufferer, clinicians want to rely on what they know from experience to be tried and true. They may at times be uneasy about relegating a given individual to a clinical trial whose outcome is unknown and will likely be unknown for a considerable period of time.

(4) The scientist tries to control all

Box 9.3 Chronic Fatigue Syndrome: A Diagnostic and Moral Enigma

Perhaps one of the most common complaints that humans suffer is tiredness. Who has not felt tired at one time or another? Most people have suffered a headache at some time. All have experienced vague discomfort or pain periodically. People who visit doctors with symptoms in patterns that include fatigue, vague bodily complaints, and aches and pains have received a variety of diagnoses over the years. They inhabit an area of anomie, of ambiguity. A century ago they might have been diagnosed with neurasthenia or chlorosis; today they may be diagnosed with chronic fatigue syndrome. In the U.S. the oldest epidemic usually included with this realm of symptom patterns is an outbreak of atypical poliomyelitis among hospital workers at Los Angeles County General Hospital in 1934. Since the early 1980s a series of case studies and epidemiological and psychosocial research have been undertaken on a disease characterized by viral-like symptoms manifest as weakness, exhaustion, and other self-defined symptoms. First officially called Epstein-Barr virus

(EBV), it appeared to be associated with serological evidence of recurrent or prolonged infection with this virus. One of the difficulties involved in this diagnosis is that most everyone has EBV antibodies because almost everyone has been exposed to the virus at one time or another. During the same years that the studies were published, the Atlanta Centers for Disease Control investigated an outbreak of a prolonged sickness in over 100 patients near Lake Tahoe. EBV antibodies were not particularly associated with this outbreak. A new consensus conference was called and the disease was renamed chronic fatigue syndrome (CFS) and new diagnostic criteria were developed (1988). All diagnostic criteria were patient-defined signs. Later, scepticism resumed in the face of the lack of patho-biological criteria, and studies calling CFS a psychiatric disorder emerged. Widespread medical legitimation of CFS is still lacking and people continue both to suffer the symptoms and, in many instances, to be refused disability insurance.

variables in the interest of objective and generalizable findings. The clinician, in contrast, is faced with a unique individual and, usually, subjectively experienced symptoms.

(5) The researcher is aware of continuous change in research findings as new hypotheses are put forward and supported or rejected. The clinician, on the other hand, attempts to practise medicine under the direction and with the support of practice standards that must have a longer life than frequently changing scientific hypotheses.

Medical Technology: The Technological Imperative

New medical technologies are continuously being developed, manufactured, distributed, and used. Among the new technologies are cardiac life support, renal dialysis, nutritional support and hydration, mechanical ventilation, organ transplantation and various other surgical procedures, pacemakers, chemotherapy, and antibiotics. The question that we ask here is, what is the relationship between medical science and this evaluation process that culminates in the use of new technologies? Available evidence suggests that practitioners consistently adopt new technologies before they are evaluated and continue to use them after evaluation indicates they are ineffective or unsafe (Rachlis and Kushner, 1989: 186).

In fact, the introduction of new medical technologies and their use patterns have been shown to be related to four social forces (Butler, 1993). These are: (1) key societal values; (2) federal government policies (while this study is based in the U.S., there is no reason to assume that Canadian legislation provides greater safeguards, and, in fact, available evidence indicates that in many ways, e.g., the thalidomide disaster of the fifties and sixties, the Canadian situation is more lenient); (3) reimbursement strategies; and (4) economic incentives.

(1) A number of social commentators have described the love affair of North Americans with high and new technology of all sorts. Enthusiastic optimism rather than realistic caution typifies our attitude to new technology. For example, while we have yet to understand all of the possible constraints to freedom and democracy created by the so-called information highway, plus other threats that may easily result from widespread electronic communication, it already exists and is in widespread use. Moreover, the development of other new and related technologies continues to precede considerations of and safeguards for social impacts.

(2) In the health area, the federal government, through the Medical Research Council, the Heart and Lung Association, and the National Cancer Institute, quietly funds biomedical research. Taxation policies, free trade agreements, support for education and science, and other federal incentives encourage the discovery of new technologies. Our national medical care system has operated to foster growth and expansion of the use of medical technologies immediately upon their development.

(3 and 4) While there are no definitive studies of the costs of new technology, a variety of studies taken together suggest that 20 to 50 per cent of the annual increases in health-care costs over the past 20 or so years are the result of progressive innovations in medical technology. A few new technologies appear to save costs – e.g., cimetidine for peptic ulcers and lithotripsy for the removal of kidney stones (*ibid.*). Most new technologies are expensive, however, and add costs to the medical care system. As long as the medical-industrial complex is even partly guided by the profit motive, the development and dissemination of medical technological innovations will result from market principles (in the short run) rather than planned, rational, and evaluated change strategies. For instance, babies weighing as little as one pound can now be "saved" at a cost of several hundreds of thousands of dollars per baby through neo-natal intensive care for a number of months, and then continuing care for those with ongoing medical and other needs. By contrast, a low-technology approach to preventing low birth-weight babies that would include feeding pregnant women nutritious diets and maintaining minimal socio-economic standards – a much more effective and efficient strategy for a

healthy citizenry – has yet to be widely implemented.

One of the best examples of the tendency for new technologies to be adopted first and then evaluated later is demonstrated by the case of electronic fetal monitoring. Designed first for use with high-risk births, it was to provide doctors information regarding the health of the fetus during labour. If a fetus showed dangerous vital signs the physician could actively intervene in the labour process by Caesarean section, for example. Electronic fetal monitoring (EFM) was initially made available in the 1960s. By the 1970s EFM and ultrasound were widely available in most hospitals. By the 1980s 30 per cent of all obstetricians had EFM in their offices to detect pre-natal problems. It rapidly became a standard monitoring device – even for low-risk situations. Its widespread use was associated with a growth in the diagnosis of pre-natal problems. The Caesarean section rate, which was 4.5 per cent in 1965, was 16.5 per cent by 1980 and 24.7 per cent by 1988. (C-sections were the most common surgical procedure in U.S. hospitals in 1989.) Studies, moreover, found that the rate varied substantially by region, race, socio-economic status, and availability of insurance for payment.

Despite the rapid growth in their use, randomized, controlled trials undertaken since 1976 have failed to demonstrate benefits of electronic fetal monitoring (EFM) in comparison with simpler methods of monitoring, such as the stethoscope. Moreover, EFM use leads to certain risks for fetus and mother. The safety of ultrasound still remains to be established. In addition, overuse of technology has been estimated at 20 per cent, 17 per cent, and 15 per cent for cardiac pacemakers, gastro-intestinal endoscopy, and coronary bypass surgery, respectively (*ibid.*). McKinlay and McKinlay (1981) provided a model – the seven stages in the career of a medical invention – they argued could be used to explain the dissemination of new medical technologies.

(1) A promising report.
(2) Professional and organizational adoption.
(3) Public acceptance and state (third party) endorsement.
(4) Standard procedure and observational reports.
(5) Randomized controlled trial.
(6) Professional denunciation.
(7) Erosion and discreditation.

The most important point is that evaluation, which is purported and believed to be the bedrock of scientifically based treatment innovations, occurs long after the introduction and widespread use of a medical invention.

Medical Science Reinforces Gender Role Stereotypes

Scientific medical knowledge is portrayed as an objective, generalizable, and true accomplishment. Yet, what is taken to be objective medical science has been shown to reflect fundamental cultural and social-structural beliefs. Normative categories of social relations, in fact, infuse medical conceptions.

Box 9.4 Gender: Menstruation and Medicalization

Medicalization has been said to weigh even more heavily on women than on men. "The essence of this feminist elaboration of the concept medicalization is that women are deprived by the medical world of their 'authentic' form of knowledge of their own female body" (Bransen, 1992: 98). Following this, it is the aim of one part of the feminist movement to recapture women's native intelligence about their own bodily experiences. Critics of this view argue that everything is so permeated by medicalization that it would be impossible for women to "gain back ownership" of their bodies. Thus women "can never form the path along which authentic knowledge of their own female body can be recovered" (*ibid.*: 99). Bransen selected one aspect of female physiology – menstruation – and asked women to talk about their experiences with and views on their cycles. She was interested in the extent to which their descriptions could be subscribed under a unified medical model. Her findings suggested that there are (at least) three different genres women use to talk about their menstruating bodies: *emancipation*, characterized by an active functioning "me"; *objective*, where the body and the menstrual cycle belong to certain experts, in particular doctors; and *natural*, where the menstruating body exemplifies a benign and undistorted nature. Each of these models conceptualizes the relationship with the doctor differently. In the emancipative genre the doctor is to be used as required at the behest of the woman. In the objective genre the doctor is there for treatment if there is some problem with the body. In the natural genre the doctor is largely superfluous because nature, given time, can sort itself out; only in extreme situations and only sometimes is a doctor consulted. By this study, Bransen raised important questions about the hegemony of the medical model.

Findlay (1993) studied the ten most widely circulating texts in obstetrics and gynecology in the 1950s in Canada, as well as a representative selection of academic articles from five major obstetrics/gynecology journals and from the *Canadian Medical Association Journal*. Her research showed how physicians' descriptions and understandings of the female body guarded and reflected family values. The publications emphasized the importance of separate spheres for men and women, of stable marriage and family life, and encouraged fertility among women (who were assumed to be white and middle-class). Findlay noted that the essence of the "normal" woman was to be always potentially fertile.

Women's bodies were described largely with respect to fluctuations in their hormones and menstrual cycles. They were described as living to reproduce. As Findlay reports, one influential obstetrician/gynecologist explained: "The desire for children by the normal woman is stronger than self interest in beauty and figure, stronger than the claims of a career, in the man it is less intense" (Jeffcoate, 1957, in Findlay, 1993). By contrast, the abnormal woman was fundamentally defined as one who had sexual or reproductive problems.

Emily Martin's work, *The Woman in the Body*, also instructs us about gender biases in medical conceptions. She demonstrates how culture shapes what biological scientists

Box 9.5 Medicalization and Demedicalization of Homosexuality

In the nineteenth century, homosexuality was defined as a disease. In the past two decades or so it has been demedicalized. The definition of homosexuality as a disease was part of a nineteenth-century trend. As science and medicine were becoming more widely legitimate and as secularization was beginning to occur, behaviours formerly defined as sin became adopted as diseases. The sick were thereby held to be largely blameless. This process was true of drug and alcohol addiction, compulsions, anxiety, and depression as well as homosexuality. As well as being a part of the general trend to humanize and secularize behaviour, disease-making in the case of homosexuality had other explanations. Hansen (1988) links the medical "discovery" of homosexuality to urbanization and the gradual development of self-conscious homosexual communities in large cities whose density of single people enabled single-sex groupings and culture. Parallel to this was the development of a new medical specialty: neurology. Neurologists tended to deal on an outpatient basis with people with a variety of "nervous" symptoms. Neurologists were a logical choice for assisting people who were ambivalent about their sexual appetites (Hansen, 1988: 104-33). It wasn't until 1869 that the first medical case describing same-sex sexual activity was published in Berlin in *Archiv für Psychiatrie und Nervenkrankheiten*. After that, more cases were published in Germany, France, Italy, Britain, and the U.S. The terms used at this time included sexual inversion, contrary sexual instinct, and sexual perversion. Hansen's analysis emphasizes the mutual benefits to doctors and homosexuals involved in the medicalization of homosexuality. The diagnosis provided a name for people and this enabled them to realize that they were not alone. Ironically, then, the labelling facilitated self-conscious acceptance among some, the development and maintenance of alternative homosexual cultures, and, eventually, the modern successful lobbying that led to the demedicalization of homosexuality.

see. One interesting illustration of her thesis is the ways in which assumptions about gender infuse descriptions in medical textbooks of the reproductive cycles and their elements such as the egg and sperm. For instance, the female menstrual cycle is described in negative terms. Menstruation is said to rid the body of waste, of debris, of dead tissue. It is described as a system gone awry. By contrast, while most sperm are also "useless" and "wasted," the life of the sperm is described as a "feat." The magnitude of the production of sperm is considered remarkable and valuable. Whereas female ovulation is described as a process where eggs sit and wait and then get old and useless, male spermatogenesis is described as continually producing fresh, active, strong, and efficient sperm. While the eggs are swept and drift down the fallopian tubes, sperm actively and in a "manly" fashion burrow and penetrate.

The Sociology of Medical Practice

Just as medical/scientific knowledge is a social product with social consequences, so, too, is the everyday practice of medicine. Have you ever left your doctor's office only

Box 9.6 Task Force on Sexual Abuse of Patients

A task force to examine sexual abuse of patients was established by the College of Physicians and Surgeons of Ontario. Its 1991 report documented the significant level of sexual abuse by physicians of patients. In all, the task force documented 303 detailed reports of sexual abuse by physicians and others in a position of trust. Sixteen reports were of the abuse of men; in the remaining 287 cases the "victims" were women. The actual incidence has been estimated in a few other studies, and it appears that 7-13 per cent of physicians have had sexual or erotic contact with patients. A survey of Ontario women found that 8 per cent of Ontario women reported sexual harassment or abuse by doctors.

The consequences of this abuse of power and sexual integrity are known to be far-reaching and include physical symptoms such as abdominal pain, pelvic pain, gastro-intestinal tract problems, and headaches, as well as eating disorders and drug and alcohol abuse. The psychological problems include intense anxiety, fear, panic, depression, suicidal feelings, loss of trust in the world, difficulty in developing and maintaining intimate and/or sexual relationships, flashbacks, nightmares, sleep disorders, and others.

In light of these findings, the task force recommended a policy of zero tolerance so that sexual abuse is never acceptable or tolerated. They also recommended education for the public about appropriate physician examinations, awareness of warning signs, education for physicians about how to prevent such behaviour in themselves and how to help patients who had been abused by another physician, and education of the College to encourage it to respond sensitively and effectively to complaints and reports of physician misbehaviour. Various other remedial and monitoring strategies were advocated as well.

The issue of penalties for physicians found guilty of sexual abuse has generated the most controversy, in particular the mandatory revocation of licence for a minimum of five years and a fine of up to $20,000.

---

to realize that you had forgotten to tell or ask him/her about something? Have you ever left the office unclear about what the doctor has said about your problem/disease, your medication, or something else? Have you ever felt that you "couldn't get a word in edgewise" in a conversation with your physician? Have you ever asked for a second opinion or been sceptical about a doctor's diagnosis?

Considerable evidence demonstrates that medical knowledge and practice are profoundly affected by social characteristics of both patients and doctors. First, with regard to patients, there is evidence that physicians tend to prefer younger patients and to hold negative images of elderly patients. Elderly patients tend to be seen as both sicker and less amenable to treatment than younger patients. The older patient, "far in excess of actual numbers, represents the negative idea of the uncooperative, intractable, and generally troublesome patient" (Clark *et al.*, 1991: 855). Elderly patients are significantly more likely to be treated with digitalis, tranquilizers, and analgesics, regardless of their actual diagnosis. Several studies have looked at the impact of gender on medical diagnosis and treatment. Their findings are contradictory (*ibid.*; see Chapter Five for a more complete discussion of age, gender, and health). Physicians' attitudes to racial characteristics reflect those of the wider socio-cultural context of which physicians are a part. For instance, a series of U.S.-based studies has demonstrated that black patients tend to be referred to specialists less often, are more often treated by doctors-in-training, are more likely to be placed on a ward, are admitted less frequently to hospital, and are more likely to be involuntarily hospitalized for mental health problems. Black patients also tend to receive less aggressive workups and intervention. There are documented differences, too, in the way that physicians treat patients of different class backgrounds. For example, patients with poorer backgrounds are likely given poorer prognosis and less "state of the art" treatment (see Chapter Six for a more detailed discussion of race, ethnicity, class, and health). The social characteristics of physicians themselves, including gender, age, professional training, education, and form of practice, have also been shown to influence their work. Some research has shown that female doctors are less likely to dominate in physician-patient discussion and that female physicians tend to spend more time with patients than male physicians (Clark *et al.*, 1991).

Cultural Variation in Medical Practice

In an intriguing study, Lynn Payer (1988), a journalist, travelled and visited doctors in several countries: the United States, England, West Germany, and France. To each doctor she presented the same symptoms.

She also examined morbidity and mortality tables and read medical journals and magazines. Using this casual and commonsensical method, Payer found strong cultural differences in diagnostic trends and patterns. They seemed to reflect fundamental differences in history and present culture. Both diagnoses and treatments varied from country to country, even under allopathic medical care. "West Germans, for instance, consume roughly six times as much cardiac glycoside, or heart stimulant, per capita, as do the French and the English, yet only about half as much antibiotic" (Payer, 1988: 38). In general, Payer found that German doctors were far more likely to diagnose heart problems than doctors in other countries. English physicians, by contrast, are characterized as parsimonious. For this reason, Payer describes the British as the accountants of the medical world. They prescribe about half the drugs that German and French doctors prescribe and perform about half the surgery of American doctors. "Overall in England one has to be sicker to be defined ill, let alone receive treatment" (*ibid.*: 41). By contrast, the Americans are spendthrift and aggressive. They have a tendency to take action even in the face of uncertainty. They do not, however, focus on a particular organ. Among the French, most ills are ultimately attributable to the liver.

Payer argued that these patterns reflected the German emphasis on the heart – on romance, in literature and music, for instance; the French focus on the pleasures of eating and drinking; the English are preoccupied with rationalizing the medicare system; and the Americans emphasize getting things done, and getting on with it. Payer's work suggests, in broad strokes, something of the relationship between culture and medical practice.

Class and Resistance to Medical Knowledge

The way that lay people interpret and resist "medical knowledge" is described in relation to a cancer education project developed for a white, working-class, inner-city area that was known as a "cancer hot spot" because of the relatively high rates of cancer mortality (Balshem, 1991). The problem was believed by the local inhabitants to be largely the consequence of air pollution from nearby chemical plants and occupational exposure. With this belief system in mind, the community resisted health education about cancer. To illustrate the resistance, Balshem, who was working as a health educator at the time, describes the aftermath of her slide show and talk about the cancer prevention possibilities of a diet that is high in fibre and low in fat. Immediately after this talk Balshem asked if there were any questions. She was met with silence. Then she raffled off a hot air popcorn maker. People responded warmly, with pleasure. After that, there was silence again. The meeting adjourned and the subtext of the silence emerged. One person talked about her old neighbour (93 years old) who ate whatever she liked and was still alive. Another teased Balshem: "you mean your husband will eat that stuff; mine sure won't." Another confessed that the people in the room liked their kielbasa (spicy sausage) too much to

eliminate it from their diets. Finally, Balshem was invited to their next church supper for some really good eating. Balshem describes the meeting finale as follows:

> Then, the social climax: I am offered a piece of cake. The offerer, and a goodly number of onlookers, can barely restrain their hilarity. Time stops. Then I accept the cake. There is a burst of teasing and laughter, the conversation becomes easier, the moment passes. We eat, pack our equipment, and leave. (*Ibid*.: 156)

While the general atmosphere of these meetings was amiable, the explicit health messages were ignored or indirectly criticized as impractical and (relatively) less important than things such as pleasure, family feeling, and "human" nature.

In order to understand the community and its responses, Balshem engaged in survey research, long open-ended interviews, and focus-group research strategies. One of the findings was that the community members have sharply contrasting attitudes toward heart disease and cancer. Table 9.1 portrays some of the findings regarding this difference with respect to the possible causes of these two diseases.

As the table demonstrates, the causes and treatments of heart disease were both fewer and were considered more responsive to lifestyle alterations. Cancer was described as the result of a horrible fate. It was seen as caused by almost everything in their environments. Many of the respondents directly denied the scientific views of cancer causation and prevention. In particular, there were

TABLE 9.1 Factors Mentioned as Preventing or Causing Cancer and Heart Disease

LIFESTYLE FACTORS	CANCER	HEART DISEASE
Diet	35	22
Smoking	16	7
Attitude	7	–
Proper exercise	5	7
Other (including sun exposure, alcohol, checkups, self-care, caffeine)	21	7
Total mentions	84	43
ENVIRONMENTAL (EXTRAPERSONAL FACTORS)		
Environmental pollution	64	1
Heredity	34	8
Other (including fate/God's will, food additives, cannot prevent, causes not known, occupational exposure, stress, everything, germs)	74	3
Total mentions	172	12

Source: Adapted from Balshem, 1991: 157.

16 direct denials of smoking and 11 direct denials of fat as cancer-causing. By contrast no one questioned the standard scientific views about the prevention and cause of heart disease. Balshem called this response resistance and explained that:

> maintaining a rebellious consciousness is part of constructing a valued self, valued community, valued life, in a subordinate class environment. Self and community, valuing and supporting each other, process myriad insults, betrayals, and frustrations. Local belief and tradition, it is asserted, are superior, as is local insight into the workings of authority and hegemony. (*Ibid.*: 166)

For an abundance of reasons, class solidarity proved to be more important than expert health knowledge.

Despite the power of "medicalization from above," there is always resistance, or as Cornwell (1984) says, "medicalization from below." Calnan and Williams (1992) demonstrate another type of resistance to medical hegemony. They studied lay evaluations of the trustworthiness of doctors with

Box 9.7 Premenstrual Syndrome: A Case of Medicalization from Above or from Below?

Recently, PMS, premenstrual syndrome, has gained credibility as a term that may provide medical and even legal legitimacy. Some women have expressed relief at having their cyclical and bodily experiences medicalized and their discomfort or unhappiness legitimized as due to a medically recognized syndrome. Other women regret the medicalization of what they consider to be a largely normal occurrence. Women who take this perspective ask questions such as how to distinguish real PMS from normal menstruation. PMS has been used as a legal defence by two women accused of murder in Britain. In each case evidence as to the woman's cyclical irrationality and aggressive impulsivity just prior to menstruation was accepted as a defence. One woman received probation and the other was acquitted.

The dubious advantages of the diagnosis of PMS, or late luteal phase dysphoric disorder, are hotly debated among feminists and physicians. The symptoms are all quite clearly subjectively defined (which, in practice, means defined as a reality or not by whomever has power in the interaction) and include: (1) marked mood swings; (2) persistent and marked anger or irritability; (3) marked anxiety or tension; (4) marked depressed mood or thoughts; (5) decreased interest in usual activities, work, friends, hobbies; (6) fatigue or marked lack of energy; (7) difficulty concentrating; (8) marked change in appetite; (9) hypersomnia or insomnia; (10) other physical symptoms such as breast tenderness, bloating, headaches, etc. (Stoppard, 1992: 123).

respect to nine specific medical care issues. In particular they asked whether or not lay persons would unquestioningly accept medical opinion with regard to the following nine interventions: (1) prescription of antibiotics; (2) hernia operation; (3) operation for bowel cancer; (4) prescription for tranquilizers; (5) hip replacement operation; (6) hysterectomy; (7) heart transplant; (8) test tube babies; and (9) vasectomy. Their findings indicated that in only one case would the majority of respondents accept medical intervention without question: 54 per cent said that they would accept antibiotics without question; 41 per cent said they would only accept the doctor's recommendation for antibiotics with an explanation. Moreover, the views of the public regarding all interventions varied according to gender, class, age, and health categories. Table 9.2 portrays the survey results for six of the nine medical interventions as well as responses to a general question concerning faith in doctors.

In making their decisions, Calnan and Williams note, respondents were guided by certain fundamental values of their own. A good intervention was characterized in the following ways: as life-saving rather than life-threatening; enhancing rather than diminishing quality of life; natural rather than unnatural; moral rather than immoral; necessary rather than unnecessary; restoring independence rather than promoting addiction/dependence; and giving good value for money rather than being a waste of money.

The lay population knows that medical knowledge does not form a consistent

TABLE 9.2 Summary of Basic Findings Regarding the Lay Evaluation of Modern Medical Technology

DOCTOR'S RECOMMENDATION	ACCEPT WITHOUT QUESTION	ACCEPT WITH EXPLANATION	NOT READILY/ NOT AT ALL
Antibiotics			
%	54	41	5
(N)	(239)	(180)	(20)
Bowel cancer operation			
%	29	60	12
(N)	(119)	(249)	(48)
Tranquilizers			
%	8	29	63
(N)	(35)	(120)	(262)
Hysterectomy			
%	20	65	16
(N)	(43)	(138)	(32)

*Numbers may not total 454 for each question due to the exclusion of those who did not answer.

Source: Adapted from Calnan and Williams, 1992: 239.

whole. Nor do the different conceptions of medical knowledge complement one another: "The medical world is a melting pot of contradictory theories and practices, controversies and inexplicable phenomena about which doctors and lay people are in constant debate" (Bransen, 1992: 99).

Medical Knowledge Becomes Popular Knowledge

When scientific information is translated into the mass media for wider public edification, the errors and limitations in the original research methodology are frequently reported inaccurately. Thus, information that is widely available may be more or less error ridden. Not only is the original science

imperfect, but its portrayal is often less than perfect. Moyer *et al.* (1994) examined the accuracy of popular accounts of breast cancer research and mammography screening over a two-year period. The researchers noted whether references to the original research (so that the public could do their own additional investigation) were available. They also compared the reported findings with the actual findings in the original research articles. Out of the 116 popular accounts investigated, the researchers found 113 references; 60 of these references were ultimately traceable to their original articles. In the 60 direct references there were 42 specific content-based inaccuracies. Of the four sources of popular accounts – newspapers, women's magazines, science magazines, and

Box 9.8 Evaluation of Medical Practice in Ontario

Frances Lankin, as Ontario's Minister of Health in the early to mid-1990s, stated that about one-third of the health care delivered in Canada is inappropriate (Rachlis and Kushner, 1994: 95). As one illustration, a study done by the Ontario Medical Association and the Ontario Ministry of Health estimated that the bill for the common cold alone was about $200 million per year in Ontario. Yet, a survey of the provincial licensing bodies for physicians, nurses, dentists, pharmacists, and optometrists revealed that none of these bodies had comprehensive quality assurance programs for the evaluation of ambulatory care. Only 8 per cent (4 out of 50) of these organizations had explicit standards for patient outcomes; 12 per cent had developed criteria for assessing the performance of practitioners. One of the organizations that had undertaken some systematic evaluation of the quality of care, the College of Physicians and Surgeons of Ontario, assessed 1,142 practitioners between 1981 and 1988. Fewer than half of the doctors were practising according to the standards of the time (Rachlis and Kushner, 1994: 105-06).

---

health magazines – women's magazines had the highest percentage of inaccuracies and the lowest percentage of traceable citations.

Doctor-Patient Communication

Doctor-patient communication reflects broader social structure and culture. Not only do physicians embody their own particular space as carriers of culture and structure, but their patients do so in turn. In a study based on ethnographic fieldwork, involving joining the surgical ward rounds at two general hospitals, Fox (1993) examined the communication strategies used by doctors to maintain authority and power in interaction with surgical patients. When patients tried to ask questions such as why they felt the way they did, how soon they would feel better, and when they could go home, the surgeons tended to ignore them. Instead, the surgeons maintained control by focusing on the success of the surgery and its specific outcome. Fox demonstrates that ward rounds can be understood as an organizational strategy entered into by surgeons to capture and maintain discursive monopoly. By keeping the discussion focused on surgeon-centred themes, patients have few opportunities to introduce their own views, concerns, or worries.

One area of social life around which there is a great deal of ambiguity and ambivalence is sexuality. On the one hand, sex is more openly discussed, portrayed, and symbolized in all of the mass media today than in the past. Acknowledgement of the pervasiveness of sex outside of the strict bounds of monogamous marriage is widespread. Accompanying the "liberalization" of sexuality, and particularly women's sexuality, is the belief that a satisfactory sex life is an important part of a satisfactory life as a whole. Yet many are still ambivalent about sex and many still believe it to be a shameful duty to be kept secret. Today, people are more likely to consult doctors when dissat-

TABLE 9.3 Doctors' Strategies for Talking about Sex to Patients

STRATEGIES	DEVICE
DELAYING	– delaying discussion of sex
	– refraining from answering "sensitive" questions
	– acting agitated in the context of delicate terms
AVOIDING	– using vague, indirect, and distant terms
	– avoiding certain delicate terms
	– using pronouns
DEPERSONALIZING	– avoiding personal references
	– using definite articles
TUNING/ADAPTING	– adopting and repeating patients' use of pronouns and their omission of delicate terms

Source: Adapted from Weigts et al., 1993: 4.

isfied with their sexual functioning (Weigts et al., 1993). Moreover, women's dissatisfaction with sex life seems to be hidden behind complaints about physical functioning, including such things as vaginal infections and pain during intercourse (Stanley and Ramage, 1984).

It is useful to understand how this ambiguity and ambivalence about sex is manifest in personal relations and in talk between doctors and their female patients. One study of doctors' and patients' talk showed constructions of sexuality were managed in the doctor's office quite "sensitively" so as to reinforce gender stereotypes about the "shame" and "mystery" surrounding female sexuality. The strategies used to discuss such "delicate" matters are best characterized as delay, avoidance, and depersonalization. Reflected in the talk and the silence is the construction of the "delicate and notorious" character of female sexuality in the context of the possible discourse with a powerful male (doctor). Table 9.3 describes the strategies used by the social actors to demonstrate their expressive caution.

Sociological discussion of talk is more than trivial. It is important both theoretically and practically. Delicacy with respect to sexuality, particularly female sexuality, is a major factor in the transmission of the AIDS virus in heterosexual populations. To the extent that women remain unable to talk clearly and confidently about their sexuality, about their genital and reproductive health, and about devices such as condoms, they may be more likely unable to refuse unwanted and/or unprotected sex.

Summary

(1) Medical knowledge is socially constructed. It reflects cultural values and social-structural locations. It has varied historically and cross-culturally.

(2) Some of the specific values of contemporary medicine include: mind/body dualism, physical reductionism, specific etiology, machine metaphor, and regimen and control. Sociological research provides a critical overview of these medical assumptions.

(3) There is a large and significant gap between the findings of biomedical research and the implementation of the consequences of these findings in medical practice. The values of medical scientists and medical practitioners are, in many ways, at odds with one another.

(4) Available evidence demonstrates that new technologies are usually adopted (even widely) before their safety and effectiveness are established.

(5) One example of the way that medical science is infused with cultural stereotypes is found in the work of Emily Martin, who contrasted the gender stereotypes observed in the descriptions of male and female reproductive systems as described in medical textbooks.

(6) Research shows that medical practice, too, is infused with cultural stereotypes, including those that pertain to age, gender, class, and race.

(7) One cross-cultural study of medical practice by Lynn Payer offers tantalizing evidence as to provocative cultural differences.

(8) That there are also significant class differences in the lay understanding and acceptance of medical knowledge is demonstrated in research by Balshem.

(9) Calnan and Williams show how lay views of medical practice vary according to gender, class, age, and specific health categories.

(10) Media information about medical knowledge is frequently inadequate or inaccurate.

(11) The discursive strategies of medical doctors as they try to keep control over their own definitions of reality in the face of patient questioning are described.

(12) Medical practice regarding sexuality issues reflects cultural practices and the associated shame and mystery.

CHAPTER TEN

Medicalization: The Medical-Moral Mix

WHAT IS THE MEDICAL response to illness, sickness, disease, and death? We have discussed the processes by which people notice signs and call them illness. What are the processes by which doctors recognize some of these signs, label them symptoms, and provide a diagnosis? How have medical diagnostic categories changed over time? What are the relationships among medicine, law, and religion? In what sense is medicine an institution of social control? To what extent have medicine's powers of social control been increasing over the past century? What is medicalization? Are women's bodies more medicalized than men's bodies? Do some people resist medicalization? What are the origins of our contemporary medical care system? To what extent is the practice of medicine a science? To what extent is it an art?

In the following chapters we turn our attention away from the social structure and the distribution, causes, and experiences of illness to the history and present state of the medical care system, ideas, organization, financing, power structures, and personnel. While medicine and illness are intertwined, they are not necessarily co-extensive. Today the definition and diagnoses of illnesses are made primarily by the medical care system. The signs or symptoms that people pay attention to, and those they ignore, are largely determined by medical definitions of illness. And the expectations people have of their bodies, and the way they sometimes communicate by being ill, also depend in part on categories of disease defined by the medical care system.

There are, however, some illnesses that resist medical definition for a number of reasons. Chronic fatigue syndrome frequently remains without a clear diagnosis for a long period of time. Multiple sclerosis is notoriously difficult to diagnose because it lacks a clear diagnostic marker and at times mimics normal though exaggerated behaviour such as periodic stumbling, slurring, and tiredness. At other times, medical diagnosis precedes a person's awareness of a physical problem. High blood pressure is, for instance, often only detected by tests, not by any physical sensations experienced by the person. The point is that sometimes what is defined as a deviant, unusual, or unacceptable feeling, behaviour, or attitude is seen as really a medical problem. Some of these problems may also fall within the realm of religion or law. For instance, AIDS is viewed as a disease by the medical care system; it is also seen as evidence of sin by some churches because the sexual behaviour of a person who has been diagnosed with AIDS is considered immoral. Because AIDS can be contagious, AIDS patients may also be subject to legal controls.

In this section of the book the theoretical paradigms are more difficult to observe. Generally speaking, very little from the symbolic interactionist tradition will be found here. In part, this is because the symbolic interactionist perspective provides for analysis on the micro level, i.e., it examines meaning as it is constructed, maintained, or changed as people interact in social situations. To the extent that sociology of medicine focuses on the medical system, it tends toward a macro level of analysis. Thus, both structural functionalism and conflict theories are more pronounced here. When classical conflict theory is evident it will usually be discussed as Marxist or feminist. For example, parts of Chapter Ten are from a conflict perspective. For the most part, however, the medical care system and its functioning are described and analysed from the more traditional structural-functional perspective.

A Brief History of Medical Practice

At the beginning of written history, medical practitioners and priests in the Tigris-Euphrates and Nile valleys were one. Illness was a spiritual problem. It was regarded as punishment for sins or for violations of such norms of society as stealing, blaspheming, or drinking from an impure vessel (Bullough and Bullough, 1972: 86-101). In Egypt under Imhotep, the Egyptian pharaoh who built the stepped pyramid, medicine began to receive some separate recognition. Even here the medical functionaries' roles were strictly curtailed by modern Western standards. Medical practitioners could only treat external maladies. Internal illnesses were firmly believed to result from and be treatable through supernatural intervention.

Modern Western medicine appears to have been derived from Greece in the fourth and fifth centuries before Christ, and from medieval Europe. In both times, the tie between the body and the spirit, the physician and the cleric, was strong. Early Greeks erected temples in honour of Hyglia, the Greek goddess of healing, and those who were ill sought treatment in these temples. Sometimes they simply slept in them in hopes of a cure. Sometimes temple priests acted as physicians, and used powers of persuasion and suggestion to heal. Early Greek physicians viewed their calling as holy or sacred.

One of the most important Greek physicians of the time, Hippocrates, perhaps best illustrates this view. He drew up the Hippocratic oath, still subscribed to by physicians today. The first sentence of the oath illustrates both the sense of calling of the physician and the dedication to the gods that medical practice involved.

I swear by Apollo Physician, by Asclepius, by Health, by Panacea and by all the gods and goddesses, making them my witnesses, that I will carry out, according to my ability and judgement, this oath and I this indenture.

The oath further states: "But I will keep pure and holy both my life and my art" (Clendening, 1960: 1, 5). The Hippocratic oath contains prohibitions against harming the patient, causing an abortion, and becoming sexually involved with a patient. It considers that things said by a patient to a doctor are to be kept confidential and treated as "holy secrets."

Hippocrates is also recognized for the introduction of detailed observational and experimental methodology. The idea of balance that dominated Hippocratic medicine in the fourth and fifth centuries before Christ continues and persists today in a variety of forms. To Hippocrates, health depended on a harmonious blend of humours – blood, phlegm, black bile, and yellow bile – which originated in the heart, brain, liver, and spleen respectively. Sickness resulted from an imbalance in any of these four humours. Symptoms reflected this lack of balance. Treatment relied largely on the healing power of nature and on the use of certain diets and medicines to return the organism to balance.

There were two types of *practitioners*, each catering to a different social class. Private physicians cared for the aristocrats. Public doctors were retained by most large towns, partly for the prestige of having a doctor and partly to serve those who needed medical care. Both the public and private physicians tended to cater to the wealthier classes. The poor and the slaves usually received an inferior quality of medical care from the physician's assistant (Rosen, 1963).

The Greek period is often thought to have culminated in the work of Galen (A.D. 130-201). His discoveries were influential for more than a thousand years after his time. Working with the principles of Aristotelian teleology, he thought that every organ had a purpose and served a special function. But Galen's greatest contributions were his anatomical and physiological works (his knowledge of anatomy was based on dissections of pigs and apes) and his systematic speculation (Freidson, 1975: 14).

When the Roman Empire collapsed, medicine and other sciences fell into decay and religious scholarship developed greater prominence. There was conflict between two modes of thought – the spiritual and philosophical, in which truth was deduced from accepted religious principles without any reference to the real world, and the empirical, in which truth could only be inferred from evidence based on observation in the real world. Medicine lost much of the scientific analysis and empirical practice developed by the Greeks. Religious dogmatism limited scientific advances by prohibiting dissection and by forbidding independent thought, experimentation, and observation. The only knowledge deemed acceptable was that found in ancient texts and approved by the Church. For medieval Christians, disease was a supernatural as well as a physical experience. Secular medical help and public health measures were criticized as signs of lack of faith. The Church, its liturgies, and its functionaries were the source of healing: sinning was the source of illness.

The Church thus influenced the practice of medicine. It also influenced its organization. Medicine was taught in the universities by rote and faith, through the memorization of the canons of Hippocrates, Aristotle, and Galen. The clergy practised medicine, but they were not allowed to engage in surgery or to use drugs. These two physical treatments were left to the lower orders. Barber-surgeons treated wounds, did other types of surgery, and cut hair. Even lower than barber-surgeons were the apothecaries, who dispensed medicines. Hospitals, too, were given up by the state and taken over by religious orders, thus coming under the control of the Church.

Medicine itself did not progress much during the medieval period. However, the epidemics of disease and death caused certain new methods of inquiry that were instrumental in the later development of scientific medicine. Faced with a horrendous death rate such as that during the bubonic plague (the Black Death is said to have taken one-third of the population of Europe in the fourteenth century), people began to raise questions about disease. In the first place, it became clear that the plague was contagious from person to person. Second, not all people succumbed to the plague. Questions about the background differences of those who did and who did not fall ill seemed relevant. The

bases for quarantine, germ theory, and the case history were laid.

By the eighteenth century scientific medicine was becoming distinguished from religious practice and folk medicine. Available medical knowledge was organized and codified. Many new medical discoveries were made. The universities, particularly in Western Europe and Scotland, became centres of exciting medical advances in both research and treatment. New tools such as forceps and the clinical thermometer were invented. New medicines such as digitalis were made available. Edward Jenner demonstrated the value of the smallpox vaccine. At the same time, the popular climate was confused by the competition among various types of healing. People visited shrines or used the services of "quacks" and a variety of other alternative healers. Medical research was inadequately financed, lacked facilities, and had no specialties; the few practising doctors were dreadfully overworked.

The modern separation of medicine and the Church is the result of a number of social processes. The secularization of the human body as an object of science is a part of this process. This happened in part because of the developing Christian doctrine of the separation of the body and the spirit, a doctrine that paralleled the philosophical discussions of Descartes. One of the results of the separation was that autopsies were allowed. If the body was no longer the house of the soul, then the wholeness of the body after death was no longer of great importance. Institutional secularization occurred, too, as the Church

Box 10.1 Christian Science

Religion and medicine are irrevocably intertwined among several religious groups who practise today. One such group is Christian Science. Christian Science was founded by Mary Baker Eddy in 1866. Born in New Hampshire in 1821 to a family of Puritan background, Mary Baker Eddy spent the first 45 years of her life poor and in bad health but committed to the self-study of various medical systems, including allopathy, homeopathy, and hydropathy. She met and was influenced by a hypnotist healer named Phineas P. Quimby.

In 1866 Mary Baker Eddy fell on a patch of ice and was said to have been told by doctors that her life was at an end. Within a week she was well and walking. She claimed that she healed herself with the aid of God and the power of the mind over the body. Overwhelmed by this experience, she told others. She trained students, the first Christian Science practitioners, in a series of twelve lessons for which she charged $100. She wrote *Science and Health With a Key to the Scriptures*, she said, under direct inspiration from God. By 1879 Mary Baker Eddy was able to found a church, the First Church of Christ, Scientist, in Boston. In 1881 she established an educational institution, Mrs. Eddy's Massachusetts Metaphysical College. The church grew quickly and by 1902 there were 24,000 church members and 105 new churches. By 1911, the year that Mary Baker Eddy died, there were 1,322 churches in Canada, Great Britain, Europe, Australia, Asia, and Africa.

Today the church is world-wide, and each church around the world follows the same lessons and readings at the same time. While there are no ministers, there are practitioners who must graduate with a Christian Science degree. The basis of the philosophy of Christian Science is that sin and sickness are not real but represent the lack of knowledge of God. According to Mary Baker Eddy,

> Sickness is part of error that truth casts out. Error will not expel error. Christian Science is the law of truth, which heals the sick on the basis of one mind on God. It can heal in no other way, since the human mortal mind, so-called, is not a healer but causes the belief in disease.
>
> Then comes the question, how do drugs, hygiene, and animal magnetism heal? It may be affirmed that they do not heal but only relieve suffering temporarily, exchanging one disease for another. We classify disease as error, which nothing but truth can heal, and this mind must be divine not human. (Eddy, 1934)

Treatment for sin and sickness involves prayer. Thinking about God and concentrating on God both lead to and constitute healing. Sickness is the result of incorrect, sinful, or ungodly thoughts.

Healing involves changed thought:

> To remove those objects of sense called sickness and disease, we must appeal to the mind to improve the subjects and objects of thought and give the body those better delineations. (*Ibid.*)

Christian Science constitutes an archetypal modern example of the tie between religion and medicine.

Charles Samuel Braden, *Christian Science Today: Power, Policy, Practice* (Dallas: Southern Methodist University Press, 1958).
Mary Baker Eddy, *Science and Health With a Key to the Scriptures* (Boston: Published by the trustees under the Will of Mary Baker Eddy, 1934).
Bryan R. Wilson, *Sects and Society: A Sociological Study of Three Religious Groups in Britain* (London: Heinemann, 1961).

FIGURE 10.1 Making Medical History

-400	-300	-100s
• Hippocrates separates medicine from religion and philosophy, treats it as a natural science	• Anatomy and physiology develop in Alexandria	• Asclepiades brings Greek medicine to Rome; bases treatment on diet, exercise, baths, massage

100s	200s	400s
• Ancient medicine culminates with Galen; his influence will last until Renaissance	• Growing Christian religion emphasizes healing by faith	• Fabiola founds first hospital in Western world at Rome

700s	800s	900s
• Arabs develop pharmacology as a science separate from medicine	• Monk-physicians treat the sick in infirmaries attached to monasteries	• Influential medical school founded at Salerno, Italy; students include women

1000s	1200s	1300s
• Arab physician Avicenna writes the *Canon*, textbook used in medieval Europe	• Thomas Aquinas describes medicine as an art, a science, and a virtue • Human dissection practised at Bologna	• Urine sample first used • Black Death kills one-third of Europe's population; medicine powerless to stop it

1500s	1600s	1700s
• First attempts to restrict right to practise to licensed and qualified doctors • Advances in anatomy and surgery as influence of Galen wanes	• W. Harvey discovers circulation of blood • Descartes conceives of body as machine and sees medicine as part of developing modern science • Hôtel-Dieu in Quebec City founded, first hospital in Canada • Use of microscope leads to new discoveries	• First successful appendectomy performed • Guild of surgeons formed in England separate from barbers, with whom they had been joined • Advances in scientific knowledge begin to be reflected in medical practice • Edward Jenner proves value of vaccination in preventing smallpox

1800s	1810s	1830s
• Medical specialties begin to develop	• René Laënnec invents stethoscope	• Theodor Schwann shows all living structures made of cells

1840s	1850s	1860s
• Inhalation anesthesia discovered • Edwin Chadwick brings about public health reforms in England	• Nurses led by Florence Nightingale save thousands in Crimean War • Dr. Elizabeth Blackwell founds New York Infirmary for Women	• International Red Cross founded • Joseph Lister introduces antiseptic surgery • Gregor Mendel develops law of heredity • Louis Pasteur shows that diseases are caused by micro-organisms

1870s	1880s	1890s
• Robert Koch discovers tubercle bacillus	• Otto von Bismarck introduces first state health insurance plan in Germany • Sigmund Freud begins to develop psychoanalytic method • Founding of Johns Hopkins medical school introduces systematic medical education in U.S.	• Malaria bacillus isolated • Wilhelm Roentgen discovers x-ray

1900s	1910s	1920s
• The hormone adrenalin is isolated • Flexner Report leads to reform of medical education	• Influenza epidemic kills millions worldwide	• Banting and Best produce insulin for use by diabetics • Iron lung invented • Alexander Fleming discovers penicillin

1930s	1940s	1950s
• Norman Bethune of Canada introduces mobile blood transfusion unit in Spain, continues work in China	• Use of antibiotics becomes widespread • World Health Organization founded	• J. André-Thomas devises heart-lung machine • Discovery that DNA molecule is a double helix provides key to genetic code • Jonas Salk develops polio vaccine • Ultrasound first used in pregnancy

1960s	1970s	1980s
• First state health insurance plan in North America successfully introduced in Saskatchewan despite doctors' strike • Michael DeBakey uses artificial heart to keep patients alive during surgery • Christiaan Barnard performs first heart transplant	• Smallpox eliminated from earth • First baby conceived outside the womb born in England	• Cyclosporin allows full-scale organ transplantation • Nuclear Magnetic Resonance makes possible more accurate diagnosis • A new disease, AIDS, kills thousands; no cure or vaccine in sight

Source: *Compass*, May, 1988, p. 10.

became separate from the state. Through this process clearer distinctions between disease, deviance, crime, and sin were forged.

The growing belief in the potential power of science and the emphasis on individual rights and freedoms contributed to the development of modern secular medicine, too. In the nineteenth century, particularly the last half, an enormous number of new discoveries occurred. Biology moved from the level of the organ to that of the cell, and both physiology and bacteriology were studied at that level. Germ theory emerged. Surgery grew in sophistication along with asepsis and anesthesia. Figure 10.1 outlines some of the most important developments and discoveries in the history of allopathic medicine.

Medicalization: The Critique of Contemporary Medicine

Medical science became influential during the period that urbanization, industrialization, bureaucratization, rationalization, and secularization developed. Medical institutions began to increase their powers as agencies of social control. More and more types of human behaviour began to be explained in medical terms. It has been argued that as the medical system's powers of social control increase, so do the religious institutions' powers of social control decrease. Behaviours that were once viewed as sinful or criminal are now more likely viewed as illnesses. Alcohol addiction is a case in point. There was a time when drinking too much, too frequently, was seen as a sign of moral weakness, in fact, a sin. Today, however, alcohol addiction is seen as a medical problem. Treatment centres for the "disease" are located in hospital settings, and treatment frequently includes medication and a variety of other medical therapies.

Medical institutions, including hospitals, extended-care establishments, pharmaceutical companies, and manufacturers of medical technology, have grown in importance. An increasing part of the gross national product is being spent on health care. Health-care expenses have grown to such an extent that, as one person quipped, by the year 2000 the federal government will be part of the Ministry of Health, rather than the reverse. One definition of medicalization, from the work of Zola (1972), is that it is a process whereby more and more of life comes to be of concern to the medical profession. Zola portrays medicalization as an expanding "attachment process," with the following four components:

(1) the expansion of what in life is deemed relevant to the good practice of medicine;
(2) the retention of absolute control by the medical profession over certain technical procedures;
(3) the retention of near absolute access to certain areas by the medical profession;
(4) the expansion of what in medicine is deemed relevant to the good practice of life.

In this view, the first area of medicalization is the change from medicine as a narrow, etiological model of disease (which has been called the medical model) to a broader concern with the physical, social, spiritual, and moral aspects of the patient's life. As well as bodily symptoms, the entire lifestyle of the patient is now considered of concern to the doctor. For example, some physicians now routinely include in their patients' case histories questions about eating habits, friendships, marital and family relationships, work satisfaction, and the like.

The second component is the retention of absolute control over a variety of technical procedures. A doctor is permitted to do things to the human body that no one else has the right to do. Doctors are responsible for surgery, prescription drugs, hospital admittance, and referral to a specialist or another doctor. Doctors are the gatekeepers to numerous associated services and provisions. The average doctor generates significant health-care costs annually through prescriptions, hospital admittance, treatment costs, and the like.

The maintenance of near absolute control over a number of formerly "normal" bodily processes, and indeed over anything that can be shown to affect the working of the body or the mind, is the third component. Zola argues that the impact of this third feature can be seen by looking at four areas: aging, drug addiction, alcoholism, and pregnancy. At one time, aging and pregnancy were viewed as normal processes, and drug addiction and alcoholism were seen as manifestations of human weakness. Now, however, medical specialties have arisen to deal with each of these. Zola illustrates this point with a discussion of the change in the view and the treatment of pregnancy.

For in the United States it was barely 70 years ago when virtually all births and their concomitants occurred outside the hospital as well as outside medical supervision . . . but with this medical claim solidified [to manage births] so too

was medicine's claim to whole hosts of related processes: not only birth but prenatal, postnatal and pediatric care; not only conception but infertility; not only the process of reproduction but the process of sexual activity itself. (Zola, 1972: 77)

The last component is the expansion of what in medicine is seen as relevant to a good life. This aspect of medicalization refers to the consideration of a variety of social problems as medical problems: homosexuality, criminality, and juvenile delinquency are among those "problems" that were once moral/religious problems and are increasingly seen as amenable to medical definition and treatment.

The Medicalization of Human Behaviour

Conrad and Schneider (1980) have charted the impact of the medicalization process in a number of areas, including mental illness, alcoholism, opiate addiction, delinquency, hyperkinesis, homosexuality, and crime. In all of these they attempt to show how medicine is increasingly an institution of social control. Their research on hyperkinesis (now called attention deficit disorder or ADD) provides one specific illustration of the process of medicalization.

Hyperkinesis is a relatively new "disease" that has been "discovered" over the past three decades. It is estimated that it affects between 3 and 10 per cent of the population of elementary school children. Although its symptoms vary a great deal from child to child, typical symptom patterns include some of the following: excess motor activity, short attention span, restlessness, mood swings, frigidity, clumsiness, impulsiveness, inability to sit still in school or comply with rules, and, finally, sleeping problems. Most of these behaviours are typical of all children at least part of the time. In fact, all these behaviours are probably typical of all people at least once in a while. Conrad and Schneider have explained the processes by which these "normal" behaviours became grouped and categorized as indicators of "disease." They argue that hyperkinesis was "discovered" for a number of sociological reasons.

The first step was the discovery in 1937 by Charles Bradley that amphetamine drugs had a powerful effect on the behaviour of children who had come to him with learning or behaviour disorders. Only later, in 1957, were the "disorders" to become a specific diagnostic category – hyperkinetic impulse disorder. A national task force in the U.S., appointed to deal with the ambiguities surrounding the diagnosis of the disorder and its treatment, offered a new name: "minimal brain dysfunction" (MBD). In 1971 Ritalin, a new drug similar to amphetamines but without the negative side effects, was approved for use with children. Soon Ritalin became the drug of choice for children with hyperkinesis or minimal brain dysfunction, which became the most common childhood psychiatric problem. Special clinics to treat hyperkinetic children were established, and substantial amounts of federal research funds were allocated. Articles appeared regularly

in the mass media periodicals in the 1970s, and many teachers developed a working clinical knowledge of the diagnosis. In short, MBD became a popular disease around which a great deal of lay knowledge and activity was organized, once the drug to treat the disorder was synthesized, produced, and made available.

Three broad social factors aided the discovery of hyperkinesis: (1) the pharmaceutical revolution, (2) trends in medical practice, and (3) government action (Conrad and Schneider, 1980: 157). First, the pharmaceutical revolution led to a great number of drug-related success stories, such as penicillin's success as a widely effective antibiotic and the results of psychoactive drugs in a variety of mental illnesses. Such successes encouraged hope for the potential value of medications in many areas of life. Second, at this time the mortality rate from infectious diseases in children decreased and the possibility of concern with less-threatening disorders emerged. Medical practice consequently began to pay more attention to the mental health of children and to child psychiatry. Third, government publications, task forces, and conferences, along with the activities of the pharmaceutical companies, reinforced the legitimizing of MBD as a new diagnostic category to be managed by the medical profession.

The points made in this analysis are several. First, the behaviour labelled hyperkinetic existed long before the diagnosis. Indeed, such behaviour was and is widespread throughout the population. Second, the popularizing of the diagnosis corresponded to its widespread recognition as a pharmacologically treatable disorder. Third, the popularizing process was aided by the entrepreneurial behaviours of government and the drug companies, along with the Association for Children with Learning Disabilities in the United States. To conclude, medicalization is seen, in this example, as a process by which a common behaviour becomes codified and defined as entailing certain symptoms that are best managed through medical interventions.

Two other new "diseases" are PMS (premenstrual syndrome) and menopause. Recently, considerable medical and then critical feminist attention has been paid to these. McCrea (1983) documents the history of the discovery of menopause as a deficiency disease and notes several parallels to Conrad and Schneider's analysis of hyperkinesis. McCrea dates the discovery of the disease to the 1960s when a prominent gynecologist was given more than $1 million in grants by the pharmaceutical industry. Very soon, this prominent gynecologist, sponsored by the pharmaceutical industry, was writing and speaking about menopause. He described it as a deficiency disease leading to a loss of femininity and "living decay." The diagnosis was coupled with a solution – ERT (estrogen replacement therapy). In a recent feminist analysis, Dickson argued that the "mounting sales of estrogen are a result of the expanding concept of menopause as pathology, or increasingly as prevention of debilitating pathology" (1990: 18). Still, many argue that menopause is not a disease. It is part of the natural aging process through which most women pass with little difficulty.

The Contemporary Physician as Moral Entrepreneur

During the nineteenth and twentieth centuries the medical model has reached its peak. Complete separation of body and the soul/mind and of church and state has occurred in these two centuries. In all of its major institutions, society has become more secularized. The modern world increasingly relies on reason as the way to truth, not faith. The rationalization of the world is seen in the spread of the money economy, capitalism, complex division of labour, bureaucratic social organization, technological development, mass production, factory organization, urbanization, and the like. Modern medicine is seen as a type of science, and the hopes people hold for the benefits of science are unbounded.

Yet there are many ways that the physician can be seen partly as a physical scientist and partly as a moral decision-maker. Not only must the doctor work out a diagnosis that is consistent with his/her understanding of both scientific and medical knowledge as well as the expectations of the patient, but the doctor must also do this within the context of his/her own religious and other values. Diagnosis involves negotiation between the patient, who presents some symptoms and not others, and the doctor, who sees as symptoms things the patient does not notice and disregards some things that the patient sees as symptoms.

A direct link between medical and moral considerations in medical decision-making is described in the work of Talcott Parsons (1951: 428-47; also see Freidson,

Box 10.2 Disease Definition

Mildred Blaxter (1978) has commented on the categorization of illness. She notes that the most generally accepted and complete list of medical categories or diagnoses, the International Statistical Classification of Diseases, Injuries, and Causes of Death, includes a great number of different models of diseases. Some classes of disease are virtually assertions about the cause, e.g., cut on finger; others are simply descriptions of visually obvious or verbally presented symptoms, e.g., high blood pressure. Some are classified by site, e.g., diseases of the stomach; some are categories of symptoms, e.g., headache; others are the names of syndromes that include the nature, symptoms, cause, and prognosis, e.g., Tay-Sachs disease. This list of categorizations could be extended. But the point is that disease diagnosis is not a straightforward and unequivocal procedure. Diseases vary fundamentally in their certainty, ranging from the best defined, e.g., major anatomical defects caused by trauma, to those with unknown etiology and variable description, e.g., multiple sclerosis. Given variability in the meaning of disease, it is not surprising that the process of diagnosis is sometimes considered to be an art rather than a science.

1975: 205-77). In his work on the sick role, Parsons argued that medicine legitimates illness through diagnosis, on the condition that the patient plays the prescribed sick role. To be exempted from social responsibilities due to illness and from responsibilities for the condition, the patient is expected (1) to want to get well; (2) to seek technically competent help; and (3) to co-operate with the "appropriate" helper in getting well. The sick role involves social evaluation and judgement along with physical anomalies.

Freidson's argument elaborates on Parsons's work. Freidson suggests that because medicine is the authority on what illness is, it creates illness as a social role. And illness, because it is generally assumed to be unwanted and people are expected to want to get well, is a type of deviance from the norms defining "normal" health. Human, and therefore social, evaluation of what is normal, proper, or desirable is as inherent in the notion of illness as it is in notions of morality. Quite unlike neutral scientific concepts like that of "virus" or "molecule," then, the concept of illness is inherently evaluational. Medicine is a moral enterprise like law and religion, seeking to uncover and control things that it considers undesirable (Freidson, 1975: 208).

Illness, in this perspective, is legitimated deviance. The physician, as the labeller of illness, can be thought of as a moral entrepreneur. Calling behaviour illness rather than sin is a moral act. The consequence, for instance, of labelling drug addiction as an illness rather than a moral weakness results in the minimizing of punishment and the avoidance of moral condemnation. The addicted person is treated with sympathy rather than with opprobrium. The choice of label is a moral act. It is an instance of what Zola (1972) calls medicalization.

The labelling of an illness is one instance of the moralizing of the physician. Other decisions that must be made by the doctor in the course of his/her work may also be seen as moral decisions. Tuckett (1976) enumerated a number of situations where decisions would have to be made between conflicting demands. Each decision is affected by religious and moral values. The first results from the conflict between the needs of one patient and the needs of a group of patients. Sometimes the adequate care of one patient may require the neglect of other patients. The need of an Alzheimer's patient for care 24 hours a day while in a nursing home or hospital may, for instance, have to be balanced against the needs of the ward nurses and other patients for order and their own ongoing care. The administration of experimental chemotherapeutic drugs may lead to suffering or the death of a cancer patient but can lead to knowledge that will benefit a large number of similar cancer patients at a later date.

A second conflict situation concerns the allocation of time, resources, and skills among individual patients. Organ transplantation may be the last resort for many patients who experience severe heart failure. It is exceedingly costly, however, and there is a limit to the number of transplant surgeons available to carry out the surgery; it also requires intensive, round-the-clock nursing. In this case the doctor may have to choose to allocate resources to one patient rather than another. Some people are likely to die and some to live as a result of the doctor's decision.

A third conflict involves the choice the doctor must make between the present and the future interests of a patient. For example, morphine might be the drug of choice for a victim of severe burns because of its pain-killing properties. However, morphine is addictive and in the long run could cause immense problems for the patient once he or she had recovered from the burns.

A fourth conflict has to do with meeting the expected needs of the patient versus the needs of the patient's family. While it may be in the patient's interests to be cared for at home, this may conflict with the interests of the family members. A schizophrenic patient may be seen by his or her family as incapable of self-care. The family may want hospitalization and the patient may reject it. In this situation a diagnosis and treatment plan must consider at least these two sets of interests.

A fifth conflict situation arises when a physician is unable to help a patient and thus cannot live up to his or her self-perception as a healer. At times a patient may present a physician with a problem that the physician does not feel is within his or her realm of understanding or expertise, e.g., difficulties with sleep, or alcohol, or with a father, mother, child, or boss. Advice concerning such problems is frequently sought from a general practitioner. Even though the doctor may not see the problem as a medical one or as within his or her official jurisdiction, he or she may, perhaps to satisfy the patient and to reinforce his or her self-image as a healer, look for biological causes and prescribe "medical" remedies.

Sixth, a doctor may experience conflict between service to the patient and service to the state or some other organization. Company doctors may be torn between the interests of a patient who wants legitimation for

Box 10.3 The Voice of Medicine and the Voice of the Life World

The clinical encounter between the doctor and the patient has been described as a struggle for the control of discourse between the "voice of medicine" and the "voice of the life world" (Mishler, 1984). A number of different studies have shown the voice of medicine tends to dominate in the patient-physician encounter. Physicians ask most questions (Mishler, 1984; Waitzkin, 1989). Patients' responses are narrowly circumscribed and when they move off the topic they are interrupted, ignored, or their discussion is redirected to fit the goals of the physician. Nevertheless, patients often persevere in trying to have their stories told (Mishler, 1984). Mishler argues that allowing and even encouraging patients' stories may have two important benefits – patient satisfaction and physiological control of disease.

---

sickness because he or she desires time off work and the interest of the employer. Or the conflict might be between the interests of an insurance company and the interests of an individual.

A seventh conflict results from a doctor's dilemma in balancing the advancement of his or her career against the interests of patients. A doctor is unlikely to enhance his status or wealth by serving in a small Inuit village, and yet the members of the village may need the services of the doctor more than do those in urban areas that are often oversupplied with doctors.

An eighth and final conflict is between the doctor's role as a doctor and his or her role as a religious person, a father, a mother, a wife, a husband, a friend, and so on. These conflicts, too, affect the decision-making of a physician (see Gerber, 1983, for a discussion of the conflicts in medical marriages).

Uncertainty and Medicalization

Uncertainty is a fundamental aspect of diagnosis, prognosis, and treatment. As many have said, medicine is an art as well as a science. While the lay person expects the physician's work to be straightforward, the physician is constantly having to make judgements in situations lacking in clarity (Burkett and Knaft, 1974: 82). When faced with an ambiguous situation or when having to choose to do something rather than nothing, the medical practitioner generally tends toward active intervention (Parsons, 1951: 466-69; Freidson, 1970: 244-77; Scheff, 1963: 97-107). This is another instance of medicalization. Scheff has called this tendency to act in a situation of uncertainty the "medical decision rule." Several studies document this decision rule. In one study, Bakwin (1945) reported on physicians who judged the advisability of tonsillectomies for 1,000 schoolchildren. Of these, 611 were judged to need, and subsequently had, their tonsils removed. Those remaining were examined by another physician, and an additional 174 were selected for tonsillectomies. Finally, 215 children remained. They were examined by another physician, and still another 99 were judged to require a tonsillectomy.

In another common treatment, antibacterial drug prescription, a similar tendency toward action in the face of uncertainty is evident. Most sore throats are not sore because of an infection due to strep bacteria (the treatment of which requires antibiotics). Yet the administration of antibiotics, which at times are known to have negative side effects, does not always depend on proof of the existence of strep bacteria.

Clifton Meador (1965) has explored this tendency and has suggested some of the social sources of medical diagnoses. One is that there is no category of illness called non-disease. Because the physician's job is to diagnose illness, not health, all diagnostic categories are for diseases. They omit the very important additional set of categories that would indicate the absence of a suspected disease. Meador suggests that there must be some prevalence of non-tuberculosis, non-brain tumour, non-influenza, and so on.

Sometimes, too, people want diagnosis or medicalization for a condition (Kohler-Reissman, 1989). The problems faced by people with any of the "new diseases" such as chronic fatigue syndrome, fibromyalgia, tight building syndrome, and total allergic reactions are not the problems that result from medicalization. Rather, they result from a lack of medical definition, research, and treatment. And so people who suffer from such diseases are likely to have to search for a physician who will provide them with a medical explanation for their symptoms. Without a diagnosis such sufferers may be without disability pensions, sick leave provisions, unemployment insurance,

and the like. Because allopathic doctors have been given the right by the state to define wellness and illness, people must depend on their signatures for compensation when they feel ill. While a naturopath or an acupuncturist might recognize an illness and even have an explanation for its cause and treatment, because of the "illegitimacy" of these practitioners (in the policies and procedures of the state and of corporations) their understandings may not be used as the basis for claims for compensation.

That diagnostic decision-making is not always in the direction of active intervention has been discussed by Szasz (1974) and Daniels (1975). Szasz examines the concept of malingering, cases in which a person's claim to be ill is not accepted by the medical diagnostician. Daniels suggests that in some settings, such as the military, a person's claim to be ill is more likely to be rejected than in others. There is, however, an ironic possibility that a person who claims to be ill, to absent her or himself, may be seen as having another special kind of illness – a psychosomatic illness. This general tendency toward "medicalization" or active intervention depends on the labels and categories of illness available, the social characteristics of physicians, the social and economic situation in which the diagnosis occurs, and the demographic characteristics of the patient. Sudnow (1967), in a study of hospital emergency rooms, has shown that the age, social background, and perceived moral character of patients affect the amount of effort that is made to attempt revival of the patient when clinical death signs are detected.

In *The Sanctity of Social Life* (1975), Crane provides additional evidence that physicians respond to social variables in treating the chronically and terminally ill. In making a prognosis, they consider the extent to which patients are able to relate to others. The "treatable" patient is one who is most capable of interacting with others. The social status of the patient, while not as important as the ability to interact, is, nevertheless, an important consideration. Crane notes considerable differences among physicians of varying specialties in terms of the types of decisions made with regard to treatability. The social status of the affiliated hospital in which the medical practitioner works apparently affects his or her judgement in predictable ways. For instance, physicians in more prestigious institutions tend toward active intervention when compared with those in less prestigious ones.

Medicalization has been described as a unilateral and non-problematic process generated by the powerful medical establishment. Lay people, however, may either resist or encourage medicalization. A growing field of study is lay epidemiology, which is beginning to demonstrate the pervasiveness of lay beliefs about symptoms, their causes, and treatments (Gabe and Calnan, 1989; Hunt *et al.*, 1989; Kaufert, 1988; Walters, 1991, 1992, 1994).

The critique of medical practice is not only at the macro level of systems. A number of sociologists have recently investigated the patient-doctor relationship by means of observation, recording, transcription, and analysis of the verbal interaction.

Waitzkin's (1989) work is one example of this type of study that tries to link the relationship between personal troubles and social issues (Mills, 1959). Observations showed how doctors interrupt, question, and in a variety of ways direct the conversation as they desire it to go. Their greater power in the interaction, as the providers of definition of the problem (diagnosis) and treatment, allowed doctors to direct the verbal interaction toward technical issues and away from social issues. When patients raise issues about their lives, doctors, Waitzkin found, tended to question and interrupt so as to redirect the attention to technical/medical solutions. Thus, what might otherwise be seen and responded to as social issues deserving and requiring social, political, and economic response became smaller matters amenable to medical intervention. In this way, the doctor forestalls political/economic analysis and critique and operates as an instrument of social control in support of prevailing social practices. Waitzkin's observations supported the following three propositions: "(i) that medical encounters tend to convey ideologic messages supportive of the current social order; (ii) that these encounters have repercussions for social control; (iii) and that medical language generally excludes a critical appraisal of the social context" (Waitzkin, 1989: 220).

Mishler's (1984) work reinforces the observations and analysis offered by Waitzkin. Again, through the analysis of detailed transcriptions of doctor-patient interaction, Mishler documents attempts by patients to raise "voices of the life world" (the everyday, largely non-technical problems that patients carry with them into the medical encounter with doctors) and doctors' tendency to respond through the "voice of medicine" (the technical topics of physiology, pathology, pharmacology, and so on). Mishler, too, noted that doctors use conversational strategies such as interruption, questioning, and topic changing to maintain control of the doctor-patient interviews (1984).

Medicalization and Demedicalization

The link between health and illness and morality is a universal phenomenon (Freidson, 1970, 1975). Religion, medicine, and morality are frequently connected. This integration may become a problem, however, in a complex industrialized society such as ours, in which the medical and the religious institutions are separate. The official perspective is that doctors deal with physiologically evident illnesses, while the clergy and the courts deal with moral concerns. The jurisdictions are believed to be distinct. Yet, the doctor is accorded a good deal more power, prestige, and influence in our society than the clergyman or average lawyer (Blishen, 1969; Clarke, 1980). This power is granted in part because the doctor's work is seen as altruistic, related to the service of others, and impartial (see Parsons, 1951, and Chapter Nine for further discussion). In fact, however, as we have demonstrated, the doctor makes moral judgements in ever-widening spheres of life (Illich, 1976; Zola, 1972; Conrad, 1975).

Physicians tend to act in the face of uncertainty, to diagnose disease and not non-disease, to consider social characteristics of patients in their diagnoses and treatments. In a variety of ways the job of the doctor is to label or "create" a definition of illness for the person who consults the doctor.

A central aspect of the work of a physician is the diagnosis of illness. Illness is not an objectively defined physiological occurrence independent of cultural meanings, but involves social, moral, and physical considerations. Various physical states do exist. Whether they are called health or illness does not depend directly on the physical states, but rather on the evaluation of the states by those who label them.

A number of people would argue that demedicalization is more characteristic of the contemporary society than medicalization (Fox, 1977). There is evidence that the power of the medical model to determine how we think about health and illness has recently declined. The dominance of the physician in the medical labour force is being challenged by other types of health-care providers. The prevalence of the medical model's way of thinking about health, illness, and treatment has been criticized frequently (Carlson, 1975; Foucault, 1973; Illich, 1976; Freidson, 1975). Increasing expenditures on the provision of medical care have not meant better health in the population. The new disease profile, which demonstrates the prevalence of chronic illness, mitigates the pervasive power of the medical model. Chapter Thirteen describes this process in more detail.

Summary

(1) The definition and diagnosis of illness today are largely the result of the labelling activities of the medical profession. "Deviant" and "normal" feelings are labelled by the medical profession, and these definitions become reality for social actors.

(2) The chapter gives a brief overview of the relationship between medicine and religion through history.

(3) Over the past two centuries, medicine has become a distinct discipline. Definitions of sin, crime, and illness have changed. Some criminal or sinful behaviours are now viewed as illnesses, e.g., alcohol addiction.

(4) Through medicalization, medicine has increasingly become an institution of social control. Medicalization is characterized by four components: movement from a narrow view of disease to a broader one; control by the medical profession over a variety of procedures; the almost exclusive access of doctors to certain areas, such as the body; and the ability of the medical profession to identify certain social problems as medical problems.

(5) The physician is not only a scientist but a moral decision-maker. Medicine can legitimate the illness it diagnoses on the condition that the patient adopts the "sick role" that is prescribed by the doctor. Medicine defines what is deviant from "health" and also how the patient is to react to that definition. Illness is legitimated deviance insofar as it has been identified by the physician and the appropriate steps are taken by the patient to get well.

(6) Doctors also make other moral decisions: decisions regarding the allocation of resources, choosing between the present and future interests of the patient, balancing the needs of the patient versus those of his family.

(7) Medicalization has also caused doctors to take action in situations of uncertainty.

(8) Medicalization can be observed in the micro-situation of verbal interaction between the patient and the physician.

CHAPTER ELEVEN

Medical Practitioners, Medicare, and the State

W**HAT IS THE HISTORY** of the medical profession in Canada? Have allopathic doctors always been dominant in the provision of medical care? What constraints has the state placed on the practice of medicine in Canada? How did medicare develop? How successful has medicare been in broadening and equalizing access to the medical care system for all Canadians? Has the change in accessibility led to changes in the overall health of the population or in the distribution of health in the population? These are among the questions that will be addressed in this chapter.

Early Canadian Medicine

The first Canadian medical "system" was composed of the various medical/religious institutions of the many groups of Native peoples. Each of these groups had its own culturally unique definitions of what constituted health and what constituted illness, its own pharmacopoeia, and its own preferred types of natural and supernatural interventions. Since these peoples handed down their traditions orally, the only written accounts are from white settlers, priests, explorers, and traders. These writers tell us that medicine men or shamans were frequently called on to diagnose and treat various types of injuries and disease. Shamans, and other Native people, too, developed a number of very effective botanical remedies such as oil of wintergreen, and physical remedies such as sweat lodges and massages.

Canada's Native peoples are known to have used over 500 different plants as medicines. Some were chewed and swallowed, some drunk in herbal teas, some were boiled and the vapours inhaled, some infused and even poured as medicine into the patient's ear. Sometimes plants were used for ritual purposes only. For example, thorny or spiny plants were used to ward off evil spirits or spirits of disease and death. Sometimes the value resided in the pharmacological effectiveness of a particular plant for a particular symptom or disease. A famous example of a plant with specific and, now, scientifically substantiated medicinal benefit was the scrapings of white bark of cedar, which is rich in vitamin C, for the treatment of scurvy.

Box 11.1 Medicine among the Native Peoples of Ontario

In the seventeenth century, at the time of the arrival of the Europeans, the Indians of Ontario were divided into two linguistic groups: the Algonquian and the Iroquoian. A study of Iroquoian bones dug from a burial mound near Kleinburg, Ontario, revealed that the Native peoples suffered from such well-known diseases as arthritis, osteomyelitis, and tumours. Healing of a trepanned hole in a skull revealed the skill of a local surgeon. Some treatments devised by the Native people, when the cause was obvious, were empirical, rational, and effective. Internal conditions of unknown cause were often attributed to supernatural origins such as (1) the breaking of a taboo, e.g., mistreating an animal or showing disrespect to a river, (2) ghosts of humans, which craved company, (3) the evil ministrations of a menstruating woman, or (4) unfulfilled dreams or desires.

The Native people believed that everything in the world had a spirit or a soul, including animals, trees, rocks, the sky, lakes, and rivers. Thus, everything in the world was to be respected. In some Native cultures, all young people, especially males, were expected in early adolescence to search for a vision or a dream as a guide through life. To achieve this goal it was customary to spend at least a week alone without food. Hungry, lonely, and full of transient concerns, the young person would generally have a vision, which was then interpreted by the medicine man or father as a help through adulthood. Medicine men were both the spiritual leaders and healers. They were able to cast out spells, predict the future, recover lost objects, diagnose and treat disease, and bring rain. Some were also magicians and jugglers.

A practical armamentarium of medicines was evolved, which included treatments for widely different medical problems, from fractures and wounds to freezing and frostbite, burns and scalding, rheumatism, arthritis, urinary problems, fevers, intestinal disorders, cancer, blood poisoning, and toothaches. Some of the herbs and plants used by the Native people continue to be used today. Clearly, the medicine of the Native people was well developed and quite complex (Holling, 1981).

This was taught to Jacques Cartier by Native people. By today's medical standards, there seem to have been pharmacologically useful treatments for a wide variety of symptoms such as wounds, skin eruptions, gastrointestinal disorders, coughs, colds, fevers, and rheumatism.

The health of the early settlers in Canada was continually assaulted. Even before they arrived, immigrants faced grave dangers from the overcrowded conditions in the boats in which they came to Canada. These densely packed quarters greatly increased susceptibility to the spread of contagious diseases. Pioneer life, too, was fraught with hazards. The winters were often extremely cold. The growing season was short and difficult. Accidents occurred frequently as people cleared the bush for roads and buildings. Accidents also occurred on the rapidly flowing waterways. The building of roads, canals, and railways was extremely dangerous. Epidemics of smallpox, influenza, measles, scarlet fever, and cholera decimated the population from time to time. Childbirth in pioneer conditions was often dangerous for both the mother and the child (Heagerty, 1928). Most medical treatment was performed by local midwives, a few doctors trained in Britain and the United States, and travelling medical salespeople; other treatments were based on folk remedies or on mail-order medicines.

The Origins of the Contemporary Medical Care System

The origins of the type of medical practice dominant today can be found in nineteenth-century Canada. At that time a wide variety of practitioners were offering their services and selling their wares in an open market. Lay healers, home remedies, folk cures, and other kinds of medicines were all medical options available to the population. Whisky, brandy, and opium were used widely as medicines, as were various types of patent medicines. Medicine shows were common in the 1830s and 1840s.

Most of the first allopathic medical practitioners in New France were either barber-surgeons from France, who had received primitive training as apprentices, or apothecaries, who acted as general practitioners dispensing available medicines. Barbering and surgery both required dexterity with a knife; they were handled by the same person because of the ubiquitous practice of bleeding as a treatment for a wide variety of ailments. Surgery was only practised on the limbs and the surface of the body. Internal surgery almost always resulted in death because it was not known how to treat bleeding and sepsis. Two other common treatments were purging and inducing vomiting. Among the first doctors in Upper Canada were army surgeons; there were also some civilian physicians. Homeopathic doctors and eclectics (practitioners who used a variety of treatments) worked alongside allopathic practitioners. Lay-trained midwives delivered babies in the home, and other lay-trained persons performed surgery and set bones. No one type of healer was predominant.

Although the first medical school was established in Canada in 1824, theories about disease were still unscientific. The first involvement of the state in this hodgepodge of medical practices was in 1832, when advance warning came of the possible arrival in Canada of immigrants with cholera. The government immediately appointed a Sanitary Commission and a Board of Health, which issued directives for the protection of the people. Infected people were quarantined. Contaminated clothing was burned, boiled, or baked. Private burials were ordered. Massive outbreaks of cholera in 1832 and 1854 necessitated the establishment of a quarantine station at Grosse Isle on the St. Lawrence for ship passengers who were infected and thus unable to enter Canada. Both Montreal and Quebec City adapted buildings to isolate disembarking passengers who had cholera. Public ordinances, which received the force of law in 1831, prevented the sale of meats from diseased animals and appointed civil authorities to inspect dwellings for their state of cleanliness. Early public health measures primarily emphasized quarantine and sanitation. Later the government became

Box 11.2 Canada's Cholera Epidemic

The world-wide epidemic of cholera reached Canada for the first time in 1832. Apparently Irish immigrants, who were escaping the potato famine and the beginnings of cholera in Ireland, were already affected when they set forth for Quebec City and Montreal. The boats they travelled on were built for 150 but carried as many as 500 passengers. Such unsanitary circumstances invited the spread of the virulent and contagious disease. It attacked apparently healthy people, who could die within a matter of hours or days. It erupted in a number of symptoms, including severe spasms and cramps, a husky voice, sunken face, a blue colour, and finally kidney failure as various bodily processes collapsed. Doctors could do virtually nothing to help their patients. Although there were various theories associating the disease with dirt and filth, its cause was not understood.

When it realized the nature of the calamity before it, the Canadian government established a Board of Health with a mandate to inspect and detain ships arriving in Canada from infected ports. A Quarantine Act was passed in February of 1832 and remained effective until February, 1833. Another Act provided a fund for medical assistance to the sick immigrants and to help them to travel to their destinations when they had sufficiently recovered. In the spring of 1832 Grosse Isle in the St. Lawrence, which was directly in the path of ships arriving from Europe, was established as a place of quarantine. All ships were stopped for inspection. Distinctions were made between ships arriving from infected and non-infected ports. Those from infected ports were required to serve a quarantine period, while the ships with ill passengers were thoroughly cleaned and those who were diseased were disembarked and treated. The dead were buried there. Some of the actions of the government were very unpopular because they restricted individuals' freedoms, and resulted in riots in various locations throughout the country. Crowds burned down some cholera hospitals.

Cholera invaded Canada at three other times – in 1834, 1852, and 1854. While statistics are not entirely reliable, it is estimated that as many as 20,000 people died in all the epidemics.

Geoffrey Bilson, *A Darkened House: Cholera in Nineteenth-Century Canada* (Toronto: University of Toronto Press, 1980).
J. Heagerty, *Four Centuries of Medical History in Canada*, Vol. 1 (Toronto: Macmillan, 1928).
Geoffrey Marks and William K. Beatty, *Epidemics* (New York: Charles Scribner's, 1976).

involved in various public health efforts, including the Public Health Act of 1882 of Ontario, which was soon adopted in the rest of the country, the Food and Drug Act, the Narcotics Control Act, the Proprietary and Patent Measures Act, and the establishment of hospitals and asylums.

The Efforts of Early Allopathic Physicians to Organize

In spite of the primitive state of their medical knowledge, numerous attempts were made by allopathic doctors, dating from 1795, to pass legislation that would (1) prohibit any but allopathic practitioners from practising; (2) provide the allopaths with licences under which they would practise; and (3) control admittance to allopathic practice. The allopathic practitioners often had high social standing in the new colony. In the English-speaking areas, they were usually British immigrants and often ex-military officers (Torrance, 1987). Generally, they moved in the highest social circles, married into important families, held political office at times, edited influential newspapers, and provided care for the wealthier classes (Gidney and Millar, 1984). In small towns outside Toronto, allopathic practitioners were often among the most important businessmen, church leaders, and town politicians. By 1852 an informal group of these men was able to establish the *Upper Canada Journal of Medical, Surgical and Physical Science.*

The allopathic doctors expressed frustration with their working conditions. They claimed that they were under-rewarded and under-esteemed as a result of the competition from "Irregulars," and because of "quackery" and the disorganized state of the medical schools. Many of the "Irregulars" were also educated, and because they did not engage in the heroic measures of the allopaths, such as the application of leeches, blood-letting, and applying purgatives, they were less likely to cause harm. But the allopathic doctors thought of them as uneducated, ignorant pretenders who ruthlessly and recklessly administered untried, untested methods with dangerous results.

In fact, even at this time, the "Irregular" doctors exhibited considerable strength and organizational skills. In 1859 homeopathy was the first profession to be legalized and to establish a board to examine and license practitioners. The eclectics were successful in consolidating their own board in 1861 (*ibid.*). Perhaps because of their relative success in organizing themselves, the homeopaths, eclectics, and other heterodox practitioners were "contemptuously lumped together by the established profession as "emperics" (Hamowy, 1984: 63). The *Upper Canada Journal* denounced homeopathy: homeopathy "is so utterly opposed to science and common sense, as well as so completely at variance with the experience of the medical profession, that it ought to be in no way practiced or countenanced by any regularly educated practitioner" (quoted *ibid.*). Opposition to the eclectics was equally passionate. They were called "spurious pretenders" and seen as embodying a continuous and strong threat to "true science."

Competition within the ranks of the allopaths, largely between "school men" (the university-trained and affiliated practitioners) and the Upper Canada practitioners (Gidney and Millar, 1984), prevented the unified stance necessary for the establishment of standards of education, practice, and licensing. In 1850 the "school men" held sway. By 1865 they had conceded some of their authority to the elected representatives of the ordinary practitioners. In the same year (1865) these ordinary practitioners succeeded in passing self-regulatory licensing legislation. This was revised in 1869 under the Ontario Medical Act to create the College of Physicians and Surgeons of Ontario. Much to the surprise of the "regular" or allopathic physicians, both homeopaths and eclectics were included under the same

Box 11.3 The Flu Epidemic of 1918

In the fall of 1918 Canada's population was about 8 million. About 60,000 Canadians had died in the First World War. Between 30,000 and 50,000 died during the fall as a result of the dreaded flu epidemic that was sweeping much of the world. Apparently the flu came to Canada on a troopship, the *Anaguayan*. One hundred seventy-five of the 763 soldiers on board took ill. The ship was quarantined at Grosse Isle. Yet the disease was passed on to Canadians. By the end of September it was clear that Canada had a serious problem. The epidemic spread quickly. It spread up and down from the U.S. into Canada and vice versa. New York and Massachusetts were hit. It also spread westward along the railways and highways. It was attributed to Spain and called the Spanish flu, not because it started there but because, as Spain was neutral in the war, it was easy to attribute it to the Spanish.

It was unique because it tended to hit young adults and often kill them. Previous flu had tended to affect both the very old and the very young. The flu epidemic attacked one in six. Canadian schools, auditoriums, and various halls were opened as temporary hospitals. Children were left without parents. Quarantines were imposed. Public meetings were forbidden. Partly because of the impact of the flu, the need to establish a federal health authority was acknowledged. The bill to institute such a department received first reading in March, 1919, and the new department became operational that fall. But quarantines did not seem to work. Many people thought that quarantine was unjust. Others just did not believe that quarantine worked because they saw people succumbing who had been very careful, while others, less careful, remained disease-free.

Because quarantine did not seem to work, other measures were introduced. Laws were passed to ensure that people wore masks. But the laws differed. In some municipalities those who were caring for the ill were required to wear masks. In others, anyone who was in contact with the public was expected to wear a mask. Alberta required that anyone outside the home had to use a mask. Yet masks proved as ineffective as quarantine. Rather than boiling and sterilizing them frequently, and certainly between wearings, people allowed the trapped germs to spread and multiply in the moist, warm environment inside the mask.

Most homes had their own trusted preventative measures and treatment. Mothballs and camphor in cotton bags worn about the neck were common. Travelling medical salespersons reaped profits from the salves and remedies they sold. Alcohol and narcotics were prescribed.

Losses to business were enormous. People were too sick to shop or were afraid to venture into the stores for fear of catching the disease. Many staff, too, were off sick. Theatres and pool and dance halls suffered heavy losses. Some 10,000 railway workers were off at one time. Ice storms, blizzards, and below-zero weather had never exacted so heavy a toll as this epidemic. Telephone companies were heavily overextended, both because people were relying on the phone rather than leaving the house and because so many employees were away sick. The insurance industry was one of those most heavily hit. Apparently there were often many flu claims. All in all, Spanish flu had an enormous effect on Canadian society.

Janice P. Dickin McGinnis, "The Impact of Epidemic Influenza: Canada, 1918-1919," Canadian Historical Association, *Historical Papers*, 1977.

John J. Heagerty, *Four Centuries of Medical History in Canada*, Vol. 1 (Toronto: Macmillan, 1928).

Eileen Pettigrew, *The Silent Enemy: Canada and the Deadly Flu of 1918* (Saskatoon: Western Producer Prairie Books, 1983).

legislation. Homeopaths continued to be represented by the College until 1960. Eclectics were excluded in 1874.

This 1869 legislation gave the practitioners, via their representatives on the College, control over the education of medical doctors. Proprietary (privately owned) medical schools were founded, but they were rapidly affiliated with universities in order to grant degrees. The power struggles between the university-based doctors (school men) and the practitioners continued.

In Lower Canada the attempt to define and limit the work of physicians was complicated by the tensions between French and English doctors. The College of Physicians and Surgeons of Lower Canada, formed to regulate practitioners in Lower Canada, was created in 1847. In 1849 legislation was passed to allow automatic incorporation of anyone who had been engaged in practice in 1847.

The competition and in-fighting that characterized much of the medical care system in the nineteenth century had receded by the beginning of the First World War (Coburn *et al.*, 1983). The year 1912, however, marked a turning point in the position of allopathic practitioners. The Canada Medical Act, which standardized licensing procedures across Canada, was passed in 1912. By this time a patient who sought the services of an allopathic practitioner had more than a 50 per cent chance of being helped by the encounter. The passage of the Canada Medical Act coincided with another important event. The momentous Flexner report, *Medical Education in the United States and Canada*,

sponsored by the Carnegie Foundation and the American Medical Association and financed by the Rockefeller philanthropies, was published in 1910. The Flexner report severely criticized the medical systems of Canada and the United States. It advised the elimination of the apprenticeship system; the standardization of entrance requirements to medical schools; and the establishment of a more rigorous scientific program of study. It recommended the closing of many medical schools, particularly the proprietary ones, because they did not meet the criteria of scientific medicine.

Flexner's report radically changed medical education in the U.S. and raised the standard of medical education in Canada. Medical education was taught as a scientifically based scholarly field under the aegis of universities. McGill University and the University of Toronto were given very high ratings. The medical schools in Halifax and London moved rapidly to become affiliated with Dalhousie University and the University of Western Ontario, respectively.

The Flexner report had a significant impact on the organization of Canadian medical education and the practice of medicine for the next half-century. The report enhanced the legitimacy of science as the basis of clinical practice. It reinforced the importance of empirical science with its emphasis on observation, experimentation, quantification, publication, replication, and revision as essential to medical-scientific research. It emphasized the importance of the hospital for the centralized instruction of doctors-to-be and the use of medical technology for standardization in diagnosis and treatment. By the 1920s the hospital-based, curatively oriented, technologically sophisticated medical care system that Canadians know today was firmly established.

One important side effect of the report was that the schools that were closed were primarily those that educated women and blacks. Thus, the closing of the proprietary schools further entrenched the training of white, middle-class males.

Universal Medical Insurance in Canada

A system of universal medical insurance was first suggested by Mackenzie King in 1919 as part of the Liberal Party platform; it was recommended regularly by organized labour after the end of World War One (Walters, 1982). Next, universal medical insurance was proposed at the Dominion-Provincial Conference on Reconstruction in 1945. At this time, the provinces opposed the federal initiative, favouring free medical health insurance. In 1957 the federal government introduced the Hospital Insurance and Diagnostic Services Act, which provided for a number of medical services associated with hospitalization and medical testing. The federal government was to pay 50 per cent of the average provincial costs.

In 1961, a Royal Commission on Health Services was appointed by the federal government. The commission, under Supreme Court Justice Emmett Hall, recommended that the government, in co-operation with the provinces, introduce a program of universal health care. The result was the Medical Care Act of 1968. Finally implemented in 1972, the new universal medical insurance scheme was to cover medical services, such as physicians' fees, that were not covered under the previous Hospital Insurance and Diagnostic Services Act. The scheme had four basic objectives. (1) *Universality*. The plan was to be available to all residents of Canada on equal terms regardless of such differences as previous health records, age, lack of income, non-membership in a group, or other considerations. The federal government stipulated that at least 95 per cent of the population was to be covered within two years of the provincial adoption of the plan. (2) *Portability*. The benefits were to be portable from province to province. (3) *Comprehensive coverage*. The benefits were to include all necessary medical services, as well as certain surgical services performed by a dental surgeon in hospital. (4) *Administration*. The plan was to be run on a non-profit basis.

The Canada Health Act added *accessibility* to make five principles. The costs for the new plan were to be shared 50/50 by the federal and provincial governments. They were also to be shared in a way that would serve to redistribute income between the poorer and the richer provinces.

The provinces soon faced increasing financial pressure as health-care costs grew. Physicians who felt they were not paid enough "extra-billed" their patients. When a critical level of frustration and complaint was reached, the government appointed Emmett Hall again to chair a committee to re-evaluate medicare. In 1980, Hall reported

Box 11.4 Time Line: The Development of State Medical Insurance

West Germany and Western Europe introduced social welfare insurance, including health insurance, in the 1880s.

New Zealand introduced social welfare insurance, including health insurance, in the early part of the twentieth century.

Great Britain introduced national health insurance in 1948.

CANADA

1919 Platform of the Liberal Party under Mackenzie King includes medicare.

1919 End of World War I: organized labour began what was to become an annual statement by the CCL concerning the importance of national health insurance.

1919-1920 Talk of medicare in the U.S.: several states passed medicare legislation that was later withdrawn.

1934 Canadian Medical Association appointed a Committee on Medical Economics, which produced a report outlining the CMA position in support of national health insurance, with several provisos.

1934 Legislation passed for provincial medical insurance in Alberta. Government lost power before it could be implemented.

1935 British Columbia introduced provincial medical insurance legislation. Despite public support via a referendum, this legislation was never implemented because of a change in governments.

1935 Employment and Social Insurance Act including a proposal for research into the viability of a national medical insurance scheme was introduced.

1942 Beveridge report on Britain's need for a National Health Service published in Great Britain – the subject of much discussion in Canada.

1945 Dominion-Provincial Conference on Reconstruction included proposals for federally supported medical insurance. Conference broke down in the wake of federal-provincial dispute.

1947 Saskatchewan implemented hospital insurance.

1951 A Canadian Sickness Survey completed. It demonstrated income differences in illness.

1958 Hospital Insurance and Diagnostic Services Act passed.

1962 Saskatchewan CCF/NDP government introduced provincial medical insurance; Saskatchewan doctors' strike.

1962 Royal commission appointed to investigate medical services.

1966 Federal legislation for state medical insurance passed.

1968 Federal legislation implemented.

1971 All provinces fully participated in medicare.

1972 Federal legislation included the Yukon and Northwest Territories.

1977 Federal government changed the funding formula with the provinces – Established Programs Financing Act (EPF).

1984 Canada Health Act. Reinforced the policy that medical care to be financed out of the public purse (penalties for hospital user fees and physician extra-billing).

1987 Ontario doctors' strike.

1990 Bill C-69 freezing EPF for three years, after which future EPF growth was to be based on GNP minus 3 per cent.

1991 Bill C-70 passed to freeze EPF for two additional years before new formula (-3 per cent) came into effect. Made it possible for federal government to withhold transfer payments from provinces contravening the Canada Health Act.

258 HEALTH, ILLNESS, AND MEDICINE IN CANADA

to the federal government that unless extra-billing was banned, Canada's universal health-care system was doomed. In response, but amidst much controversy, the Canada Health Act (1984), which limited the provinces' right to permit extra-billing, was passed.

The Canadian Medical Association then filed a lawsuit against the federal government, claiming: (1) that the Canada Health Act went beyond the constitutional authority of the federal government and (2) that it contravened the Charter of Rights and Freedoms in that it prohibited doctors from establishing private contracts with their patients. The Ontario Medical Association, which is the largest and most vocal provincial group and represents about 17,000 physicians, also challenged the Act in court. The Ontario government passed the Health Care Accessibility Act (Bill 94) in June, 1986, in response to the federal legislation, thus making extra-billing illegal. The Ontario Medical Association staged a 25-day strike (the longest in Canadian history) in protest, but eventually had to back down when the government refused to capitulate. Public opinion was decidedly against the strike action, and many doctors trickled back to work even before it was officially over.

Factors in the Development of Medicare in Canada

The development of medicare over this half-century in Canada was influenced positively or negatively by a number of significant social forces. The most important of these are: (1) the widespread movement in Western industrialized societies toward rationalized bureaucratic social organization and monopoly capitalism; (2) the spread in Western Europe and beyond of social welfare legislation in public education, old age pensions, family allowances, unemployment benefits, and medical care insurance; (3) the interests of the medical profession in maintaining fee-for-service, cure-oriented, hospital and technologically based medical practice; (4) the interests of the life and health insurance companies in perpetuating their share of a profitable market; (5) the interests of the drug, medical, and hospital supply companies in continuing to develop their increasingly profitable markets; (6) the interests of the urban labour unions and farm co-operatives in social welfare benefits for their members; and (7) the charismatic qualities and dedication of individuals such as Tommy Douglas, who had both a position of power at the right time in Saskatchewan and a commitment to universal medical care. Each of these will be discussed in turn.

First, the movement toward state medical insurance in Canada must be seen as part of a widespread trend in Western industrialized nations toward rationalization and bureaucratization in the context of monopoly capitalism. It was not until this century that the work of the physician came to be widely regarded as the most effective form of medical care. With the development of antibiotics to treat bacterial infections and psychoactive drugs to treat the psychoses associated with various forms of emotional despair and mental illness, the efficacy of the doctor became firmly established in the public mind. Along with the increase in legitimacy of allopathic medicine, medical practice itself began to alter as hospitals changed from being institutions for the dying and the indigent into being the doctor's place of work. There was an increase in medical specialization and a growth in medical and paramedical occupations. All of this growth and development served to enmesh the physician within enormous bureaucratic structures.

> The once-familiar physician with his little black bag is being replaced by a complex 'health-delivery system' centred in a proliferating number of large, urban-based bureaucratic settings, such as university medical centres, hospitals, community health centres and health maintenance organizations. (McKinlay, 1982: 39-40)

Bureaucratic organization is technically efficient. It is also suited to capitalist development because it assists in organizing the expansion and maintenance of control over profit necessary for capitalist accumulation.

Second, the introduction of medicare in Canada must be seen in the context of the spread of similar policies in the Western industrialized world. The first legislation was passed in Germany in the 1880s. By the time Britain first introduced universal medical insurance in 1912 (to be formally established as the universal National Health Service in 1948), much of Western Europe and New Zealand had already pursued this course. Talk of medicare and the passage of legislation in support of medicare (which was later withdrawn)

occurred in the United States by 1919-20. From 1919 until the introduction of comprehensive state-sponsored medical care in Canada a half-century later, both the federal and provincial governments made various attempts to draft legislation to implement medicare.

Third, the medical profession has had an impact on the timing and the nature of medicare. In 1934, the Committee on Medical Economics of the Canadian Medical Association completed a report that described the characteristics of the ideal medical insurance scheme. At this time the CMA expressed support for state medical insurance provided that: (1) it was administered by a non-political body of whom the majority would be medical practitioners; (2) it guaranteed free choice by physicians of their method of payment; (3) it provided for medical control over fee scheduling; and (4) it allowed for compulsory coverage up to certain levels of income (Torrance, 1987). Following the Second World War, after the return of economic growth when most patients were able to pay their bills either independently or through private medical insurance, the majority of the profession opposed medicare. The strongest statement of the opposition of the doctors to state medical insurance was the 23-day strike by about 800 of Saskatchewan's 900 physicians.

Fourth, the life and health insurance companies, through the Life Insurance Officers Association, opposed medicare. They argued that the role of the state was to provide the infrastructure for the development of such things as transportation and communication. State-sponsored

Box 11.5 Tommy Douglas

Tommy Douglas is one of the most important figures in the development of a universal state-supported medical care system. He was born in Scotland in 1904 and immigrated with his family to Winnipeg, Canada, when he was just a boy. An incident in his own history stands out because it was often said to have provided the motivation for his determined fight for medicare. Before coming to Canada, Douglas injured his knee in a fall. As a result of the injury he developed osteomyelitis and was forced to undergo a series of painful operations. Despite these operations, osteomyelitis recurred while he was living in Winnipeg. The doctors in Canada felt that amputation was necessary. However, while Douglas was at the Children's Hospital in Winnipeg in 1913, a famous orthopedic surgeon, Dr. R.J. Smith, became interested in his condition. Dr. Smith took over the case and saved his leg. He did not charge the far-from-wealthy Douglas family. As Douglas says of this experience, "I always felt a great debt of gratitude to him, but it left me with this feeling that if I hadn't been so fortunate as to have this doctor offer his services gratis, I would probably have lost my leg."

In addition to this personal experience, Douglas was moved by the social conditions in Winnipeg in the early twentieth century. Unemployment was high. People lived with a great deal of uncertainty and were able to afford only the barest of necessities. These conditions made a lasting impact on Douglas. He was also sensitive to and strongly opposed to the discrimination and prejudice based on ethnic and racial differences that he saw around him in Winnipeg.

He acted out his heartfelt commitment by first becoming a Baptist minister in 1930 and then by becoming a parliamentarian. In 1932 he joined the Farmer-Labour Party. In 1935 he was nominated as the CCF candidate for the Weyburn federal riding in Saskatchewan and was elected to the House of Commons. He was re-elected in 1940. From 1944 until 1961 Tommy Douglas was Premier of Saskatchewan, the leader of the first socialist government in Canada. He worked toward economic security for the farmer, full employment for the urban worker, and the development of natural resources. Tommy Douglas promised that "socialism would provide free health and social services and lift the burden of taxation from the shoulders of the people and place it upon the fleshy backs of the rich corporations."

By January, 1945, free medical, hospital, and dental care was provided in Saskatchewan for "blue card" pensioners and indigent people. Treatment for mental disorders and polio was made free to all. A school of medicine was opened at the University of Saskatchewan to increase the supply of doctors. Geriatric centres were built to provide care for the chronically ill. Canada's first district-wide medical insurance program was established at Swift Current, where 40 doctors served a population of 50,000 people. This was financed by a family payment of $48 per year and a land tax. Thus Swift Current became a testing ground for a provincial program.

By January 1, 1947, Douglas introduced the Hospital Insurance Plan. The premium at the time was $5 per person and $10 per family. At the end of 1947, the *American Journal of Public Health* stated that 93 per cent of the population of Saskatchewan was covered by the new scheme. The only exceptions were some remote northern communities.

In 1959 Douglas announced the introduction of medicare, which was to be based on five basic principles. (1) Medical bills would be prepaid and patients would never see a doctor's bill. (2) The plan would be available to everyone regardless of age or physical disability. (3) The plan would accompany

Box 11.5 continued

improvements in all areas of health service. (4) The plan was to operate under public control. (5) The legislation was to satisfy both patients and physicians before it went into effect.

A few days after Douglas stepped down as Premier of Saskatchewan in 1961, the medical bill passed. In 1962 the Saskatchewan doctors' opposition to the government medical scheme intensified. A strike followed the announcement that the Act was in force. The strike lasted 23 days, during which time the government kept medical services in operation by flying doctors over from Britain.

The Saskatchewan plan soon became the basis for a Canada-wide plan. In 1964 Prime Minister Pearson announced that the federal government would give financial aid to any province that had a satisfactory medicare system. By 1969 most of the provinces were participating in the plan in which 50 per cent of the costs were covered by the federal government.

Tommy Douglas re-entered federal politics and remained a member of the House of Commons, first as the leader of the New Democratic Party, which had been formed in 1961 by a coalition between labour and the CCF, from 1961 to 1979.

Thomas H. McLeod and Ian McLeod, *Tommy Douglas: The Road to Jerusalem* (Edmonton: Hurtig, 1987).
Doris French Shackleton, *Tommy Douglas* (Toronto: McClelland and Stewart, 1975).
Lewis H. Thomas, ed., *The Making of a Socialist: The Recollections of T.C. Douglas* (Edmonton: University of Alberta Press, 1982).
Robert Tyre, *Douglas in Saskatchewan: The Story of a Socialist Experiment* (Vancouver: Mitchell Press, 1982).

medical insurance was to be limited to those who could not pay for their own policies and to assist in payment for catastrophic illnesses. The insurance companies were already making a profit and wanted to continue to do so.

Fifth, the drug, medical, and hospital supply companies and their representative body, the Canadian Manufacturers' Association, agreed that to maintain their position of increasing profit they should oppose government intervention in medical financing.

Sixth, the labour unions and farm co-operatives long advocated national medical insurance. Except in the case of Saskatchewan, however, they did little actively to bring about state medicare.

Seventh, the importance of individuals such as Tommy Douglas cannot be overlooked. He was especially dedicated to the principle that medical care be accessible and available on a universal basis. He owed his well-being and perhaps his life to the fact that a physician operated on him without fee and saved his leg from amputation when he was a boy. Partly as a result of this personal experience, and also, of course, because of his socialist ideology, Tommy Douglas made medicare a fundamental plank in his CCF platform.

The Impact of Medicare on the Health of Canadians

The most important goal of medicare was the provision of universally accessible medical care to all Canadians regardless of class, region, educational level, religious background, or gender. To what extent has

Box 11.6 Impediments to the Development of State-Sponsored Medical Care in Canada

1. The traditional division, early in the century, between the urban industrial labour force and the farmers, and the consequent inability of these two working-class groups to form a unified labour lobby in favour of medicare.
2. The strengthening and unification of the Canadian Medical Association in 1920 with the appointment of Dr. T.C. Routley as leader, which gave the CMA a unified voice with respect to the conditions under which medicare might be acceptable.
3. The extensive representation of medical doctors in local, provincial, and federal politics and departments of health gave physicians the power to voice their individual views regarding medicare.
4. Through the British North America Act, 1867, the responsibility for health rested with the provinces.
5. Opposition of Canadian Roman Catholics to state medical insurance.
6. Regional, ethnic, occupational (labour/farmer) heterogeneity of Canadian society retarded the development of social democratic policies.
7. Steady growth of private insurance companies, especially during the fifties and sixties. Medical insurance became an increasingly common fringe benefit in collective bargaining agreements.
8. Repeated opposition of the Canadian Life Insurance Officers Association, the Canadian Chamber of Commerce, and the Canadian Manufacturers' Association to state medical insurance. (Walters, 1982)

this goal been reached? Before the introduction of medicare there was a positive and clear relationship between income level and use of medical services. Low-income groups visited the doctor over 10 per cent less often than did high-income groups (see Tables 11.1 and 11.2). After medicare this pattern was reversed: lower-income groups are now almost twice as likely to visit the doctor as are those from the highest income group (28 per cent versus 14.8 per cent).

In spite of the equal availability of health care to people of all classes and the disproportionately greater use of medical care facilities by the poorer classes, health continues to vary by class. People of the lower classes still live shorter lives and have more days of disability during these shorter lives. The health of those of lower income and

education in Canadian society continues to be poorer than the health of those of higher socio-economic status (see Chapter Four for a detailed examination of this statement). Obviously, health results from more than accessibility to medical services.

The Impact of Medicare on Medical Practice

According to Naylor (1982), before the Great Depression, from approximately 1918 to 1929, doctors made more than four times the average Canadian salary. The depression had a devastating effect on all incomes; in the period of recovery in the 1940s, doctors' salaries averaged about three times the national average. By 1989, with various private insurance schemes in place, doctors

Table 11.1 The Use of Medical Care Services before Medicare (1950-51) by Income Level

Rate per 1,000 Population

	ADULTS	CHILDREN	ALL PERSONS
Low Income	508	368	474
Medium Income	524	536	540
High Income	528	581	544
Higher Income	571	663	588

Source: *Medicare: The Public Good and Private Practice. A Report by the National Council of Welfare on Canada's Health Insurance System*, 1982: 10, Table 1, Catalogue 82-3267. Reproduced with permission of Minister of Supply and Services, Canada, 1989.

TABLE 11.2 Distribution of Insured Health Services among Canadians in Different Income Groups after Medicare (1974)

INCOME QUINTILE	MEDICAL CARE PROGRAM %	HOSPITAL INSURANCE PROGRAM %	BOTH PROGRAMS %
Lowest	28.8	33.1	31.7
Second	22.0	23.9	23.3
Middle	18.0	17.2	17.4
Fourth	16.4	13.4	14.4
Highest	14.8	12.4	13.2

Source: *Medicare: The Public Good and Private Practice. A Report by the National Council of Welfare on Canada's Health Insurance System*, 1982: 23, Table 4, Catalogue 82-3267. Reproduced with permission of Minister of Supply and Services, Canada, 1989.

made about 3.7 times the average Canadian salary. Initially medicare gave a dramatic boost to doctors' salaries. In the early 1970s doctors' salaries were approximately 5.2 to 5.5 times the national average. Wage and price controls, which were introduced in the mid-1970s, retarded the expansion of physicians' salaries during that period. The number of doctors in Canada grew much more quickly than the population and continues to do so. Between 1968 and 1978 the number of doctors grew by 50 per cent, the population by 13 per cent. As Table 11.3 indicates, the physician-to-population ratio

TABLE 11.3 Number of Physicians and Ratio of Physicians to Population, 1871 to 1992

YEAR	NUMBER OF PHYSICIANS	RATIO (PHYSICIANS TO POPULATION)
1871	2,792	1:1,248
1881	3,507	1:1,223
1891	4,448	1:1,087
1901	5,442	1: 987
1911	7,411	1: 970
1921	8,706	1:1,008
1931	10,020	1:1,034
1941	10,723	1:1,072
1951	14,343	1: 977
1961	21,290	1: 857
1971	32,942	1: 659
1980	44,274	1: 544
1992	53,836	1: 514

Note: 1941 figures do not include 1,150 physicians serving in the armed forces. After 1941 active civilian physicians only. Data since 1947 include interns and residents.

Sources: Ronald Hamowy, *Canadian Medicine: A Study in Restricted Entry* (Vancouver: The Fraser Institute, 1984), p. 270; Statistics Canada, *Health Reports*, 4, 2 (1992).

has been narrowing for more than a century. In 1871, there were 1,248 citizens to every physician; by 1951 the ratio was 1:977; and by 1980 the ratio was 1:544. As Table 11.4 shows, the growth rate in the number of doctors varies substantially from province to province. Compare, for instance, the low rates of growth of less than 1 per cent in Saskatchewan and Newfoundland with the almost 5 per cent increase in the Northwest Territories and the Yukon. Table 11.5 shows that from 1980 to 1992 the country's population as a whole was growing approximately 1 per cent per year whereas the number of doctors was growing 3 per cent per year.

As the physician-to-population ratio decreases, some areas and regions may experience an excess of doctors and others an insufficient number. An oversupply of doctors could be a factor in the hypothesized decline in doctors' salaries and in a decrease in power and prestige. It could also be the occasion of excess rates of surgery, prescriptions, or other unnecessary medical intervention. Even successful treatments may be accompanied by harmful side effects. An undersupply of physicians, by contrast, can lead to unnecessary suffering and lengthy travel or waits to enable a visit with a physician.

The involvement of the state in the practice of medicine has resulted in a number of

TABLE 11.4 Average Annual Growth Rate of Active Civilian Physicians for Canada and the Provinces, 1986-1992

| | NUMBER OF PHYSICIANS | | AVERAGE ANNUAL GROWTH |
	1986	1992	1986-1992(%)
Newfoundland	847	892	0.87
Prince Edward Island	174	173	–
Nova Scotia	1,536	1,759	2.29
New Brunswick	853	1,024	3.09
Quebec	12,564	14,534	2.46
Ontario	16,881	20,473	3.27
Manitoba	1,856	1,995	1.21
Saskatchewan	1,423	1,493	0.80
Alberta	3,650	4,441	3.32
British Columbia	5,736	6,953	3.26
Yukon	29	38	4.61
Northwest Territories	46	61	4.82
Canada	45,595	53,836	2.81

Source: Statistics Canada, *Health Reports*, 4, 2 (1992).

TABLE 11.5 Growth Rate of Population and Physicians for Canada

	POPULATION	PHYSICIANS
Growth Rate 1980-86	0.9%	3.4%
Growth Rate 1986-92	1.4%	2.8%
Growth Rate 1980-92	1.1%	3.1%

Source: Statistics Canada, *Health Reports*, 4, 2 (1992).

changes in the actual work of the doctor. These include changes in (1) working conditions, (2) the degree of control over clients and over other occupations in the medical field, (3) self-regulation in education, licensing, and discipline, and (4) the actual content of the work (Coburn *et al.*, 1983). Charles (1976) has documented the ways that universal medical care insurance has altered the medical profession through increased administrative, economic, political, and social constraints.

With respect to administrative constraints, Charles was chiefly referring to the increased bureaucratic surveillance of the actual practice of individual physicians through the computerized account systems detailing the number of patients seen and the medical problems diagnosed in a given period of time. The government, through this bureaucratic control, obtained the ability to investigate individual doctors whose practices varied substantially from the norms. The privacy of the independent entrepreneur was effectively eliminated through the powers of this bureaucratic surveillance.

Certain financial constraints resulted. While the salaries of physicians continued to increase under universal medical care, it was no longer possible for individual doctors to reach salary levels they might desire. The Quality Service Payment Formula dictated the maximum number of weekly "units" of service that could be provided by an individual physician without jeopardizing the quality of patient care. Doctors whose weekly incomes were regularly higher than the designated quota were then subject to closer examination by a Medical Review Committee.

The question of what proportion of the resources of a society should be dedicated to medical care, essentially a political concern, is the third area of constraint noted by Charles. One effect of universal medical care insurance is that physicians' fees are now negotiated between the government and the doctor. The government needs to keep down medical care costs and to ensure availability of service across the nation. It must also seek to minimize inequity among different medical care personnel within a global budget that includes hospitals and extended care facilities, drugs, and expensive new technology. All of this is then balanced against the desire of physicians for a certain standard of living.

The final constraint noted by Charles is social. One such example is that in order to provide equitable medical care to people throughout the country, even in remote and isolated areas, governments may implement quota systems that would allow doctors to practise in a given area only if their services were needed as determined by a standardized physician-to-population ratio.

The Impact of Medicare on Health-Care Costs

Clearly, health-care costs as a portion of the GNP have increased during the period 1960-1991. As Table 11.6 indicates, the increase has been from 5.6 to 9.9 per cent of GNP. In fact, Canadian health-care costs today are a greater proportion of the GNP than that of any country in the world except the United States. In 1991 Canadians spent about $67 billion on private and public health care, an average expenditure of $2,474 for every citizen (*Canada Year Book*, 1994: 128). The provinces and territories manage their own health-care systems, educational programs, and health personnel and certification. Provinces vary in respect to the numbers of programs in addition to the basic medicare services offered. Some, for instance, provide additional benefits, including psychologists, dentists, optometrists, chiropractors, podiatrists, as well as home care, drugs, and general preventative services. Many Canadians belong to private insurance plans for additional services. About 43.5 per cent of the population were covered by dental services in 1989. The Canada Assistance Plan is available for uninsured medical services (*Canada Year Book*, 1994: 144).

What, then, is the impact of universal insurance on physician use in Canada as compared with the U.S.? Overall, Canadians of all classes are more likely to visit a physician than Americans are. This difference is minimized but not eliminated among the elderly, who in the U.S. have medicare available to them. However, there are no significant differences between Canada and the U.S. with respect to hospital admission rates or length of stay. In addition, there are greater between-class disparities in the U.S. than in Canada – the poor in the U.S. visit the doctor less often, yet once they have made an initial physician visit they tend to have a higher number of visits and to stay in the hospital, once admitted, longer than those of the higher class levels (Hamilton and Hamilton, 1993).

Table 11.7 illustrates the fact that the fastest-growing portion of the medical care dollar in the years between 1960 and 1990 was that portion devoted to expenditures on medical institutions. The growth in this one area has comprised 10 per cent of the total budget. By comparison, the proportion of funds spent on physician services has declined from 16.6 per cent to 15.2 per cent. While policies supporting

TABLE 11.6 Total Health Care Costs, as a Percentage of Gross National Product (GNP), Canada, 1960-1991 and Selected U.S. Figures

YEAR	HEALTH CARE COSTS AS % OF GNP U.S.A.	HEALTH CARE COSTS AS % OF GNP CANADA
1960		5.6
1961		6.0
1962		6.0
1963		6.1
1964		6.1
1965		6.2
1966		6.2
1967		6.5
1968		6.8
1969		6.9
1970	7.5	7.3
1971	7.7	7.5
1972	7.9	7.4
1973	7.8	7.1
1974	8.1	7.0
1975	8.6	7.5
1976	8.7	7.4
1977	8.8	7.4
1978	8.8	7.4
1979	8.9	7.2
1980	9.5	7.5
1981*	9.7	7.6
1982*	10.5	8.4
1985	10.5	8.5
1987	10.9	8.8
1988	11.1	8.7
1989	11.9	8.9
1990	12.2	9.4
1991	13.2	9.9

*Provisional.

Sources: Health and Welfare Canada, *National Health Expenditures in Canada 1970-1982* (Ottawa, 1982); *ibid.* (Ottawa, 1984), p. 9, figure 3 (Cat. 84-2094); Policy, Planning, and Information Board.

TABLE 11.7 Percentage Distribution of Health Expenditures, Public and Private

	Hospitals	Other institutions	Physicians	Dentists	Other professional services	Drugs and appliances	All other health costs
1960	37.9	5.7	16.6	5.1	2.3	14.2	18.2
1965	42.7	6.1	16.0	4.7	1.9	13.3	15.3
1970	45.0	7.2	16.6	4.2	1.5	12.5	13.0
1975	44.4	9.7	15.7	4.9	1.1	10.7	13.5
1980	40.9	11.6	15.2	5.8	1.1	10.8	14.6
1981	41.2	11.4	15.0	5.6	1.2	11.0	14.6
1986	39.6	10.3	15.8	5.5	1.2	13.4	14.2
1987	39.2	10.3	16.0	5.4	1.4	13.9	13.8
1988	38.9	10.5	15.7	5.5	1.4		
1989	38.6	10.6	15.5	5.5	1.4		
1990	38.2	10.7	15.2	5.5	1.5		

Source: *Canada Year Book*, 1994.

deinstitutionalization have been introduced, the rate of growth of institutions has not declined proportionately. Drugs and other appliance costs declined from 1960 to 1982, but the growth in drugs alone was the largest of all areas in the period from 1985 to 1991. It is, however, still not at as high a rate as in 1960. As the population ages, the need for institutional care may increase again, particularly among the very elderly. Controlling health-care costs in the future will demand the creative development of a wide variety of home-care services that will allow the elderly and the sick to remain in their own homes for a longer period of time and at the same time not increase the burden on the household and family caregivers.

The cost of medical care has been, along with the provision of other social services, front and centre in political debates during the years since the implementation of medicare. It is fair to say that the issue is now critical – and has been dramatized by physician and nurse strikes, the closing of numerous hospitals through regionalization and debt, government caps on funding, the political focus on the level of deficit in the government financing, the free trade agreement, and a variety of other initiatives designed to restructure the provision of medical care. Using a model of analysis that focuses on the inherent contradictions in a capitalist and welfare state, Burke and Stevenson (1993) provide a critique of contemporary Canadian strate-

Box 11.7 The Canadian Medical Care System: A Qualified Success

Mhatra and Deber (1992) state that "the Canadian health care system can be considered a qualified success." Canadians are by and large satisfied with and even proud of their system; it provides universal and comprehensive coverage to good effect and at a cost comparable to those of other advanced industrial nations. Among the qualifications to success, Mhatra and Deber point to a growth in physician supply that is in excess of population growth; limited innovation in the hospital sector; possible overuse of certain drugs (especially in the elderly), some types of tests, and some surgical procedures; inadequate provision for quality control of drugs and for independent physician practice.

Today, claim Mhatra and Deber, there are two major challenges to Canada's health-care system: access to technology and increasingly broad understandings of the causes of health and illness. Today, for example, the federal government recognizes the importance of income, education, and social support in health and illness.

First, with respect to technology, there is evidence that certain high-technology procedures (e.g., magnetic resonance imaging) are more easily available elsewhere, particularly in the U.S., than in Canada. This claim is debatable, however, because there is some evidence both of overuse in the U.S. and of an appropriate level of use in Canada. Second, health today is seen in broader terms than equitable access to medical care. Many health determinants are outside the broad areas of the provision of medical care and include such important issues as education, housing, racial and gender equity, environmental protection, and traffic safety. Canada's ability to satisfactorily address these challenges to health equity depends on the Canadian economic trend, demographic predictions, federal-provincial disputes, and ideological disputes about medical services. Until and unless these concerns are dealt with, equitable access to health in Canada will continue to be elusive.

---

gies for controlling costs and redesigning health services. They argue that the political economy of health care must be studied in relation to the wider national political economy. Furthermore, they suggest, current policies suffer from several biases, including (1) "therapeutic nihilism" (a critique of the medical model that is used to cut back on the various essential medical services without provision for an alternative), (2) "healthism" (a model of health that attaches health not only to medical services but also to various other issues such as lifestyle), which can be used by the state to provide it the discursive space to claim that almost every expenditure has a health enhancing effect and thus "medical" care is less important, and (3) "the discourse of health promotion," which can easily be "captured by neoconservative forces to further an ideology favouring decentralization, flexibility, the intrusive state, individual responsibility, and the sanctity of the family." Such privatization, however, leads to inequality (*ibid.*: 70). All of these forces are in fact evident in the federal government's health emphases over the past decade and a half.

There are various proposals for cutting costs. The first set of proposals includes user

fees, extra-billing, coinsurance, and deduc-
tibles (suggested usually by the medical
profession and often with state support).
However, a number of researchers have
found that such strategies do not work to
control costs but rather serve to decrease
equality of access to care, deter use by the
poor, and redistribute the burden of paying
for health care from taxpayers to sick people
(Barer, Evans, and Stoddart, 1979; Stoddart
and Labelle, 1985). The second set of proposals concerns controls over the supply of
physicians and hospitals through limits on
immigration of foreign physicians, enrol-
ments in medical schools, residency positions, monitoring of physician billing,
hospital and bed rationalization and closings, decreasing budgets, and contracting
out of services to the private sector. The
problems associated with such provisions
are that they are largely *ad hoc* and will,
without explicit planning, inadvertently
result in the dismantling of universal
medical care. The third set of proposals
includes increase in alternative models
(homeopathy, nurse practitioners, chiropractic) of health care, the introduction of
alternate physician payment schemes such
as capitation (per patient annual payment
to a physician) and salary, health promotion education, and publicly financed
competition between types of health-care
practitioners.

Summary

(1) The chapter provides an overview of
the history of medicine in Canada. Native
peoples had medicine men and shamans
who used a wide variety of effective herbal
remedies. Early Canadian settlers relied on
a variety of treatments provided by midwives, local doctors, travelling salesmen,
and mail order dispensers of remedies.
During the nineteenth century, the government started to institute public health measures to deal with cholera and to license
allopaths, homeopaths, and others.

(2) The implementation of universal
medical insurance in Canada, which was
completed in 1972, was a long, uphill battle.

(3) There are several explanations for
the development of medicare in Canada.
Two of the most important are: movement
in Western industrialized societies toward
rationalized bureaucratic social organization and monopoly capitalism and the
spread in Western Europe and beyond of
social welfare legislation.

(4) The main goal of medicare in Canada
was to provide equal health care to all
Canadians, thus enhancing the general
health of the population. To some extent,
this goal has been reached; people in low-
income groups visit physicians more often.
Health, however, continues to vary by class.

(5) Medicare has also had an impact on
medical practice. Doctors' salaries increased
dramatically but have since levelled off. The
ratio of doctors to population in Canada
continues to grow. Working conditions, control over other occupations in the medical
field and over clients, self-regulation in
education, licensing, and discipline, and the
actual content of the work have all undergone significant alterations. As well, the
medical profession is now under increased

administrative, economic, political, and social constraints.

(6) Health-care costs, since the introduction of medicare, have increased as a portion of Canada's GNP, but medicare seems to have controlled potentially spiralling costs of medical care. The greatest growth in medical expenditures has been in institutions to 1982 and drugs to 1990. In spite of some deinstitutionalization policies, the growth in this sector is likely to continue as the population ages. Moreover, the elderly disproportionately consume pharmaceuticals. Again, as the older population expands, so, too, can drug costs be expected to grow. The cost of medical care is currently front and centre in a number of political debates regarding the provision of a universal social safety net.

CHAPTER TWELVE

The Medical Profession

"SOCIETY EXPECTS A formidable array of virtues and abilities in its doctors: technical competence, mastery of medical knowledge, sensitivity to the 'whole patient,' communicative ease and skill, wise judgement, compassion and professional integrity" (Gallagher and Searle, 1989).

Why is the occupation of the physician thought to be a profession? What is a profession? How do physicians fit into the whole medical care system? Are doctors losing or gaining power, prestige, and income? Does membership in a professional occupation guarantee a high moral calling and a dedication to service? How do doctors learn all that must be learned in medical school? How do doctors handle mistakes or conflicts? These are among the questions that interest medical sociologists. The purpose of this chapter is to analyse the work of the physician as a profession and to describe the process of medical education in the context of Canadian society, both today and in the past. The chapter will examine the division of labour and medical practice, the issues of social control and autonomy within the profession, ideas about physicians' networks and their cultural environment, their handling of mistakes, and medical education.

The Profession of Medicine

The idea that medicine is a profession is relatively new. Yet in the last half-century or so, the medical doctor has become the archetypal professional. Medicine is generally thought to be the model profession, not only by theorists of occupations and professions but by the lay public and by those other occupational groups who are aspiring to reach the "heights" of professional status.

There are three distinctly different ways to think about professions and professionalization. First, professions can be considered occupations that have certain specific characteristics or traits. Second, professions can be viewed as aspects of processes of occupational change over time. Third, it is possible to consider the notion of the profession as ideology.

Profession as Occupation

The view of the profession as a particular type of occupation is called the trait approach. William Goode (1956, 1960) thought that professions were special occupations that embodied two basic characteristics: (1) prolonged training in a body of specialized, abstract knowledge; and (2) a service orientation. These two chief characteristics were broken down further by Goode (1960). According to Goode, the characteristics of a profession include the following:

(1) It determines its own standards for education and training.

(2) There are stringent educational requirements.

(3) Practice often involves legal recognition through some form of licence.

(4) Licensing and admission standards, procedures, and profession are determined and managed by members of the profession.

(5) Most legislation with respect to the practice of the profession in question is shaped by the profession.

(6) The profession is characterized by relatively high power, prestige, and income.

(7) The professional practitioner is relatively free of lay control and evaluation.

(8) The norms of practice enforced by the profession are more stringent than legal controls.

(9) Members are more strongly identified and affiliated with their profession than members of other occupational groups.

(10) Members usually stay in the profession for life.

The characteristics of the medical profession in Canada today fit very closely those suggested by Goode. Physicians themselves determine what constitutes appropriate subject matter for the study and practical experience of physicians-in-training. Two recent reviews and revisions of medical education are important examples. The first is the *General Professional Education of the Physician and College Preparation for Medicine* report published by the Association of American Medical Colleges in 1984. The second is the recently completed *Educating Future Physicians for Ontario*, which was begun in 1980 by faculty at Ontario's five medical schools. Medical schools have some of the most stringent admittance criteria of any educational programs. In addition to grades, many medical schools consider the personal qualities of the applicants, such as their ethical views, their behaviour, self-presentation in an interview, ability to work with others as a "team" player, their leadership abilities (Clarke, 1980), and, increasingly, whole-person communication skills (Kendall and Reader, 1988). Licensing and admittance standards, too, are determined by the profession. Medicine is characterized by the relatively high power, prestige, and income of its professional practitioners. Physicians are largely controlled via various committees and organizations, such as the College of Physicians and Surgeons, which are composed primarily of practising doctors. Their own norms, incorporated in the Code of

Box 12.1 The Formal Code of Ethics of the Canadian Medical Association

---

PRINCIPLES OF ETHICAL BEHAVIOUR FOR ALL PHYSICIANS, INCLUDING THOSE WHO MAY NOT BE ENGAGED DIRECTLY IN CLINICAL PRACTICE

I Consider first the well-being of the patient.

II Honour your profession and its traditions.

III Recognize your limitations and the special skills of others in the prevention and treatment of disease.

IV Protect the patient's secrets.

V Teach and be taught.

VI Remember that integrity and professional ability should be your advertisement.

VII Be responsible in setting a value on your services.

Canadian Medical Association, *Code of Ethics*, 1989.

Ethics (see above), are more exacting although not necessarily strictly enforced. A doctor's identity as a person is tied to his or her occupation. This affects all other aspects of his/her life, including family life. (For a study of the impact of the occupation on family relationships, see Gerber, 1983.) Generally, a career in medicine is an ultimate and therefore lifetime career.

Goode's (1956, 1960) view of professions has a number of limitations. First, it is based on an acceptance of the viewpoint or ideology put forward by the professional group itself. The ideology of a particular occupation cannot help but be self-serving. Second, it ignores the fundamental importance of the power of the professional group. It is this power that enables it to maintain the prevailing ideology both internally and in society at large. Third, its analysis does not take into account the changes in the profession over time; it assumes that the peak of professionalism is portrayed by the traits of the profession as they are described by members of professions. Fourth, it ignores the relationship between the profession and the rest of society – the state, the economic and political structure, and the social organization of the society. In brief, the trait approach ignores the historical development of the profession and its place in society.

Profession as Process

The second view of professions is as a process by which various occupational groups may progress or hope and aspire to progress over time under various conditions (see Table 12.1). Within this perspective there are those who accept the ideological view promulgated by the profession; they see the traits such as those described by Goode (1960) as the end points of a pattern of forward growth. Wilensky's (1964) work is representative of

Box 12.2 Selections from "Responsibilities to the Patient"

AN ETHICAL PHYSICIAN:

STANDARD OF CARE

1. will practise the art and science of medicine to the best of his ability;
2. will continue his education to improve his standards of medical care;

RESPECT FOR PATIENT

3. will ensure that his conduct in the practice of his profession is above reproach, and that he will take neither physical, emotional, nor financial advantage of his patient;

PATIENT'S RIGHT

4. will recognize his limitations and, when indicated, recommend to the patient that additional options and services be obtained;
5. will recognize that the patient has the right to accept or reject any physician and any medical care recommended to him. The patient, having chosen his physician, has the right to request of that physician opinions from other physicians of the patient's choice;
6. will keep in confidence information derived from his patient, or from a colleague, regarding a patient and divulge it only with the permission of the patient except when the law requires him to do so;
7. when acting on behalf of a third party will assure himself that the patient understands the physician's legal responsibility to the third party before proceeding with the examination;
8. will recommend only those diagnostic procedures which he believes necessary to assist him in the care of the patient and therapy which he believes necessary for the well-being of the patient. He will recognize his responsibility in advising the patient of his findings and recommendations and will exchange such information with the patient as is necessary for the patient to make his decision;
9. will on the patient's request assist him by supplying the information required to enable the patient to receive any benefits to which the patient may be entitled;
10. will be considerate of the anxiety of the patient's next-of-kin and co-operate with him in the patient's interest.

Canadian Medical Association, *Code of Ethics*, 1989.

such an approach. The steps to becoming a professional in this perspective are as follows: first, the members of the occupation engage in full-time work; second, they establish a relationship with a training/education program; third, they establish an association; fourth, they gain legal status; and fifth, they construct a code of ethics. This approach brings an historical perspective to the concept of the profession. It also has the advantage of showing that occupations can be seen as falling along

TABLE 12.1 Profession as Process

NON-PROFESSION		PROFESSION
Very little, very simple concrete education or training not associated with a university.	EDUCATION	Complex, abstract, esoteric, lengthy education associated with university.
Anyone can call themselves a	LICENSING	Profession itself determines eligibility and continuously evaluates ongoing suitability to maintain professional status.
Each person for himself or herself and to his or her own benefit.	CODE OF ETHICS	Altruism, higher calling, collectivity orientation and identity.
Each person for himself or herself and to his or her own benefit.	ASSOCIATION	Organized group encompassing most people actively pursuing the profession.
State-based laws to control individual behaviour and discipline.	PEER CONTROL	Professional association determines standards of education, professional activity, and disciplines members in violation.

continua with respect to the degree of professionalism they incorporate. Underlying this model, as well as the trait approach, is an acceptance of the view that the profession is a more desirable and higher calling than other occupations.

Johnson's (1972, 1977, 1982) model of the professions bridges the process and the ideology approaches. Johnson describes professions with respect to process, too, but he focuses on the way that occupational groups become professions as they increase in power. He discusses the theoretical explanations for the power of professions. He sees power as the ability of the professional practitioners to define "reality" in an increasingly broad way, and even to define the "good life" for its clients.

From this viewpoint the fundamental characteristic of a profession is the ability of the group to impose its perspective and the necessity of its services upon its clients. Professional power arises out of the uncertainty in the relationship between the client and the professional; this uncertainty stems from the social distance between the two parties. Three crucial variables determine the degree of power held by a professional group: (1) the more esoteric the knowledge base of professional practice and the less

Box 12.3 Specialization

Another characteristic of the contemporary medical care system is extensive specialization. Recently, the *Globe and Mail* advertised vacancies in all the following position, many of which did not exist only a few decades ago. These were all advertised under the heading *HOSPITAL, MEDICAL AND SOCIAL SERVICES*.

Psychologist

Ultrasound Technologist

Pharmacist

Staff Occupational Therapist

Cardio-Pulmonary Perfusionist

Speech Pathologist

Manager of Nursing Practice

Executive Assistant (to the President)

Independent Community Living Worker

Administrative Coordinator, Regional Geriatric
   Program

Histology Manager

Clinical Nurse Specialist

Information Systems Coordinator

Research Assistant

Registered Technologist

Health Science Librarian

Physiotherapist

Director of Plant and Engineering Maintenance

Nurse Manager

Supervisor, Accounts Payable

Psychiatrist

Executive Assistant

Medical Officer

Coordinator, Community Mental Health Program

Director of Pediatrics and Obstetric Nursing

Kinesiologist

Nuclear Medicine Technician

Social Worker

Manager, Nursing Practice

Chaplain

Ultrasonographer

Director, Sexual Assault Care Centre

Audiologist

---

accessible to the lay public, the greater the power of the profession; (2) the greater the social distance between the client and the professional when the professional is of a higher income and social class than the client, the greater the power of the profession; and (3) the greater the homogeneity of the professional group in contrast with the heterogeneity of the client group, the greater the power of the profession.

Esoteric knowledge is important because the less accessible the knowledge to a wide spectrum of the population, the greater the power of the professional group compared to the potential client group. Social distance refers to the relative prestige of the occupation within the labour force, the income of the practitioners, and even the relative class background from which the practitioners are typically drawn as compared to average societal members. The more heterogeneous the consumer group, the more likely they are to be disorganized and unable to work together to protect their interests. On the other hand, the more homogeneous the professional group, the more they are able

Box 12.4 Norman Bethune

Norman Bethune, born in 1890 in a small Ontario town called Gravenhurst, became a prominent Canadian thoracic surgeon who served in three wars and distinguished himself as a surgeon on three continents.

Bethune was the offspring of a family with high standards and widely recognized achievement. His great-great-grandfather, Reverend John Bethune (1751-1815), built Canada's first Presbyterian church in Montreal. One of his sons, John Bethune (1791-1872), became Rector of Montreal and Principal of McGill University. Two other sons, Alexander and Angus, also achieved prominence in their chosen professions: Alexander became the second Anglican Bishop of Toronto; Angus became a successful businessman, helping head the two great Canadian fur-trading companies in Canada, the North West Company and the Hudson's Bay Company. Angus's son, Norman (Bethune's grandfather), was educated in Toronto and learned surgery in Edinburgh before returning to Toronto to become one of the co-founders of Trinity Medical School in 1850. Norman himself was one of three children of Elizabeth Ann Goodwin and Malcolm Bethune.

Between 1893 to 1907 the family moved frequently. Bethune worked in a number of jobs, as a lumberjack and as a teacher, teaching immigrants how to read. From 1909 to 1911 he was registered in pre-medical studies at the University of Toronto. Just before his third year of medicine, Bethune joined the Royal Canadian Army Medical Corps. After he was wounded at Ypres he returned to Canada to complete his medical degree by 1916. In 1923 he married Francis Penney Campbell, with whom he had stormy and ambivalent relationship. They were divorced and later married again. In 1927-28 Bethune succumbed to tuberculosis and was treated at Trudeau Sanatorium in New York. As a result of his own experience of illness, Bethune developed an interest in thoracic surgery and went into practice with Dr. Edward Archibald, a renowned thoracic surgeon at the Royal Victoria Hospital in Montreal. Here Bethune engaged in research on tuberculosis. During this time he invented a variety of surgical instruments that were widely used in North America.

Bethune is, however, best known for his political activities. Sixty years ago, in 1936, Bethune tried to establish a state-supported medicare system for Canadians. He organized a group of doctors, nurses, and social workers, called the Montreal Group for the Security of People's Health. Outraged at the refusal of the Canadian Medical Association to support plans for socialized medicine, he sailed for Spain at the outbreak of civil war in October, 1936, to aid in the struggle against the forces of Mussolini and Hitler. In Spain he devised a method to transport blood to the wounded on the battlefields.

As a professed Communist, it was almost impossible for Bethune to remain in Canada. When war broke out in China, Bethune decided to help in the fight for communism. Medical treatment was needed. With funds donated by the China Aid Council and also from the League for Peace and Democracy, Bethune purchased much-needed medical supplies. He arrived in Yannarv in 1938 and quickly moved to the front lines in the mountains of East Shanxi. Travelling with the Eighth Route Army, he devoted his remaining years to fighting fascism and saving the lives of the soldiers who stood for the cause he, too, believed in. His fame reached China's political leaders. Summoned to meet Chairman Mao Tse-Tung, Bethune convinced the leader of the need for surgical knowledge in the front lines of the Chin-Cha'a-Border Region.

Bethune founded a medical school and model hospital that would later be renamed the Norman Bethune International Peace Hospital. He had placed his mark in history's pages as a great surgeon and humanitarian.

Bethune died in China of septicaemia on November 12, 1939.

Sidney Gordon and Ted Allan, *The Scalpel, The Sword* (Toronto: McClelland and Stewart; New York: Monthly Review Press, 1952).
D.A. Shephard, ed., *Norman Bethune – his times and his legacy* (Ottawa: Canadian Public Health Association, 1982).
Roderick Stewart, *The Mind of Norman Bethune* (Westport, Conn.: Lawrence Hill, 1977).

to organize to attain their ends. The power differential is greatest when a heterogeneous client group must deal with a relatively homogeneous professional group.

Johnson's model overlaps part of the process model in that he articulates some of the bases upon which professional power varies. His work is also related to the trait approach to professions in that the power of the profession is in its ability to persuade others of its viewpoint (or ideology).

Subordination, Limitation, and Exclusion in the Medical Labour Force

The medical profession successfully consolidated its position as the provider of medical care by restricting the scope of the work of other types of practitioners. According to Willis (1983), whose research was carried out in Australia, there are three distinct processes by which the allopathic medical profession achieved the level of dominance it has come to enjoy in the twentieth century. The first is subordination. This refers to the process whereby potentially or actually competing professions come to work under the direct control of the doctors (1983). Nursing is an example of this process, because the practice of nursing has been severely restricted and constrained so that nurses' only legal position today is subordinate to the medical profession (see Chapter Fourteen).

The second process is called limitation and is illustrated by such occupations as dentistry, optometry, and pharmacy. Such occupations are not directly under the control of the allopathic practitioners but are indirectly controlled through legal restrictions. The process by which pharmacy came to be subordinate to medicine is a case in point. In the nineteenth century, pharmacists operated as primary caregivers. They prescribed medicines to patients who came to them for help, particularly to the poor and doctorless. In many ways the roles of the pharmacists and the doctors overlapped at this time because most doctors made and dispensed their own drugs. When the pharmacists became self-regulating in 1870-71, they agreed on an informal compromise with the doctors. Pharmacists gave up prescribing and doctors gave up dispensing drugs. Today the profession of pharmacy is attempting to expand its care into the community, especially through educating and communicating with the public. One of its particular mandates is the reduction in drug-related problems (over-medication, drug overdose, drug interactions, and side effects) among the elderly (Battershell, 1994; Muzzin *et al.*, 1993).

The third process, exclusion, is the process whereby certain occupations that are not licensed and therefore are denied official legitimacy come to be considered "alternative" medicines. The Australian example given by Willis (1983) is that of chiropractors. However, chiropractors in Canada have an ambiguous position today (Coburn and Biggs, 1987). (See Chapter Fourteen for a discussion of this issue.) Perhaps a better Canadian example would be that of naturopaths. Because naturopathic medicine is based on a different model of science than traditional medicine

Box 12.5 Detachment, Efficiency, Authority: A Doctor's Perspective

The following quotation, from an excellent book by a physician (Hilfiker, 1985), exemplifies the personal and reflective nature of his autobiographical account of the life of the modern doctor. In this brief section he offers a bit of an explanation for some of the impersonality that is so often, and usually negatively, experienced by patients.

Nevertheless, physicians do deal with these stresses. We have found ways to continue practising in the face of this pressure. How do we respond? The language we use hints at the answer. We begin by learning to detach ourselves from the chaos of the situation. Mr. Smith with all his fears and insecurities becomes "the stroke in room 8," thus allowing us to concentrate on his physical disability; Mrs. Jones, who regularly disrupts the office with her incessant demands, becomes "a problem patient" whom we manage with behavior modification. We next learn the principles of efficiency and productivity. "Patient management" is the buzz word, and "productivity incentives" are standard practice at many large clinics to encourage greater efficiency in office and hospital. Then the physician discovers the protective coat of prestige and authority. When I told Mrs. Murphy that we had done everything we could to treat her husband's cardiac arrest, I was using the power of my position to close off questions, to protect myself from her implied criticisms. Any continued questioning would have been a direct challenge to me, a step only the most assertive sort of person might take. In addition, we physicians protect ourselves emotionally by keeping ourselves at the top of the medical hierarchy. How often is the physician "Dr. Hilfiker" while the nurse is "Maryanne"? Finally – though no language hints at this – there is our wealth to comfort us. If we must suffer outrageous stress, at least, so the unspoken theory goes, we will be well compensated financially.

These responses – clinical detachment, efficiency and productivity, prestige and authority, hierarchy, and wealth – are not intrinsic to the practice of medicine. Although the structure of modern medical practice may encourage them, we physicians also choose to endorse and accept them, in part as a way to relieve the inordinate pressure of our work. They seem to offer a kind of escape. Unfortunately, the escape is only illusory.

and is not yet the subject of sufficient (traditional scientific) research confirming its efficiency, it is not considered a mainstream medical theory nor are its practitioners funded under medicare.

There was a significant change in the legal status of complementary and alternative health-care practice in Ontario in December, 1993. On this date, the Regulated Health Practitioners Act came into effect. This landmark legislation restricts allopathic monopoly by delicensing all health-care professions. Consumers can now choose which form of health care they desire. Alternative practitioners are allowed to advertise. By this Act medicine is limited to clearly defined or "controlled" activities (such as surgery or prescription drugs). Twenty-one health professions are regulated under the Act, including audiology, speech pathology, chiropody, chiropractic, dental hygienics, dental technology, dentistry, denturism, dietetics, massage therapy, medical laboratory technology, medical radiation technology, medicine, midwifery, nursing, occupational therapy, opticians, optometry, pharmacy, physiotherapy, psychology, and respiratory technology. All others, such as homeopaths, acupuncturists, herbalists, and reflexologists, are unregulated. Both regulated and unregulated professions have the right to practise. Regulated professions have their own administrative colleges with independent self-governing powers and are responsible to the Ontario health ministry. Regulated practitioners may do only what their colleges permit. Unregulated practitioners can do anything, as long as it is legal.

Profession as Ideology

An ideology is a group of descriptive and prescriptive beliefs that explain and legitimize certain practices and viewpoints. The notion that professionalism is an ideological concept was raised first in the writings of Everett C. Hughes (1971). He criticized the trait approach, which was the standard of the time, because it was based on the assumption that what the medical profession said about itself – its ideal image – was an accurate representation of the profession. Hughes noted that this was a biased representation because physicians, as well as offering a valuable service to others, were rewarded generously both by their status and by their income.

Parsons (1951) adopted the prevailing ideology of professionals when he argued that a profession has the following characteristics:

(1) universalism, (2) functional specificity, (3) affective neutrality, and (4) a collectivity orientation. Each of these will be explained. A physician is expected to apply universalistic, scientifically based standards to all patients. Gender, education, income level, and race of the patients, for instance, are not expected to affect the way in which the physician deals with a patient. All patients are to be provided with the same level of care, for the same level of disease, according to universally applied standards. The work of the physician is also functionally specific to the malfunctioning of the human body and mind. Functional specificity requires that the physician not offer advice to a patient on non-medical matters such as real estate or choice of a career but restrict advice to the precise area of health and disease. The doctor is expected to refrain from emotional involvement, to be, in other words, affectively neutral. When medical issues are involved (or are believed to be involved), the physician has extraordinary access to intimate knowledge of the social and emotional life, as well as the body, of the patient. The physician is permitted special access to the private sphere of the life of the patient on the assumption that his or her duties will be performed in an objective and emotionally detached manner. The doctor is expected, on the one hand, to express concern and exhibit a "good bedside manner," while, because of his or her special knowledge of the body and life of the patient, he or she is expected to be impartial and unemotional in judgement and treatment. Finally, the physician is expected to exhibit a collectivity orientation,

Box 12.6 The Longest Doctors' Strike: Ontario Doctors Walk Out for 25 Days

June 11, 1986, was a red-letter day for about 11,000 of the approximately 17,000 doctors practising in the province of Ontario. It was the day that they began what became a 25-day strike against the Liberal government of David Peterson and the people of Ontario. They were protesting the impending passage in the Ontario legislature of Bill 94 – the Health Care Accessibility Act. The 250-member committee of the Ontario Medical Association, after approximately five months of intense lobbying in a losing battle with the Ontario government, had voted to strike. All but 6,000 of the doctors had apparently agreed and had joined the strike, closing offices, hospital wards, and numerous medical services. Most of the province's approximately 9.2 million residents were left with only emergency care.

The minority Liberal government, in a coalition with the New Democratic Party, supported the Health Care Accessibility Act. The federal government's Canada Health Act of 1984 had stipulated that for every $100 that the provinces allowed their physicians to extra-bill patients, an equal amount would be withdrawn from the transfer payments made by Ottawa to the provincial government under the stipulations of medicare policy. If extra-billing was not stopped by April, 1987, the provincial government stood to lose $100 million in federal health-care payments. Under these conditions the provincial government held firm.

The doctors opposed the ban on extra-billing not because of the potential financial losses but because they feared the encroachment of government into the way they practised medicine. In their view, banning extra-billing was the first step by government to restrict and control the work of physicians. The doctors said that they did not want to be civil servants; they wanted to reserve the right to practise where, with whom, and under whatever conditions they themselves, as a self-governing profession, saw fit. They argued that the doctor-patient relationship was sacred and that no outside body had any right to interfere in or affect the nature of this special relationship. They maintained that they were fighting for a principle and not for money.

But the doctors were forced back to work. The strike was not a success for them. They lost legislatively and they lost in more subtle ways. They lost a good deal of public respect and admiration. Nine days after the strike officially began, the Health Care Accessibility Act was passed (June 20, 1986). It stipulated that extra-billing was illegal, and that doctors who persisted could be jailed or fined up to $10,000. Some doctors retaliated, not by extra-billing, but by billing for services that they had previously provided without cost, such as prescribing via the telephone, transferring clinical records, and charging administrative fees to patients who wanted to be considered on their caseload. The OMA challenged the legality of Bill 94, claiming that it violated their rights under the Canadian Charter of Rights and Freedoms.

in other words, to provide service to others out of a sense of calling to the profession and an altruistic desire to serve others. Self-interest, whether expressed in terms of financial rewards or other aspects of working conditions, is not appropriate within the ideology of medical practice.

Freidson views professionalism as ideology, as is suggested by the title of his book, *The Profession of Medicine* (1975). The profession of medicine refers both to the occupational group, which is composed of medical practitioners, and to the statements made by doctors as they "profess" their work. In Freidson's model, medicine's claim to professional status rests on three assertions made by physicians: first, medical knowledge is complex, detailed, and difficult; second, medical work is based on the findings of objective science; and third, as professionals, doctors can be trusted to put the welfare of the public ahead of their own welfare.

This view is ideological – and it has served the medical profession well. Through the perpetuation of such a view, the public has been led to believe that doctors are "morally superior" individuals who deserve to be trusted, to dominate the practice of medicine, to hold a monopoly in the construction, maintenance, and spread of what is taken to be official "medical knowledge," to control standards for their training, and to discipline their errant members themselves.

This ideological position is widely challenged today in the wake of increasing costs without equal and parallel improvements in life expectancy and health quality, mounting evidence of doctors' mistakes, and the rising rates of malpractice suits. Moreover, in the wake of the debt crisis the future growth of the heavily professionalized and increasingly specialized occupation of medicine is likely to be curtailed. During the late eighties and early nineties hospitals were closed across the country and home health care expanded. The federal government appears to be increasingly supportive of what has come to be called health promotion and disease prevention (see Chapter Ten). Alternative health-care providers such as midwives have become legitimized with their own college in Ontario. Moreover, midwives have recently been granted a degree of professional autonomy. Challenges to the exclusive rights of allopathic doctors to medical care are evident in a variety of ways. The first blow to the Canadian public belief in the altruism of the medical profession arose at the time of the doctors' strike in Saskatchewan. A more recent and equally dramatic demonstration was the strike over extra-billing by many Ontario doctors in 1986.

A Brief History of Medical Education in North America

After 1800 a growing number of proprietary medical schools (profit-making institutions that were generally owned by doctors who also served as teachers) began to open in North America. The quality of the education they offered was poor. They were usually ill-equipped for teaching, but then medical theory was still very simple (Wertz and Wertz, 1986: 137). Bleeding, the application of leeches, and the ingestion of purgatives to cause vomiting were still the treatments of choice. At this time people were as likely to be harmed as helped by the prevalent and accepted methods of cure.

The most advanced medical training of the time was in Europe, especially France and Germany, where medical research was well supported. Louis Pasteur's germ theory, proposed in the mid-nineteenth century, had a remarkable effect on medicine and provided the basis for the discovery, classification, and treatment of numerous diseases. By the end of the century German physicians had made significant discoveries in the world of medical science, too. Rudolf Virchow described a general model of disease development based on cellular pathology (1858), and Robert Koch discovered the bacilli for anthrax (1876), tuberculosis (1882), and cholera (1883). Because of these exciting developments in European medical science, American doctors began to cross the ocean for their education and training. As many as 15,000 American doctors studied in German-speaking universities from approximately 1870 to 1914.

European training soon became a symbol of status among Americans. Those with such training were able to establish more prestigious and specialized practices than American-trained doctors. By the beginning of the twentieth century, medical research was well funded by the Carnegie and Rockefeller families. The United States began to move ahead as a leader in the development of scientific and medical resources.

Medical education, too, was radically revised and upgraded at the beginning of the twentieth century. The Flexner report, published in 1910, reviewed medical education in the U.S. and Canada. Sponsored by the Carnegie Foundation, Abraham Flexner visited every medical school in the two countries. Only three American medical schools, Harvard, Johns Hopkins, and Western Reserve, were given approval.

The others were severely criticized and were characterized by Flexner as "plague spots, utterly wretched." Flexner's observations about Canadian medical schools were very similar to those regarding the American schools. As he said:

> In the matter of medical schools, Canada reproduces the United States on a greatly reduced scale. Western University (London) is as bad as anything to be found on this side of the line; Laval and Halifax Medical Colleges are feeble; Winnipeg and Kingston represent a distinct effort toward higher ideals; McGill and Toronto are excellent. (Flexner, 1910: 325)

According to the report, 90 per cent of the doctors practising then lacked a college education, and the vast majority had attended inadequate medical schools (Wertz and Wertz, 1986: 138). Flexner recommended that medical schools consist of full-time, highly educated faculty and be affiliated with a university. He suggested that laboratory and hospital facilities be associated with universities. He recommended the establishment of admission standards. Medical education, he argued, should be graduate school level. However, the effects of the Flexner report were not as devastating to Canadian schools as to the American schools because Canadian medical schools had already been limited and controlled in their growth. In addition, there were far fewer proprietary schools in Canada. Proprietary schools were gradually closed, and medical education was slowly upgraded. Funding was donated by the large American foundations to Canadian medicine, and its status as a scientifically based practice became well established.

By the middle of the 1920s the medical profession in America had clearly established itself as a leading profession. Its standards for training had been improved and were considered excellent. Medical research had reached great heights. The power of the doctor as a healer and as a scientist was widely assumed.

Medical Education in Canada Today

The first Canadian medical school, established in 1824 with 25 students, was at the Montreal Medical Institution, which became the faculty of medicine of McGill University in 1829 (Hamowy, 1984). By the turn of the century six other university medical schools had opened their doors – Toronto, Laval (Montreal and Quebec City campuses), Queen's, Dalhousie, Western, and Manitoba. Today, approximately 1,700-1,800 students graduate annually in Canada (*Canadian Encyclopedia*, 1988: 1323). As indicated in Chapter Eleven, the number of doctors is growing faster than the population as a whole (see Tables 11.3, 11.4, 11.5).

The process of becoming a doctor requires three different steps: an undergraduate education in science and/or arts, graduate study leading to the M.D. degree, and a minimum one-year internship in which the graduate doctor works in a hospital or clinic under the supervision of practising doctors. During the internship year students write qualifying exams through the Medical Council of Canada, after which a licence to practise medicine is issued by any one of the provincial medical licensing bodies. Many doctors continue after the internship to specialize in family practice or any of a number of the 42 other different specialties (24 clinical, 6 laboratory, 12 surgical) recognized by the Royal College of Physicians and Surgeons of Canada. Today there are seventeen Canadian universities that grant an M.D. degree.

Medical students have been and continue to be drawn from middle- and upper middle-class family backgrounds (Kirk, 1994). But the rapid increase in the proportion of women enrolled in and graduating from medical school reflects a broadening of the formerly more exclusionary entrance requirements for medical school (see Table 12.2). However, financial concerns are still among the top three reasons mentioned by students who considered and then rejected medical school education (Colquitt and Killian, 1991).

In fact, as Table 12.3 shows, this class-based association has been heightened over the last ten to fifteen years. In 1983-84, 60.6 per cent of the students enrolled in medical schools had fathers who had post-secondary education. In 1968-69, 55.4 per cent of the students enrolled had fathers who had post-secondary education.

The Process of Becoming a Doctor

The most influential and complete sociological studies of medical education were done in the 1950s and 1960s. The two most

TABLE 12.2 Trends in Female Enrolment and Graduates in Canadian Medical Schools, 1957-58 to 1989-90

Year	Enrolment 1st year (%)	Total (%)	Graduates* (%)
1957-58	8.5	7.0	5.3
1958-59	9.2	8.0	5.6
1959-60	9.0	8.6	7.6
1960-61	10.3	9.4	7.9
1961-62	11.5	10.2	10.1
1962-63	11.3	10.1	8.0
1963-64	13.4	11.3	10.2
1964-65	10.9	11.1	9.5
1965-66	12.7	11.4	11.3
1966-67	12.8	12.1	11.3
1967-68	14.5	12.5	10.9
1968-69	17.6	14.3	13.3
1969-70	18.7	15.7	12.0
1970-71	20.2	17.8	13.5
1971-72	22.4	19.9	17.3
1972-73	25.2	22.0	17.1
1973-74	27.4	24.1	20.0
1974-75	29.0	26.1	22.3
1975-76	33.0	28.3	24.7
1976-77	32.7	30.3	27.1
1977-78	33.0	31.6	29.4
1978-79	36.4	33.3	30.9
1979-80	36.8	34.5	32.3
1980-81	40.0	36.2	33.4
1981-82	39.2	37.5	35.6
1982-83	43.0	39.2	36.6
1983-84	43.5	40.8	36.9
1984-85	42.0	41.8	40.4
1985-86	43.9	42.6	41.6
1986-87	43.3	42.8	42.0
1987-88	46.7	43.7	40.5
1988-89	43.4	44.4	44.2
1989-90	43.1	44.0	44.0

*"Graduates" refers to calendar year (summer and fall of year) graduations. For example, year 1960-61 graduates includes the graduates of the Spring/Summer of 1961 with the Fall graduates of 1961.

Source: B. Blishen, *Doctors in Canada* (Toronto: University of Toronto Press, 1991), p. 84, from *Canadian Medical Education Statistics 1990*, vol. 12 (Ottawa: Association of Canadian Medical Colleges, 1990).

notable were the studies done at the University of Chicago by Howard S. Becker and others, *Boys in White: Student Culture in Medical School* (1961), and at Columbia University by Robert K. Merton and his colleagues, *The Student Physician: Introductory Studies in the Sociology of Medical Education* (1957). Each in its own way contributed to our knowledge of the socialization of the physician. Each explained some of the processes whereby young men (and some young women) pass through one of the most rigorous, busy, lengthy, and pressured educational systems and come to adopt the values, norms, skills, and knowledge expected of a medical doctor.

Becker *et al.* observed and spent time with medical students during their years of medical education at the University of Kansas Medical School. These researchers noted that the major consequences of medical school were that physicians-in-training became aware of the importance of two dominant values: "clinical experience" and "medical responsibility." These dominant values then guided and directed the strategies used by the medical students to manage the potentially infinite workload involved in learning all that had to be known before graduation. "Clinical experience" essentially refers to the belief that much of medical practice is actually based on the "art" of determining, from complex and subtle interpersonal cues and in interaction with the patient, the nature of the disease and the appropriate treatment. More important than either abstract knowledge based on medical school lectures or book or general scientific knowledge,

TABLE 12.3 Enrolment by Father's Education and Program, 1968-69 and 1983-84

| | 1968-69 | | 1983-84 | |
	Total Enrolment	% Post-Sec Ed.*	Total Enrolment	% Post-Sec Ed.*
Architecture	1,997	67.5	2,271	71.0
Pharmacy	2,004	44.4	2,871	63.9
Medicine	5,045	55.4	7,519	60.6
Law	4,096	52.1	7,934	54.6
Dentistry	1,574	47.5	1,858	53.2
Arts & Science	112,207	47.5	161,909	50.2
Engineering	18,046	43.9	34,875	49.1
Agriculture	4,024	41.1	5,033	43.5
Commerce	16,315	40.9	48,123	43.1
Education	21,257	36.8	34,748	43.0
Nursing	3,721	37.6	7,156	41.9
Other	13,836	43.5	22,466	49.0

*Per cent of students enrolled whose father had post-secondary education.

Source: "Accessibility to higher education – new trend data," *CAUT Bulletin*, June, 1988, p. 15.

"clinical experience" was considered a fundamental aspect of good doctoring.

Along with "clinical experience," doctors-in-training were impressed by the notion of "medical responsibility." By this Becker and his colleagues stressed the "enormous" moral responsibility of the life-and-death decisions that frequently confront doctors in practice.

These two values were the most important aspects of learning. They served to enable the medical students to choose, from the masses of detail presented in books and in lectures, what actually had to be learned. The notion of clinical experience guided the selection of facts that had to be memorized

to pass the examination. It tended to lead the students to disregard basic science and focus on classes where they were provided with practical information of the sort that was typically not found in medical textbooks. Furthermore, the focus on medical responsibility tended to result in an emphasis by the student on interesting cases that involved life-and-death medical judgements rather than on the more common and mundane diseases. It also tended to guide the choice of specialties. The most desired specialties frequently were those that had the potential of saving patients (or killing them), such as surgery and internal medicine.

Merton *et al.* described medical education and socialization as a continuous process by which medical students learned to think of themselves as doctors and in so doing absorbed sufficient knowledge to feel comfortable in their new role. Accordingly, two basic traits were developed: the ability to remain emotionally detached from the patient in the face of life-and-death emergencies, great sorrow or joy, and sexuality; and the ability to deal with the inevitable and constant uncertainty. Fox (1957) observed three sources of uncertainty. The first occurred because it was impossible to learn everything there was to learn about medicine and its practice. The second stemmed from the awareness that, even if the student was able to learn all the available medical knowledge, there would still be gaps because medical knowledge was itself incomplete. The third and final source of uncertainty arose from the first two: the uncertainty resulting from distinguishing lack of knowledge on the part of the student and inadequacy in the store of medical knowledge.

Over time, disease and death ceased to be frightening, emotional issues and came to be seen as medical problems. The researcher elucidated the processes by which the students learned, in the face of enormous amounts of information, what they would be quizzed about and thus what they had to know to pass their courses and to graduate. *Boys in White* also described the conversion of the early idealism of the new student to the cynical stance held by students in later years. When graduation approached, the original commitment to helping patients and practising good medicine tended to return.

Getting Doctored

Shapiro (1978), in one of a number of autobiographical accounts of medical school, has critically evaluated his own experiences of becoming a physician in a book entitled *Getting Doctored*. In this analysis the two most important features of medical education are the concepts of alienation and the authoritarian personality. Alienation is evident in the relationships of medical students to one another, in the relationships of doctors, interns, and residents toward one another, and in the approach of the medical student to studies, medical school, and pharmaceutical companies and other related institutions. The authoritarian personality, which breeds alienation, is experienced at most stages of medical education and later in medical practice. Karl Marx described four types of alienation – the alienation of labour or productive activity, alienation from the product of labour, alienation of people's relationships with others around them, and alienation from life.

Alienation of labour is seen throughout the medical care system, from the workers at the top down to those at the bottom. Shapiro provides numerous examples of alienated labour in medical school. These include: fierce competition between medical students, an enormous workload, and little direction regarding what out of the mountains of information is most relevant. "Students, especially in the early stages, spend extraordinarily long hours at study without any assurance that they have mastered what is necessary" (Shapiro, 1978: 29). Shapiro uses the clever phrase "doc around the clock" to encapsulate the doctor's experience of being tied to medical work.

One tool for survival used by medical students is mnemonics to aid memorization, e.g.,

"Lovely French Tart Sitting Naked in Anticipation" (the first letters of each word in this phrase represent, in ascending order, the seven major branches of the external carotid artery).
"Thick-Thighed Ladies Live in Place Ville-Marie." The eight essential amino acids. (*Ibid.*: 72)

Aside from the obvious sexism in these two mnemonics, they illustrate the alienation of the student from meaningful work because they do not aid comprehension or enable the student to use this information. The major value is that they aid in the ability of the student to regurgitate the information for the examinations.

Alienation from the product of labour exists when the worker does not feel that the product of his or her labour is a true reflection of his or her self and values. The way that doctors learn to talk about patients and their illnesses provides some illustrations of their lack of identification with patients.

Crock: This is a term applied to persons whose complaints are not "organic" and to some who have chronic diseases about which nothing or very little can be done.

Piss poor protoplasm: This term is used to describe patients who have disease or failure of several organs or organ systems.
This patient has two neurons: Neurons are nerve cells, and this insulting remark is a house officer's [staff doctor's] assessment of a patient's intelligence. It is often applied to a patient who is unco-operative in diagnosis or therapy. (*Ibid.*: 168)

Alienation of relationships, too, is repeatedly evident in Shapiro's description of medical school. One set of interpersonal strategies will be used to illustrate this. Competition is a central feature of the interpersonal interaction of physicians-in-training. There is widespread fear of failure. Students compete for rank in class. Students vie for the best internships and for the position of chief resident. Shapiro describes the games students play with one another to avoid authentic human relationships.

I know more than you do.
I know something you don't know.
I can study longer than you can. (*Ibid.*: 40-41)

Shapiro's personal critique of medicine may be idiosyncratic but it is not unique to him.

Organization of the Medical Profession: Autonomy and Social Control

Physicians are self-regulating. This means that through their organizations they decide what constitutes good medical practice, determine the requirements for training a physician, set standards of practice, and discipline colleagues who depart from these standards. The way that practice is organized, its locus of activity, and the payment modality all affect the regulatory power.

The profession in Canada has established two major control bodies, the College of Physicians and Surgeons and the Canadian Medical Association. Both these organizations are recognized by the federal and provincial governments. Both attempt to define expectations of medical behaviour, to improve standards of performance, and to protect the status and economic security of their members.

The Canadian Medical Association is an amalgamation of the provincial medical associations. Physicians with membership at the provincial level also have membership at the national level. The CMA represents physicians as a national lobby group. The provincial bodies negotiate with the respective medical care plans for fee scales and other matters of relevance to the practice of medicine. The situation is somewhat different in Quebec. Here two organizations – the Federation of General Practitioners and the Federation of Medical Specialists of Quebec – created to defend the "social, moral and financial" interests of their members, negotiate collective agreements with the provincial government (Soderstrom, 1978: 93). The College of Physicians and Surgeons has bodies in each province whose responsibility it is to oversee the practice of medicine in the

province in the interests of protecting the public. The Medical Act describes the powers and responsibilities of the College. These include such things as the designation of the qualifications required for medical practice, the certification of specialists, and the investigation of professional incompetence or misconduct (ibid.).

The Medical Act requires that the Medical Council of Canada be responsible for the licensing of qualified medical practitioners, for supervising what these practitioners do, and for preventing unqualified practitioners from practising. Thus the provincial colleges have the power to eliminate from the register all those who are convicted of certain offences. And while practitioners may appeal the judgements of the Medical Council to the appeal courts, the judgement of the Council is almost invariably upheld. In this way the courts have in fact declared that the profession is the proper judge of the actions of its colleagues (Blishen, 1969: 81).

Autonomy and power are crucial characteristics of an occupational group that claims to be a profession. As Freidson (1975: 71) says, "The most strategic distinction [between a profession and any other occupation] lies in legitimate organized autonomy – that a profession is distinct from other occupations in that it has been given the right to control its own work." Doctors themselves determine the terms (1) of admittance to medical school, and (2) of licence to practise once medical school, internship, and residency are completed. As professionals they hold sacred their right to control their own work without outside interference.

Freidson (1975) described how the self-regulating doctors' institutions depend on the type of practice in which the doctors are engaged. Freidson distinguished among doctors practising alone, colleague networks, large group practices, and university clinics. Those who practice alone enjoy the greatest freedom from outside control. Pure forms of solo practice are generally quite rare. Solo practice is, however, still lauded as the ideal model of the entrepreneurial professional who is free from control: the doctor treats the patient privately and confidentially, without the interference of colleagues, insurance companies, governments, or any other outsiders. This patient-doctor relationship is described as a "sacred trust." Assurance of a competence by the physician in private practice is based primarily on the assumption of adequate recruitment policies, educational programs, and licensing procedures. The ongoing day-to-day control of practice ultimately rests with the individual practitioner.

Various kinds of colleague networks, including the very common network of independent practitioners who share "on-call" times, are more tightly controlled and are potentially more vulnerable to continual scrutiny of colleagues. Those who work in group practices and clinics are probably most subject to collegial surveillance, as well as observation by administrators and paramedical practitioners. But even in this situation, most of the day-to-day practice of medicine in the office is primarily under the control of the practitioner and his or her colleagues.

One of the primary sanctions against behaviour deemed inappropriate by professional colleagues is ostracism by not referring patients or by denying certain privileges, such as hospital privileges. This is only useful to the extent that the particular medical practitioner is dependent on colleagues and hospitals for practice. Usually, specialists are colleague-dependent to a greater extent than are general practitioners because access to these specialists by patients can only occur through the medical referral system. Ostracism is only of limited effectiveness in actually changing the inappropriate behaviour because the more the practitioner is isolated, the less is his/her behaviour under surveillance (Blishen, 1969); as Freidson argues, "observability of performance is a structural prerequisite for regulation" (Freidson, 1975: 157).

Hall's (1948) early studies of the medical profession showed how important the network of personal contacts was to professional control. For instance, the prestige of the hospital at which the internship is carried out significantly affects the future practice and the subsequent level of income and prestige of the physician. The first appointment was particularly symbolic because it was "a distinctive badge" and "one of the most enduring criteria in the evaluation of his status" (Hall, 1948: 330). Hall noted also how the major hospitals were organized in a hierarchy of status: intern, resident, staff member, and associate staff member. Each is a step up in prestige and power. There were also differences in status between hospitals due to, among other things, such social considerations as the class and ethnicity of the patients.

Box 12.7 Sir Frederick G. Banting and Charles H. Best

Banting and Best are known for their discovery of how to extract the hormone insulin from the pancreas. This made possible the treatment of diabetes milletus, a disease in which an abnormal build-up of glucose occurs in the body. In 1923 Banting was a co-recipient of the Nobel Prize for physiology and medicine for his research and the development of insulin.

Banting's interest in medicine was aroused as the result of a childhood incident. One day, while on his way home from school, he stopped to look at two men who had just begun the first row of shingles on the roof of a new house. As he watched, the scaffolding on which they stood suddenly broke. The two men fell to the ground and were badly injured. Banting ran for the doctor, who arrived in a matter of minutes. "I watched every movement of those skilful hands as he examined the injured men and tended to cuts, bruises, and broken bones. In those tense minutes I thought that the greatest service in life is that of the medical profession. From that day it was my greatest ambition to become a doctor."

He studied at the University of Toronto and became committed to his idea that insulin, a hormone secreted by certain cells within the pancreas, would cure diabetes. However, Banting's qualifications as an investigator of carbohydrate metabolism were limited. He needed an assistant.

Realizing Banting's dilemma, Professor J.J. Macleod at the University of Toronto mentioned to his senior class in physiology that a young surgeon was carrying out research on the pancreatic islets and the isolation of the antidiabetic hormone. Two students, C.H. Best and E.C. Noble, had previously been engaged in experimental studies of diabetes and had an understanding of carbohydrate metabolism as well as the specific skills required to perform the necessary tests. Both were chosen assistants for a period of four weeks

each. They tossed a coin to decide who would work the first four weeks and Best won. Mr. Noble never did return to assist Banting. Best became his permanent assistant. Banting and Best made a good team. Their skills complemented each other. The younger man's knowledge of the latest biochemical procedures complemented the surgical skills of the older man.

On May 16, 1921, Banting and Best were given ten dogs and the use of a laboratory for eight weeks. Their first task was to ligate the pancreatic ducts of a number of dogs. The next step involved trying to produce experimental diabetes. Many animal rights' activists argued against the use of dogs in Banting and Best's experiments. However, Banting adopted a very caring attitude toward these "assistants." He made sure the dogs were always spared unnecessary pain.

By July 21, 1921, a depancreatized dog and duct-tied dog with a pancreas were available. Banting opened the abdominal cavity of the latter dog, removed the shrivelled pancreas, and chopped it into small pieces. This mass was ground up and saline was added. Banting and Best administered 5cc intravenously to the depancreatized dog. Samples of blood were taken at half-hour intervals and analysed for sugar content. The blood sugar fell and the clinical condition of the dog improved.

On January 11, 1922, after months of research to perfect the extract, Banting and Best were ready to experiment with a patient. The first patient was a 14-year-old boy suffering from juvenile diabetes. When he was admitted to the Toronto General Hospital he was poorly nourished, pale, weighed a meagre 65 pounds, and his hair was falling out. A test for sugar was strongly positive. He received daily injections of the extract. An immediate improvement occurred. The boy excreted less sugar, and he became brighter, more active, and felt stronger. This 14-year-old boy was the

Box 12.7 continued

first of many children who were helped by the discoveries of Banting and Best.

In 1923, Banting and Professor Macleod jointly received the Nobel Prize in physiology and medicine for their investigations into and research on insulin, the active principle of the Islands of Langerhans of the pancreas and regulator of the sugar level in the blood. Macleod received this joint recognition for sharing his laboratory facilities and for finding Best as a collaborator for Banting. Macleod insisted on the verification of the initial work and the repetition of certain control experiments. However, Macleod did not create the serum. Banting was somewhat annoyed that Best did not receive any recognition or award for his work, so he divided equally his share of the award money ($40,000) with Best. Furthermore, Banting always assigned "equal credit" to the discovery of insulin to his friend and assistant, Best.

Michael Bliss, *The Discovery of Insulin* (Toronto: McClelland & Stewart, 1982).
Michael Bliss, *Banting: A Biography* (Toronto: McClelland & Stewart, 1984).
Arturo Castiglioni, *A History of Medicine* (New York: Knopf, 1941).
Lloyd Stevenson, *Sir Frederick Banting* (Toronto: Ryerson Press, 1946).
Academic American Encyclopedia, vol.3 (Princeton, N.J.: Arete Publishing, 1980), p. 71.
Colliers Encyclopedia, vol.3 (New York: Crowell-Collier, 1973), p. 602
Encyclopedia Britannica, vol.3 (Chicago: William Brenton, 1976), p. 134.

Many of the clinical decisions made by doctors are explainable in the context of the network of medical contacts. Hall (1946) found that the class, ethnicity, and educational background of the doctor affected the type of colleague networks and hospitals of which he/she was part, and these in turn affected the type of practice taken up. Practice is also known to be affected by a number of physician characteristics including gender, age, area of specialization, quality and recency of training, ongoing involvement in the professional community, and practice setting (see Clark *et al.*, 1991, for a review of various studies describing the effects of these issues on variations in physician practice). Coleman *et al.* (1957) noticed how the adoption of new drugs was dependent on the position of the doctor within a network of colleagues: the most closely integrated physicians were most likely to prescribe the new drugs first.

The practice of medicine is observed and regulated within the setting of the hospital. But access to the hospital is not equally available to all medical practitioners. This limited access could lead to the enforcement of high standards of medical practice. However, considerations other than medical expertise are known to affect doctors' hospital privileges. Hall has shown how networks based on friendship, ethnicity, religion, and social class influence which students enter certain hospitals as interns and residents and who obtains hospital privileges (Hall, 1946).

Appointments are not made on the basis of technical superiority. Those appointed must be technically proficient, but after that level of competence is reached, other factors take precedence over sheer proficiency. At this level *personal* factors play a part in determining who will be accepted. (Hall, 1948: 332)

Most hospitals have several committees that guard standards of medical practice. The credentials committee scrutinizes the qualifications and experience of the physicians who want to practise in the hospital and specifies the kinds of professional activities they are allowed to undertake. The tissue committee studies the tissue removed by surgeons and evaluates whether or not particular operations were warranted. The medical audit committee reviews medical records and pre- and post-operative diagnoses. And the medical records committee evaluates the records kept by physicians (Blishen, 1969: 78). Roman Catholic hospitals also have medico-moral committees to ensure that Catholic morality prevails, particularly with respect to matters concerning the "sanctity of life."

Recent analyses suggest that today the power and autonomy of the physician within the hospital is severely restricted. Doctors used to be the only gatekeepers to the hospital system: they determined which patients, medical colleagues, and various para-professionals would be admitted. Now hospital administrators, many of whom are graduates of university-based hospital administration programs, are the pivotal figures in organizing the hospitals (Wahn, 1987). Today, doctors function as the middle managers whose work is likely to be constrained, organized, and directed by the hospital administrators.

This new occupational group of hospital administrators is generally responsible to the government for containing costs, as well as to the board of directors of the hospital. The board of directors is composed of representatives of the doctors, the nurses, and others who work primarily in the hospital, along with members of the community. Provincial hospital commissions employ large numbers of financial experts and hospital administrators, include a few physicians, then set the budget and monitor the performance of hospitals under their jurisdiction. At times hospitals have adopted measures recommended by the commission that have been contrary to the interests of the medical personnel (*ibid.*: 427).

Hospital administrators are guided by different primary goals than doctors are. They are chiefly concerned about managing the organization in a rational and efficient manner and eliminating unnecessary services and costs so as to balance the hospital budget. This, of course, does involve satisfying the goals of promoting and maintaining good, safe, and efficient medical care. But these "medical" goals must be met in the context of rationalizing services and balancing the interests of many specialty groups who work within the hospital.

Wahn explains some of the specific processes that affect the work of the doctors in the hospital. He mentions one hospital in which the demands for financial cutbacks were used to force doctors in the emergency ward to accept salaries despite their opposition, because the fee-for-service expenses were too high. He discusses another situation in which the neurosurgeons at a particular hospital were ordered to schedule surgery every day rather than simply on one day of the week (which had been the norm) in order to maintain the department's services (nursing and so on) at a consistent level. This was opposed by the doctors because it meant, among other things, that they had to make hospital calls on the weekends. The doctors' wishes were ignored for the "greater good" – the efficient running of the hospital.

Despite the potential importance of hospital committees for the maintenance of health-care standards, they do not always do their job as well as might be expected. There is some evidence from an American study that smaller hospitals are not as likely to control standards as effectively as large hospitals (Blishen, 1969: 77). Informal controls actually operate in all hospitals, so that doctors' reputations may be spread by word of mouth by any member of the medical, paramedical, or administrative staff. The effectiveness of gossip as social control is probably greater in a small hospital in a small community than in a large hospital in a large community, and yet the effectiveness of this control has not been widely documented.

The Management of Mistakes

Mistakes happen to all of us. They happen both on and off the job. They are an inevitable part of life. Which student has ever received 100 per cent on all tests, essays, and exams? Which teacher has never made an error in adding or recording marks? Doctors are no exception: they also make mistakes. But when doctors make mistakes the results can be devastating – they can result in death or serious disability for the patient. One question Marcia Millman asks in her book, *The Unkindest Cut* (1977), is how do doctors handle their mistakes? Millman carried out her research for two years at three university-affiliated hospitals. She was a participant-observer working with doctors at all levels, from interns to residents. Because these hospitals were teaching hospitals attached to universities, their standards can probably be considered above average.

Millman's work highlights three basic points. First, the definition of what constitutes a mistake is variable. In fact, the old joke that the operation was a success but the patient died has a basis in fact. Results that patients interpret as mistakes are not necessarily considered mistakes by doctors. And doctors in different specialties may have different understandings of what constitutes a mistake. Actions that would be considered reprehensible in an attending physician are considered permissible in an intern, who is, after all, "just learning." The point here is that there is no universal standard of perfect or imperfect medical practice. Norms are worked out in practice.

Second, while the designation among doctors of what constitutes a mistake is problematic, some results of medical practice are considered undesirable enough to warrant investigation. One example would be an unexpected death during surgery.

When such an outcome occurs, Millman notes, doctors tend to use two mechanisms for dealing with it: (1) neutralization, and (2) collective rationalization. As Millman says,

> By neutralization of medical mistakes I mean the various processes by which medical mistakes are systematically ignored, justified or made to appear unimportant or inconsequential by the doctors who have made them or those who have noticed that they have been made. (Millman, 1977: 91)

Thus, an action that results in harm to a patient can be ignored, or it can be justified because of the intricate nature of the surgery involved or because it was the patient's fault for not informing the physician of a drug reaction. In justifying themselves, doctors may emphasize unusual or misleading clues, or they may focus on the patient's unco-operative attitude or neurotic behaviour. Even though "doctors may have differences and rivalries among themselves with regard to defining, blaming, and acting on mistakes, all doctors will join hands and close ranks against patients and the public." The assumption made is, "there but for the grace of God go I" (*ibid.*: 93). In the face of mistakes, doctors will band together to support one another and to explain the behaviour leading up to the mistake as blameless or at least as being as logical as possible given the circumstances.

The third point of Millman's work is that hospitals have instituted formal mechanisms for dealing with mistakes, for using errors as a source of education, and for investigating culpability. The Medical Mortality Review Committee, one such mechanism, meets monthly to discuss deaths within the hospital and to review each death in which there may have been some possibility of error or general mismanagement. The fundamental yet unspoken rule that governs the operations of the Medical Mortality Review is that it is to be "a cordial affair." Even though the atmosphere may sometimes be slightly strained and the individual physician whose decisions are being reviewed may be embarrassed, rules of etiquette and sociability are stringently enforced. All members of staff who had a part in the mistake use the committee meeting as an opportunity to explain to the others how each was individually led to the same conclusion. The discomfort level is also minimized by emphasizing the educational aspect of the event rather than its legitimate investigatory nature.

Technical and Moral Mistakes

Bosk's (1979) work on mistakes among doctors adds another dimension to the analysis already discussed. Bosk studied the ways surgeons and surgeons-in-training at a major university medical centre regulated themselves. He noted that the medical centre's hierarchy is structured and managed so that those at the bottom levels are more likely to have to bear the responsibility for mistakes than those at the top.

There were, he noted, two different kinds of mistakes – technical and moral. Technical mistakes are to be expected in the

Box 12.8 Medical Practice Is Inevitably Error-Ridden

An interview study of forty doctors' interpretations of what constitute mistakes in their work demonstrated the "anguish of clinical action and the moral ambiguity of being a clinician" (Paget, 1988). Mistakes, the interviewed doctors noted, are absolutely inevitable in the work of the doctor. Mistakes are intrinsic to medical decision-making: they are essential to the experience of doctoring. Clinical practice involves risks because it requires the application of finite knowledge to specific situations with their own limitations of responses, settings, and patient-doctor communication. Doctors are often not "to blame" or incompetent. Medicine by its very nature is an essentially "error ridden activity." Because of the method of interpretative understanding based on interview data, Paget was able to describe medical practice with an empathy for doctors. As she says, "mistakes are complex sorrows of action gone wrong" (*ibid.*: 131).

In fact, "medical mistakes are an intrinsic feature of medical work" (*ibid.*: 5). Medical work is inevitably error-ridden as the doctor knows that she/he is making decisions based on some but not all of the information available. Moreover, even if the doctor has all the available information, it may not be enough to diagnose or treat a disease. Medical practice inherently encapsulates a degree of uncertainty. Yet it usually results in, even requires, action. Paget's work enables us to see the definition of mistakes from the perspective of the practising clinician. Based on open-ended interviews and using a phenomenological approach, she shows the everyday struggles of doctors who make decisions in an error-ridden climate. The following quotations illustrate how doctors talk about mistakes.

Well, all mistakes are relative. They're relative to the setting in which they are made, and they're relative to the intent of the physician. I think mistakes

about, for example, placing an individual on oral hypoglycemics with maturity onset diabetes – that wasn't a mistake five years ago, and it might not be a mistake now, but it's certainly suspect. A mistaken diagnosis: the threat, the constant threat, the constant nightmare of many people. The commission or omission kinds of errors; there are conflicts there too, whether you . . . whether you resected the wrong breast for carcinoma. That . . . that certainly is a different order of – excuse me – a different level or dimension of error as opposed to putting an individual on the wrong medication or giving him an inadequate dosage or something like that.

I think, dealing with mistakes . . . I think, we see mistakes all the time. But the errors are errors now, but weren't errors then. If someone comes in with an obvious primary atypical pneumonia, and he's being treated with Ampicillin, you can hardly fault someone for that, because the . . . because the fact that it was a mistake was brought to light after the therapeutic procedure was effected. Similarly, in preventive medicine, you destroy the evidence of your efforts by being successful – so is it a mistake too, to put an individual on a low cholesterol diet after you first suspect that he had – that he has – hyperlipoproteinemia of one kind or another? You don't know whether that's a mistake; some people will think it's a mistake because you deprived him of all those steaks. (*Ibid.*: 54)

Neither patients nor doctors adequately acknowledge the experimental character of clinical work.

Physicians work under the peculiar burden of having to believe in their conduct, even while it is experimental, and having to mask many primitive feelings of fear

Box 12.8 continued

and anxiety, in both themselves and their patients, in order to execute, as it were, the work. (*Ibid.*)

This characteristic of medical work was less problematic a half-century ago. Today, however, when treatments can be dangerous the potential for deleterious results of mistakes has grown enormously. Irreparable mistakes are inevitable: they occur, in spite of good intentions, good education, and care.

---

practice of medicine, and colleagues are normally forgiven them. They are viewed as an inevitable part of medical work and are seen as having definite value in that they often motivate improvements in practice. People are expected to learn from their technical mistakes. Moral mistakes, however, are severely reprehensible. They constitute evidence that the physician or physician-to-be does not belong as a medical colleague (as a member of the team, or a team player). Moral errors that demonstrate an unco-operative attitude, unreliability, or a lack of responsibility to patients are more serious. Moral errors are not as easily forgotten or forgiven because the person who made them is believed to be unsuitable as a doctor.

Practice Norms and Variations

A recent study by Wennberg (1984) showed that different colleague networks develop different sets of beliefs, norms, and practices regarding specific diseases and the appropriate treatment for them. Wennberg compared hospitalization and rates of surgery in the states of Iowa, Vermont, and Massachusetts, and found tremendous variations from state to state and within states. He then divided the different areas into what he called "hospital markets" or hospital service areas (populations using one hospital or a group of hospitals). He found that different hospital service areas had highly variable rates of all measures of hospital and surgery utilization (see Table 12.1). For instance, he noted that the rate of hysterectomy (among all women under 70) varied from 20 per cent in one market to 70 per cent in another. The rate of prostatectomy varied from 15 per cent in one market to 60 per cent in another. Tonsillectomy rates varied from 8 per cent to 70 per cent in two different markets. In trying to explain these wide differences, he examined the roles played by (1) illness rates, (2) insurance coverage, (3) access to medical services, (4) age distribution, (5) per capita hospital bed ratios, and (6) physician-population ratios. None of these factors made a significant impact on the use of hospitals or surgery. Wennberg concluded that the differences must be due to subjective variations in the practice styles of physicians in each area.

Rachlis and Kushner have illustrated how the productivity of doctors varies substantially from area to area. For example, they note that radiation oncologists in Toronto see 50 per cent fewer patients than those in Hamilton and that those in Hamilton see 50 per cent fewer patients than those in Halifax. Only part of the difference is due to research responsibilities and part is due to the expansionist role of radiation oncologists – from the technical (as is common in Britain) to whole patient management (as is common in the U.S.) (Rachlis and Kushner, 1994: 5). Within the Canadian medical care system there is considerable treatment variation from place to place. Rachlis and Kushner note the following findings: (1) a person is three times as likely to receive a tonsillectomy in Saskatchewan as in Quebec; (2) county by county rates for coronary bypass surgery vary $2^{1/2}$ times throughout Ontario; (3) the length of stay for hospitalized heart attack patients ranges from 6.6 to 12.9 days in Ontario.

According to John P. Bunker (1985), such variation occurs because medicine is an art and not a science, and therefore a great deal of uncertainty is involved in the practice of medicine. Most surgery, possibly 85 to 90 per cent, is not for life-threatening problems but for conditions of discomfort, disability, or disfigurement. Moreover, in most cases the benefits just about equal the risks. Surgery is just as likely to improve the health, comfort, and life span of the patient as it is to cause an increase in morbidity. There are no clear differences in mortality rates in different hospital markets. Hospitalization or surgery rates are not correlated with declines in either the mortality or morbidity rates.

Wennberg, Bunker, and Barnes (1980) studied seven different types of surgery,

including hysterectomy, tonsillectomy, gall bladder surgery, prostatectomy, and hemorroidectomy, but in this case they compared Canada, the United Kingdom, and different locations in the United States. Generally, the rates were highest in Canada and lowest in the United Kingdom. The authors suggest that while such differences have enormous implications for costs and for morbidity and mortality rates, they are not the result of malpractice but occur because diagnostic and treatment procedures are not perfect. Norms of practice develop out of colleague and network interaction. Hospital and clinic routines play a part in determining styles of practice. As Horowitz states:

> In medicine there are few certainties. There are more questions than answers. There are few truths. And everything is changing. Most of the time, the best we can hope for is the best judgement based on the best available facts and the particular circumstances. (Horowitz, 1988: 29)

Malpractice

One of the fastest growing types of social control of medical practice is the malpractice suit. Malpractice insurance policies are now taken for granted as a necessity, and fees for malpractice insurance are growing substantially. One Canadian study indicated that malpractice suits have tripled over the past fifteen years and that the amount of awards has quadrupled. Twice as many suits against doctors are successful as were successful fifteen years ago. Most complaints originate in hospitals or other

FIGURE 12.1 Predicting Hospital Market Differentials

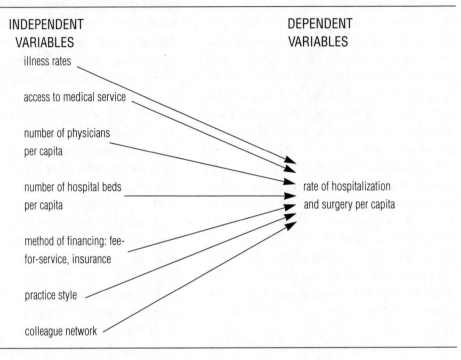

Source: Adapted from Wennberg, 1984.

health-care institutions. As a consequence of these trends, doctors are now spending $200 million annually on insurance for claims of injury alone. Family doctors paid an average of $1,500 in 1994, while those in the high-risk specialties of anesthetics and obstetrics paid $17,000 in 1994. This rate is still, however, not at all close to the insurance malpractice rate in the U.S. The Canadian doctors' insurance company, the Canadian Medical Protective Association, has grown financially from a base of $24.5 million in assets in 1983 to $614 million at the end of 1992. Growing at about $100 million a year, membership fees are running at about 30 per cent above total expenses (Rachlis and Kushner, 1994: 305).

Summary

(1) The medical profession can be thought of as an occupation, a process, or an ideology. There are certain strengths and limitations to each of these descriptions and it is possible for them to overlap.

(2) As an occupation, medicine has two basic characteristics: prolonged training in a body of specialized, abstract knowledge and a service orientation. However, the

view of profession as occupation accepts the profession at its own valuation and ignores the power of the group, changes over time, and the relationship between the occupation and the rest of society.

(3) As a process, the medical profession can be seen as developing over time. Members engage in full-time work, establish a training/education program, belong to an association, gain legal status, and construct a code of ethics. The members of the occupational group must have esoteric knowledge and social distance from the rest of the population and be more or less homogeneous.

(4) As an ideology, the medical profession holds descriptive and prescriptive beliefs that explain and legitimate certain practices and viewpoints. The ideology includes: universalism, functional specificity, effective neutrality, and a collectivity orientation. An effective ideology allows physicians to dominate and monopolize the field.

(5) Today, there are three steps to becoming a doctor: an undergraduate education, graduate study leading to the M.D. degree, and a minimum one-year internship. Medical students generally come from middle- and upper middle-class family backgrounds. Physicians-in-training become aware of two dominant values: "clinical experience" and "medical responsibility."

Medical students learn emotional detachment and the ability to deal with inevitable and constant uncertainty; they also encounter alienation and may develop authoritarian personalities.

(6) Power and autonomy of the medical profession have been somewhat curtailed by the emergence of hospital administrators who direct and organize doctors' activities. Physicians are also governed by the College of Physicians and Surgeons and the Canadian Medical Association.

(7) Doctors make mistakes. The definition of what constitutes a mistake is variable; doctors use two mechanisms for dealing with undesirable mistakes: neutralization and collective rationalization. The formal mechanism hospitals have instituted to deal with mistakes is the Medical Mortality Review Committee. There are two kinds of mistakes: technical, which are to be expected and are normally forgiven, and moral, which may indicate that the person who made them is believed somehow to be unsuitable as a doctor.

(8) Different colleague networks within the medical field develop different sets of beliefs, norms, and practices regarding specific diseases and their appropriate treatment. Variations in medical practices indicate that medicine's diagnostic and treatment procedures are not perfect but are sometimes subject to error.

CHAPTER THIRTEEN

The Medical Care System: Critical Issues

WHAT IS WRONG with the medical care system today? Is it under-funded? Is it poorly organized? Is it too "high-tech"? Will there be enough money for medicare to support the aging Canadian population? Do we overemphasize treatment and neglect prevention? Is there any justification to the charge that medicine is a sexist institution? These are the sorts of questions that will be addressed in this chapter.

This chapter will focus on two main critiques of the medical care system. The first centres on the notion that the contemporary medical care system is dominated by the "medical" model of health and illness (see Table 13.1). We will discuss here the various characteristics of this model, the value of alternatives to the medical model, such as environmental and lifestyle models, and the adequacy of the medical model with respect to the changing demographic profile of Canadian society and the epidemiological data on the causes of diseases. The second critique concerns sexism in medicine and in the experiences of health and illness.

The Medical Model

The medical model is based on the assumption that disease is an objectively measurable pathology of the physical body that results from the malfunctioning of parts of the body. All diseases are eventually explainable through a close analysis of the biological components of specific individual human beings. The first implication of this model is that the most significant advances against disease are through scientific investigations, in the laboratory and the clinic, of the individual human body. The second is that the search for causal mechanisms should focus primarily on the pathological alterations within the cells of the body. The third is that disease is considered undesirable and abnormal. The fourth is that disease, being undesirable, is to be eliminated as quickly and completely as possible with as powerful means as necessary.

The limitations of the medical model are best clarified through comparison with two competing models – the environmental/social-structural model and the lifestyle model. The environmental/social-structural

TABLE 13.1 Models of Disease Causation

MEDICAL MODEL	ENVIRONMENTAL/SOCIAL-STRUCTURAL MODEL	LIFESTYLE MODEL
Disease is an objectively measurable pathology of the physical body, which is the result of the malfunctioning of parts of the body. Cure is through chemotherapeutic, surgical, or other "heroic" means. Hospitals, as places for the practice of high-tech medicine, are of primary importance.	Disease is best understood as the result of social-structural inequalities in class, gender, race, ethnicity, and environmental conditions such as air and water pollution, dangerous and stressful work, harmful organization of major societal institutions such as the family and education.	Disease is best understood as the result of individual behaviours based on decisions about such things as 1. exercise 2. stress management 3. diet 4. smoking 5. substance use and abuse 6. sexual habits 7. seat belt use, speed limit observance, and the like

TYPICAL "HEALTH" PROBLEMS IN EACH MODEL

Coronary heart disease	Poverty	Smoking
Respiratory disorders	Unemployment	Alcohol addiction and abuse
Cancer	Pollution	Drug addiction and abuse
Sexually transmitted diseases	Family violence	Unsafe sex
AIDS	Sexual abuse	Poor dietary habits
Arthritis	Unsafe and unhealthy working conditions	Lack of exercise
Mental illness	tions	Poor stress management
Hypertension	Homelessness	Unsafe driving habits
Multiple sclerosis	Poor wages	Equipment or appliances that are unsafe
Diabetes	Racism, sexism	
Colitis	Underemployment	
Alzheimer's disease	Lack of adequate child care	
Burns	Lack of readily available and safe birth	
Influenza	control	
Pneumonia		
Broken limbs		

model focuses on causes of disease that lie outside the individual organism. In this model, disease is best understood as the result of a complex of social-structural inequalities revealed by the relationships between disease and class, gender, race,

Box 13.1 A Problem in the Provision of Medical Care to Native People

Native people are not only plagued by the effects of the alcohol they use themselves, they also suffer the consequences of the misuse of alcohol in the medical care system available to them.

The Nimpkish Indian Reserve lies midway between Vancouver Island and the British Columbia mainland. In 1979 the practising doctor on the reserve was Dr. Harold J. Pickup. On January 18, 1979, Renee Smith, a 15-year-old Native girl, was taken to Dr. Pickup complaining of severe abdominal pain, nausea, and diarrhea. The doctor prescribed medicine for her pain and she was sent home. After a very painful night, Renee was taken back to the hospital the next morning with the same symptoms. This time she was admitted.

On January 19, Renee's mother was told by the doctor that she had a severe case of the flu but that she was improving. Renee didn't seem any better. On January 20, her mother was told that it was her appendix but that an operation was not necessary. When her symptoms did not disappear, x-rays were taken of her stomach. This was on January 22, four days after she first asked for help. Dr. Pickup said the x-rays showed that she was full of gas (this was his reasoning for the abdominal pain) and an enema was ordered.

That night Renee had trouble breathing and after two phone calls the doctor arrived. However, he was too late and she died. The coroner's report showed that she died of severe generalized peritonitis resulting from a ruptured appendix.

Later it was determined that Dr. Pickup was under the influence of alcohol while carrying out his duties as a physician. It was known to many that the doctor had a drinking problem. Another woman had died under suspicious circumstances after having a hysterectomy. She was also under the care of Dr. Pickup.

At the inquest, the lawyers for the doctor tried to plant "seeds of doubt" in the minds of the jury. They blamed the problem on specific stereotypical characteristics that they attributed to Native patients. It was suggested that Renee presented herself too late, after her appendix had already ruptured. It was said that she masked her symptoms because of her high tolerance of pain and that she had been unco-operative and uncommunicative. The doctor's lawyers were speaking to a receptive audience. In spite of evidence to the contrary, the verdict at the inquest concluded the death to be natural, and it was classified as being accidental due to medical negligence.

After the jury's verdict Dr. Pickup continued to practise in the community. The College of Physicians and Surgeons of British Columbia found the doctor competent and thus his licence was not suspended.

Dora Speck Culane, *An Error in Judgement – The Politics of Medical Care in an Indian/White Community* (Vancouver: Talon Books, 1987).

ethnicity, environmental pollutants and contaminants, dangers in the workplace, and stress. These inequalities are also evident in the personal difficulties resulting from the ways in which social institutions and practices such as the family, sexuality, and education engender human suffering leading to illness. Whereas this model is not able to cure disease in the individual, at least not as quickly or heroically as allopathic medicine, and certainly not with the "magic bullet" of a pill or surgery, it can be used as a basis for planning programs of disease prevention or health promotion, programs that in the long run should decrease the burden of disease on the population.

The third model is the lifestyle model. In this model the individual is not just a body, as in the medical model, but is body, mind, and spirit, and must be considered in the context of his/her whole life. In this model disease is understood as the result of individual actions based on personal decisions regarding a style of life such as (1) exercise, (2) stress management, (3) diet, (4) smoking, (5) substance use and abuse, (6) sexual behaviour, and (7) other behaviours such as using a seat belt or observing the speed limit. As in the medical model, the individual is the source of both the problem and the cure. Unlike the medical model, however, individual choice and free will rather than biological malfunctioning are viewed as the immediate causes of illness. Both treatment and prevention require a change in lifestyles.

No one of the three competing models is in itself sufficient to explain or prevent all disease. All three models are useful and valuable. Each addresses a necessary problem. A complete medical care system would include some aspects of all three models. A basic problem with the contemporary Canadian medical care system is the over-reliance on the medical model at the expense of the other two. Disease prevention, in spite of increasing formal recognition, has received relatively little financial support. The largest expenditure of the medical care system is on hospitals – 60.0 per cent of the whole budget when all institutions, including homes for special care and others, are included. The second largest component of

the health-care dollar – 21.6 per cent – is expended for physician services. Drugs (prescribed out of hospital) and appliances comprise 5.3 per cent of the health-care budget. All these expenditures are for items primarily within the medical model.

Only 5.7 per cent of the total health expenditures lie outside these areas for public health and home care (Rachlis and Kushner, 1994: 48) and could be used for disease prevention and health promotion, or interventions that would fit within the environmental and lifestyle models. Financially, then, the system is currently dominated by the medical model.

Specific data are available from the United States. (Their precise relevance to Canada needs to be determined.) In 1979 the total portion of the U.S. federal budget spent on medicine was 8.8 per cent. Only 2 per cent of this total was spent on public health or disease prevention (Freeland and Schendler, 1981; Terris, 1980). As well, according to Fisher (1986: 93), the funds allocated to prevention are vulnerable to budget cuts. "In the fiscal year 1980 – a year when the United States was not at war – the budget for public health activities was cut by about a third, while the military budget was increased by over half."

This reliance on the medical model is particularly problematic because of the limited relevance of the notions of cure to the most significant diseases in modern society, namely the degenerative, chronic diseases. As discussed, the chief causes of death today are heart disease, strokes, and cancer. All these are diseases of middle and old age. All tend to be chronic rather than

Box 13.2 Limitations of the Medical, Environmental/Social-Structural, and Lifestyle Models

Each model has its own limitations and problems. The medical model has had spectacular success in the diagnosis and treatment of organ-specific diseases. In fact, the success has been so great that organs can even be removed and replaced by others – either artificial or from human donors. Its limitations are most evident in its inability to cure the ever-expanding range of chronic and multi-system diseases. The environmental model may direct us to cleaning up the organochlorides in the environment. It neglects, however, the basic research necessary to understand genetically inherited diseases, such as Huntington's, for instance. The lifestyle model is useful in that it provides information that can be used by a population to reduce the chances of certain diseases such as lung cancer, by a reduction in cigarette smoking. Its chief failure is its inadvertent capacity to blame the victim – for having caused his or her own illness. A number of social commentators have pointed to the tension between individual rights and freedoms, on the one hand, and those of the collectivity, on the other hand. As Stone (1986: 671) has put it, "the politics of preventive medicine . . . manifest a strong individualism . . . deeply ingrained in American politics in general. Philosophically, the whole idea of identifying high-risk individuals and high-risk individual behavior locates the source of misfortune in the individual rather than in the social structure and economic opportunity." On the other hand, societal interventions, such as those against smoking in public places, risk infringing on the rights and freedoms of the individual for the good of the collective. Health promotion actions, whether through education (e.g., public health television), engineering (e.g., fluoridation of the water supply), or enforcement (e.g., seat belt legislation), are always embedded in this tension between the individual and the collectivity (Alonzo, 1992).

---

acute. As the Canadian population ages, these and other chronic conditions will become more prevalent and the relevance of the medical model, and the cure that it seeks, will decrease. On the other hand, the importance of long-term care and patient education increases.

In addition, medical intervention has enabled many people to live who would previously have died. Premature and underweight babies are kept alive for many months in hospital by surgery, drugs, complex technology, constant care, and intravenous feeding. Young people who have suffered from motor vehicle accidents are kept alive as quadriplegics. Middle-aged or older people, having suffered a disease such as a stroke, are sometimes kept alive with artificial respiration and tube feeding. Some advances in medical care have meant longer life for many severely disabled people who can only be kept alive with intensive nursing care but who cannot be cured by medicine. Such medical interventions may lead to a decrement in actual quality of life.

The Decline of the Medical Model – Demedicalization

In spite of the many successes of modern medical practice in individual cases

FIGURE 13.1 A Framework for Health Promotion

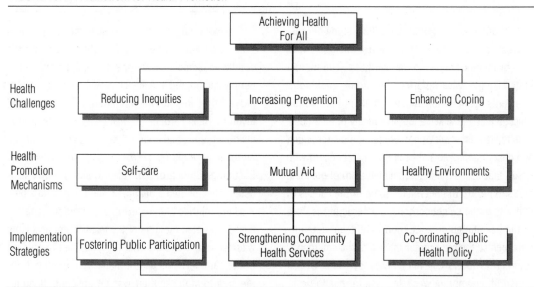

Source: Jake Epp, *Achieving Health for All: A Framework for Health Promotion* (Ottawa: National Health and Welfare, 1986). Reproduced with permission of Minister of Supply and Services, Canada, 1989.

involving such miracles as heart transplants, heart-lung transplants, artificial heart transplants, in vitro fertilization, and the like, there is widespread evidence that the biomedical model of illness is rapidly losing support. As a result of the publication of his *A New Perspective on the Health of Canadians* (1974), former federal minister Marc Lalonde has convinced many countries in the rest of the world to consider the ways morbidity and mortality are the result of a complex of factors: biological host factors such as genetic heritage; environmental conditions, including both physical and social-economic environments; and "self-imposed" factors such as cigarette smoking and excessive alcohol and drug consumption. Lalonde's work

has made an important impact on Canadian government policy and has also influenced the World Health Organization and numerous nation-states.

This new approach is called "health promotion" and is defined as the process whereby people can increase control over and improve their health (see Table 13.1). According to former Health Minister Jake Epp's 1986 report, *Achieving Health for All: A Framework for Health Promotion*, the mechanisms for addressing health problems are (1) self-care, or the decisions individuals take in the interests of their own health; (2) mutual aid, or the actions people take to help one another; and (3) the development of healthy environments. The report suggests three ways to implement this new approach:

Box 13.3 Treatment or Prevention?

Rachlis and Kushner (1989) describe a study of great interest to those involved in the treatment–prevention–promotion debate. The study, a community undertaking, was reported by John McKnight. Residents from a low-income housing project in Chicago met and talked about the quality of their community life – unemployment, crime, and overcrowded and decaying houses and health. In their concern with health they asked McKnight (and his team of graduate students) to try to understand some of the causes of ill health in the neighbourhood and some of the uses of the hospital emergency room. The team carefully documented the reasons for the use of the emergency room. They found that the emergency room, over a period of one month, was used less for health problems and more for community problems. Following is the list, in order of prevalence, of problems that came to the community emergency department:

1. traffic accidents
2. other kinds of accidents
3. interpersonal violence
4. respiratory problems
5. drug problems
6. alcohol
7. animal bites

The community decided to tackle the last problem first – dog bites. It seemed straightforward and the solution seemed clear. They established a five-dollar bounty and offered it to any citizen who reported sighting a stray dog in the community. The kids got involved. Riding around on their bikes, neighbourhood children rounded up dogs – 200 in one week. Soon the number of emergency room dog bites declined. Encouraged, the group decided to tackle traffic accidents. They plotted a neighbourhood map with a blue X where injuries had occurred and a red X for fatalities. They noted that one of the locations of accidents was near a fire hydrant that the kids used to cool off in the summer. This action made it very difficult for the cars to stop. Another was an unsafe parking lot entrance. The group was successful in changing the location of the fire hydrant and also in closing the unsafe parking lot entrance. It succeeded in diminishing accidents and emergency room use by a community-based (essentially transferring tools to the powerless) response. Other examples of community-based initiatives include community gardens, group kitchens, babysitting exchanges, job banks, and the like. Communities can help themselves and can improve their health.

---

fostering public participation, strengthening community health services, and co-ordinating public health policy.

Strategies for implementing this approach are diverse; they include media promotion, health education, health advocacy, community development, and community economic development (*Health Promotion*, 1987: 7-10). Media promotion includes the production and distribution of all sorts of materials such as films, slideshows, pamphlets, posters, and the like; these are designed to be used in community centres, schools, libraries, and hospitals and at health fairs to improve the public's general awareness. When one Toronto community identified unemployment as a significant health problem, a group undertook the publication of a brochure, which was sent around to local doctors suggesting that they refer unemployed patients to self-help, political action groups and employment counselling agencies.

Among its targets, health education includes schools, communities, and health professionals and involves many different types of popular education techniques, partly modelled on the work of Paulo Freire. Freire and his colleagues approached Brazilian peasants wherever they gathered – on the streets, at market, in their homes. Freire taught the people empowerment through literacy. So, too, the goal of health education is empowerment through knowledge.

Health advocacy involves lobbying for social policies and laws that promote health. Legislation, for example, covers smoking and non-smoking areas, lead-emission controls in automobiles, and workers' right to know the names and effects of potentially hazardous products with which they work. Health advocacy means taking a stand on health issues: seeking controls on pollutants that cause acid rain would be a case in point.

Community development and community economic development work to mobilize a community around an issue so that the members of the community work together to reach the goals they have defined for themselves. Credit unions, community banking schemes, and cooperatives are examples of community economic developments.

Canadian society has changed a great deal over the past century, as has the rest of the Western industrialized world. Most Canadians now live in cities; in fact, 30 per cent live in the three largest Canadian cities. Over 90 per cent of all Canadians in the work force work for others, the majority in service, trade, and financial institutions. The medical care system has become a medical-industrial complex and now includes the pharmaceutical industry, hospitals and long-term care institutions, and the medical supplies (e.g., CAT scanners) industry. The system is administered by a large bureaucracy: medical and hospital administrators, managers and planners, government health ministries, and local health units with jurisdiction regarding reportable and communicable diseases. In addition, there are diverse medical care workers, including nurses, pharmacists, dentists, chiropractors, allopathic doctors, and others. In the midst of such complexity, the role of the medical practitioner as the guardian and definer of health and illness has diminished. Fiscal restraint, competition for alternate and lay healers, and unionization among nurses and other medical care practitioners have each in their own way also served to diminish the role of the doctor as the high priest of the medical system, and in so doing these societal changes have diluted the force of medicine as an ubiquitous and persuasive institution of social control.

Changes in government policy reflect wider trends in Canadian society. One important trend is the women's movement, which has led to the radical alteration of many aspects of contemporary life – increased numbers of women in the labour force, equal pay for equal work and equal

Box 13.4 Waiting for a Miracle

What medical miracles are in store for us in the decades ahead? In *RX 2000: Breakthroughs in Health, Medicine, and Longevity by the Year 2000 and Beyond* (Simon & Schuster, 1992), clinical pathologist Dr. Jeffrey A. Fisher offers a fascinating look at a wide range of breakthrough cures, diagnostic procedures, and preventive techniques that will change our lives in the next 40 years. Here are some of his predictions:

1997:   First AIDS vaccine approved.

1999:   Cleaning our arteries as an outpatient.

2000:   Simple blood test identifies those with lung cancer predisposition.

2001:   Birth control pill for males finally arrives.

2002:   AIDS epidemic halted.

2005:   Surgery for breast cancer obsolete.

2007:   Drug addiction conquered, smokeless society finally achieved.

2009:   At last! Colds can be prevented.

2010:   Biological clocks reset with new drugs.

2011:   Autoimmune diseases completely controllable.

2014:   Parents can now create designer children before birth.

2015:   The ultimate treatment for Alzheimer's disease: prevention.

2025:   Average life expectancy at birth – 100 years.

2030:   Coronary disease, cancer effectively wiped out.

Healthwatch, Fall, 1993, p. 17. Reprinted by permission of Simon & Schuster.

pay for work of equal value, increased divorce rates, decreased fertility rates, and so on. Of particular relevance to demedicalization is the examination of issues in women's health.

Among the earliest thrusts of the movement was the demand for knowledge and information about women's bodies. And among the most important publications in the area of women's health was the contribution of the Boston Women's Health Collective, *Our Bodies, Our Selves* (1971). With the women's health movement, women were primed to ask questions and demand second opinions. They became aware of the deleterious effects that medicine had had on both the mental and physical health of women. They documented the prevalence of extensive and unnecessary hysterectomies and excessive and sometimes debilitating prescriptions of mood-altering drugs and tranquilizers. They became aware of the devastating, long-term side effects of using drugs that had been inadequately tested, such as thalidomide and DES (diethylstilbestrol), and the defects of various birth-control devices. Some women sued major manufacturing and pharmaceutical companies as well as their doctors. Millions of women called for natural childbirth, for the Leboyer method of immersing the baby in warm water immediately after birth, for home births, for various types of non-medicated, non-interventionist birth procedures. They taught one another how to do internal exams. They formed self-help and consciousness-raising groups to enable them to be more powerful and to have more autonomy in their individual dealings with doctors. In many ways the women's

Box 13.5 Public Health Debate: Cigarette Smoking vs. Oral Contraceptives

It has been argued that there is a strong case to be made that oral contraceptives should be available in vending machines and cigarettes available by prescription. Cigarettes are known to be directly implicated in the deaths of people from lung cancer, heart disease, and emphysema. Estimates suggest that over 100 Canadians die a day as a result of cigarette smoking. In contrast, oral contraceptives prevent unwanted pregnancies and even protect women from endometrial and ovarian cancer. Large-scale epidemiological studies of oral contraceptives over thirty years suggest that the "pill" is relatively safe. Cardiovascular disease appears to be unrelated to low-dose pills (as long as women with a family history of the disease are screened out). Most research shows no overall effect on breast cancer incidence. Nevertheless, this disease is the most important risk factor identified and more research is necessary before confidence about the benign effect of the pill can be conclusive. In fact, Aspirin appears to be more lethal than the pill. Four people died accidentally and 202 women suicided from analgesics, antipyretics, and antirheumatics in 1988. Thus, it is argued that oral contraceptives should be cheaply and easily available with package-inserted information that would allow women to screen themselves for contra-indications and educate them to use oral contraceptives safely and successfully (Grimes, 1993; Trussell *et al.*, 1993).

movement challenged the power of medicine to speak with authority about most aspects of women's lives. The recent legitimization of midwifery in Ontario is just one of the successes of the several decades of the women's movement. The Ontario report on physician sexual harassment must also be seen as a consequence of the empowerment of women and the increasing recognition of their voices and experience.

A second important trend is the changing disease profile. In the past most illnesses were either minor and self-limiting or acute and fatal. Much illness today is chronic. Chronic illness is expected to last a long time, even over a lifetime, with intermittent remissions and exacerbations of symptoms. The power of medicine to cure chronic illness is virtually non-existent. One consequence of the changing disease profile is the high rate of long-term institutional care. According to Wilkins and Adams (1983), the annual number of days of care in long-stay hospitals has more than doubled over the decade. Chronic illnesses require management by the patient as well as care, at times, by various others, including family members, friends, and nurses. While medical research continues to play an important role in chronic illness, actual medical practice plays only a minor role.

Chronic illness is predicted to be an increasingly significant component of the overall disease profile (Angus, 1984). According to Fries (1980), who has studied the changing disease profile since 1900, by the year 2050, if present trends continue, the average life expectancy will be 85, plus or minus 12 years. Fries calls this maximum life expectancy the biological wall. By the year 2021 those over 65 are expected to comprise 17 per cent, and by 2031, 20 per cent of the population. The most elderly (over 85) population is projected to grow even more rapidly than the total over-65 age group. This is particularly significant because the older the age group, the greater the incidence of chronic illness and associated frailty.

The following are the major expected health problems: (1) reproduction; (2) accidents; (3) chronic diseases – cancer, cardiovascular disorders, arthritis; (4) disorders of old age; (5) mental health problems (Mustard, 1987).

According to Mustard (1987: 26), "the main concerns of health care services will be caring for an older population that is losing, at a varying rate, the vitality of its organ systems, and caring for individuals with the mental health problems that affect all age groups." In this situation the most important role for the medical system "will be to provide efficient, effective and supportive care to an aging population" (*ibid*.: 27). Canada is making slow progress in a shift away from hospital-based services and curative medicine to community-based primary care and health promotion.

A third important trend is the growing knowledge expressing the importance of prevention in minimizing and/or eliminating the major "killers" in Canada today. A review of mortality trends in various age groups in Canada documented that the most frequent causes of mortality – heart disease, cancer, respiratory disease, accidents, suicide, and violence – were amenable to reduction through social, structural,

environmental, and lifestyle changes (Trovato and Grindstaff, 1994).

A fourth significant component is the trend to demedicalization based on finances. Health-care costs have grown. As a percentage of gross national product in 1990, per capita health expenditures were $2,266 – a total of $60 billion – and accounted for almost 10 per cent of the GNP. Aside from the U.S., which spends 12 per cent of its GNP on health care, Canadian expenditures are the highest in the world. Yet there are several countries with better health records, in spite of less expenditure and older population (Manga, 1993). The demand for health care, in fact, can grow expansively and almost endlessly because the end product of medical care – good health – is always elusive. The Economic Council of Canada projected in 1970 that if the then current rate of growth in the education and health-care sectors were to continue, by the year 2000 these two sectors would consume the whole of Canada's GNP.

There is evidence that the changing demographic profile will significantly increase the financial costs of medical care. In this context future predictions are that there will be an increasingly significant decline in the transfer of federal funds for health to the provinces. The funds available for this aging population are thus likely to decrease in the future. Gross and Schwenger (1981) suggest that by the year 2026 the elderly will consume 56.6 per cent of the total medical care budget. It is important to note that a good proportion of this increasing expenditure will be dedicated to institutional care, if current trends continue. On the other hand, the benefits of medicine are not seen as increasing as rapidly as the costs. Social policy researchers have made us aware that the great benefits to health that occurred at the end of the past century and at the beginning of this century were largely the result of public health measures, not medical measures. The present government emphasis on lifestyle changes, health promotion, and disease prevention harkens back to the success of the early public health movement.

The new focus on health promotion and the social and environmental causes of illness is not without negative implications. The very articulation and mobilization of support for a model of illness that would be broader than the medical model could ultimately be used to delegitimize the differences between health and sickness and thus some treatments. It could then be used to eliminate certain "discretionary" treatments from national or provincial medical insurance. If health becomes too broad it can no longer be addressed in specific social/health policies. Moreover, the themes elaborated as aspects of the newer health promotion trends such as "disease prevention, health promotion, iatrogenesis, individual and community empowerment, social networks, and family and home care, can as easily become ideological justifications for the privatization and deregulation of health services, with all that implies for the quality and equality of care" (Burke and Stevenson, 1993: 54).

Sexism and Patriarchy in Medicine

Both sexism and patriarchy are prevalent in the medical care system. Sexism is the tendency to construct and act upon stereotypes based on gender. Thus, for example, women who suffer from migraine headaches are likely to be seen by the medical profession as neurotic, whereas men with the same set of symptoms are likely to be seen as victims of severe occupational demands (MacIntyre and Oldman, 1984). Yet recent research has shown that doctors may be even more "blinded" by the sex of men than of women and that they are now six times as likely to make stereotypical assumptions about men than about women as patients (Groce, 1991). Such different views are partly the result of sexist attitudes held by doctors (Fisher, 1986), which are reinforced by pharmaceutical advertisements (Rochon-Ford, 1986). Patriarchy is male dominance over women. Examples can be seen throughout society, in the dominant positions held by men in the family, the schools, in sexual relations, in work, and elsewhere. Patriarchy is also evident in the dominant position held by men in the medical labour force and in the power of male health providers over (largely) female patients.

Sex and the Medical Hierarchy: A Brief History

That there have always been women healers in society is documented on stone tablets, carved in relief on ancient walls, and told in legends of ancient Egypt, Sumeria, Greece, and Rome. The notion that only men can be doctors, and that women healers must be different and subservient is a relatively recent idea. The drive to exclude women from positions of rank in the medical labour force is coincidental with the drive toward professionalization.

At first women and men were both involved in the practice of medicine. In a number of European countries women from powerful families often ran convents (Mumford, 1983). In most large religious institutions the infirmarian offered her services as physician, pharmacist, and health teacher. Herbs for medicinal purposes were grown and cared for in the gardens of the religious. Often, particularly in large convents and abbeys, nuns provided nursing services.

The famous medieval medical school in Salerno in the eleventh century accepted both men and women as medical students and as teachers. In fact, one of the best-known medical school teachers in the eleventh century was a female, Trotula. She taught medicine and wrote an important gynecological-obstetrical treatise that became a major reference work for centuries. In addition, she co-authored medical books with her husband and son. In the fourteenth century, Roman women from important families could remain single and work as doctors or professors. This was true also in the other major schools in Italy and in the Muslim world – Cairo, Baghdad, Cordova, Toledo, and Constantinople (Mumford, 1983: 27). Jewish women physicians were widely accepted in Italy and France, and

one woman ran a large medical school in Montpellier in the 1320s. Women doctors were even more common, yet no less important, in Germany. There is evidence, too, that women were practising in the Netherlands.

In the middle of the fourteenth century the Plague devastated the European countryside. Men and women survivors escaped to the cities. The feudal system based on family, community, religion, and land was threatened. The marriage rate declined, and uprooted, isolated women without families or husbands were left to support themselves.

Attempts to exclude women from practising medicine apparently began in earnest in Britain in 1412, when Cambridge and Oxford universities sent a petition to Parliament outlining the dangers of allowing "ignorant and unskilled persons" to practise medicine and surgery. The result was the near exclusion of women from the practice of medicine. At that time there were very few educated male physicians either. In 1512 Henry VIII set out to centralize state power. One strategy was the passage of the medical act of 1512, which prohibited undesirable and uneducated persons from practising medicine. Women were one of the categories of undesirables – they were believed to be causing grievous harm in the name of healing. Medical practice was restricted to those who were licensed by bishops. This effectively excluded women. However, Henry VIII's dictum and the influence of the bishops were limited to London. In the countryside men and women continued to work side by side as

Box 13.6 Ayurvedic Medicine

Documented as being the oldest medical system on earth, Ayurveda, translated from Sanskrit, means literally life knowledge (Ayu = life, Veda = knowledge). It is a holistic form of health care that originated in India approximately 5,000 years ago and is still used today by an estimated billion and a half people world-wide. The popularity of this form of health care in North America may be due to the increasing number of people turning to natural methods of disease prevention. Dr. Surendra Tripathi, one of North America's most knowledgeable practitioners, states that any health problem can benefit from Ayurvedic treatment. Problems may range from colds and depression to cancer and AIDS. Ayurveda follows a philosophy that believes perfect health can be achieved by everyone. This goal may be accomplished by balancing every dimension of one's being, i.e., mental, physical, spiritual, emotional, social, and environmental, to restore harmony and vitality for a healthy life. The basis of the Ayurvedic practice is eliminating and preventing the causes of disease. Practitioners are skilled in many traditional areas of healing, such as nutrition, herbs, medicinal plants, homeopathy, mineral therapy, massage, yoga, fasting, oils, meditation, hydrotherapy, and pancha karma, a complete cleansing therapy.

Around 480 B.C., Ayurvedic physicians established the world's first hospital, in which they performed 119 types of surgery, including brain surgery. The ancient medical system included many familiar and traditional specialties found in modern hospitals. These included geriatrics, surgery, pediatrics, gynecology, toxicology, rejuvenation, and genetics. Ayurvedic physicians were the first to perform plastic surgery, which they then introduced to the Western world in the eighteenth century.

Sometimes heralded as the mother of all healing, Ayurvedic medicine provides the structure to the essential ideas of the contemporary holistic health movement. The British pharmacopoeia continues to list 560 Ayurvedic medicines while 320 can be found in the United States equivalent.

Ayurvedic medicine identifies three main causes of disease. First, ignoring the body's wisdom (i.e., playing with animals when severely allergic); second, misusing any of the five senses (i.e., not protecting your eyes against UVA rays); and third, shifting environmental conditions (need to alter diets and habits along with seasonal weather changes).

The Ayurvedic system maintains that five "humours" control the body – air, water, ether, heat, and earth. These humours help to classify body type, i.e., vata (air), pitta (bile), and kapha (water). One humour tends to predominate although people may be deficient or excessive in any number. Humours are seen as having symbolic and literal significance. For example, a hyperactive child has too much vata (air), as does anyone with stomach gas. Determining the humour is vital to designing a lifestyle that gives one optimal health and longevity. Diagnoses and medical guidelines are determined by careful questioning, physical examinations, and pulse-taking. Pulse variations classify the humours and reveal imbalances.

The practice of Ayurvedic medicine respects the connection between the mind and the body. It remains as the most available form of health care in India. North American practitioners are often not quite qualified. These physicians should have studied for a minimum of six years and have earned a Bachelor of Ayurvedic Medicine and Surgery, which is available only in India. Future plans for Ayurvedic medicine include a proposed plan for an Ayurvedic hospital to be opened in Ontario (Imbert, 1993).

surgeons, physicians, home nurses, and midwives.

The Roman Catholic Church was particularly alarmed at the declining birth rate and the "breakdown" of the family. Seeking to bring these trends under control, the Church fathers began to write about witchcraft and its potential for spreading sin and evil. The most important work on witchcraft was the official Catholic text written by two German monks, Kramer and Sprenger – *Malleus Maleficarum* or *The Hammer of Witches* (Ehrenreich and English, 1973, 1978). Many of the people who were named witches, persecuted, and put to death were female healers and midwives, especially those of the peasant population. The *Malleus Maleficarum* declared that women were particularly likely to be witches because they were especially vulnerable to consorting with the devil in a sexual way. Lust was believed to be insatiable in women. Entry into a coven was said to involve sexual intercourse with the devil. In addition, helping and healing were in themselves evidence of witchcraft because they demonstrated a lack of faith. The Christian was supposed to accept sorrow and pain as a visitation from God and not seek to be healed.

Witch hunting in Europe spanned approximately four centuries; between 1479 and 1735, approximately 300,000 people were put to death as witches. Most of these were women. Some have argued that the campaign against witches was primarily a campaign against women (Ehrenreich and English, 1973). And most of their crimes were related to their roles as healers and midwives.

Midwives were particularly vulnerable to attack because they had access to what were considered to be prize witchcraft materials – newborn babies, placentae, and umbilical cords. In addition, it was thought particularly important to prohibit women from attending childbirth: no baby should be allowed to die unchristened, but midwives could not conduct a christening. For a time, women healers either "went underground" or left the field entirely. Later they resurfaced quietly, primarily as nurses attached to monasteries and hospitals (see Chapter Fourteen).

Then, in the middle of the nineteenth century in the United States, one woman, after numerous unsuccessful attempts, was granted admittance, as the result of a student referendum, to the University of Geneva in New York state. Elizabeth Blackwell graduated as a medical doctor in 1849 with honours in all courses, except, ironically, in the one course in which she was forbidden to enroll: "Women's Diseases." Later she established the New York Infirmary for women and children.

By the 1850s Canadian women, too, began to ask for admittance to medical schools. As late as the 1880s, the few women practising medicine in Canada had received their training outside Canada. In 1883 women's medical colleges affiliated with Queen's University and the University of Toronto were established to provide medical instruction to women. By 1895 women could take examinations at medical schools of their choice (Roland, 1988).

TABLE 13.2 Undergraduate and Professional Enrolment by Sex and Program, 1968-69, 1974-75, and 1983-84

	1968-69		1974-75		1983-84	
	Female %	Total (Male and Female)	Female %	Total (Male and Female)	Female %	Total (Male and Female)
Medicine	25.5	5,773	28.1	6,136	41.8	7,560
Nursing	95.5	4,037	97.9	5,190	97.7	7,447

Source: *CAUT Bulletin*, June, 1988, p. 16.

The Medical Labour Force in Canada Today

Both sexism and patriarchy prevail in the Canadian medical labour force at all levels of care: administrators, physicians, nurses, nurses' assistants, clerical workers, and housekeeping staff (Navarro, 1975). The higher the prestige, income, and power of the occupation, the smaller the percentage of females working in it. The medical profession is largely male; the nursing profession is largely female (see Table 13.2). Below these two levels is the service sector, made up of various technicians, orderlies, nurses' aides, nurses' assistants, and domestics. This sector is also largely female and very poorly paid. Put in simplest terms, males predominate in the most prestigious, powerful, and highest paying professions and have relegated females to the less prestigious, less powerful, and less remunerative semi-professions. This situation is illustrative of patriarchy.

It is worth adding that the proportion of visible minorities is highest in the lowest ranks of the health-care system. As I write this, the news on the CBC includes a piece about the proposal that new immigrant doctors (often, today, members of visible minorities) will be admitted to Canada and allowed to practise for five years without a licence, provided they work in under-serviced areas. At the end of the five-year period, they are to be licensed to practise as they wish. This could be seen as an example of racism. It may also reflect contempt for the nether regions – who can have doctors assigned who may not be good enough for the mainstream, particularly, urban dweller.

The Medical Profession

In the twenty years in Canada from 1969-70 to 1989-90, the proportion of female medical graduates grew from 12 per cent to 44 per cent (see Table 12.2). It may be argued, however, that during this same period the prestige and power of Canadian doctors compared to other workers in the health-care field has declined. The growing proportion of women in the medical schools and in subsequent medical practice cannot, therefore, necessarily be taken as indication of a decline in sexism.

Within the medical profession women continue to be concentrated in certain "female" specialty areas (Lorber, 1984) such as pediatrics, family medicine, and gynecology (Kirk, 1994). In general, these offer the lowest levels of financial reward and the least prestige (Ackerman-Ross and Sochat, 1980). Women are also underrepresented in the upper echelons of medicine, in medical schools, and within the hierarchy of medical schools (*ibid.*). The average salary of male doctors in Canada in 1985 (Statistics Canada, 1985) was $90,562, while the average salary of female doctors was $57,126. Similar differences by sex are evident among other health-care providers such as osteopaths and chiropractors (men made $58,645 and women made $35,680), and in optometry (men made $61,625 and women $33,600) (Statistics Canada, 1985). Moreover, women physicians, as other women workers, are vulnerable to sexual harassment in medical school, in practice with patients, and with colleagues. In fact, a recent study published in the *New England Journal of Medicine* and co-authored by Dr. Susan Phillips of Queen's University found that 40 per cent (422 of 1,064) of Ontario female family physicians have experienced sexual harassment, primarily from male patients (Graham, 1994: 14). Women physicians, the study states, are particularly vulnerable in private practice because they must examine patients physically in their offices.

Nursing

The professional status of doctors has been achieved and is continued through the dominance of male physicians. Medicalization is a process that has been largely invented and perpetuated by male thinkers, scientists, and practitioners. The dominance of males in the medical care system has been maintained through the relegation of female medical care workers, such as nurses and nurses' assistants, to relatively powerless and poorly remunerated positions.

Nursing was and is a female job ghetto, as can be seen from the figures in Table 13.2. Moreover, within nursing, males are relatively more likely to become administrators and those in positions of power. The percentage of male nurses with a higher education than the basic diploma is only very slightly higher than the percentage of female nurses with higher education. Yet, although men only have a slight edge in educational background, they have a higher percentage in the upper echelons of nursing than women do (from head nurse up to directors). (These figures give a conservative estimate of male power and potential power in nursing because the entry of males into nursing in any numbers is fairly recent; thus, male nurses will likely tend to be younger and to have had less time in which to be promoted on average than female nurses.)

In 1981 there were 45,542 physicians and 206,184 professional nurses practising in Canada. Yet, expenditures for physicians comprised 14.7 per cent of the medical care budget, while 7.1 per cent was spent on all

Box 13.7 The Registered Nurses Association of Ontario Responds to the Events at the Hospital for Sick Children and the Grange Inquiry

"The Grange inquiry was the highest priced, tax-supported, sexual harassment exercise I've ever encountered," said Alice Baumgart, the Dean of Nursing at Queen's University. The Registered Nurses Association of Ontario agreed, and in a little booklet called *The RNAO Responds* they explain why. This next section will summarize its arguments.

The arrest of Susan Nelles exemplified sexist bias for a number of reasons.

(1) The police and others jumped to the conclusion that the unprecedented number of deaths, 36 between July 1, 1980, and March 25, 1981, on wards 4A and 4B, were the result of murder. They neglected to investigate the possibility that the theory of digoxin overdose was questionable because digoxin is notoriously difficult to measure. It is normal to find some digoxin after death, and there is some evidence that digoxin levels may increase after death. No control groups were used for the baseline data.

(2) Once the police and hospital officials decided that the deaths were due to murder, they neglected to examine systematically all the possible sources of digoxin "overdoses." They failed to consider that the drugs might have been tampered with in the hospital pharmacy; or in the manufacturing or distributing branches of the pharmaceutical companies; or administered secretly by any of a number of other hospital personnel who had regular access to wards such as physicians, residents, interns, dieticians, lab technologists, or even a member of the general public who might unobtrusively have entered the hospital on a regular basis. Instead, the focus of suspicion was immediately placed on the most powerless people in the system – the nurses. Because, according to the first analysis, Susan Nelles appeared to have been the only nurse on duty for a number of suspicious deaths,

she was questioned. When she asked to see a lawyer before answering questions, the police assumed that she was guilty and arrested her. The case against her was strengthened because she apparently had not cried when the babies died. Both aspects of her behaviour – asking to see a lawyer, which was, of course, responsible adult behaviour, and her failure to cry – violated sex stereotypes; thus her behaviour was taken as evidence that Nelles must be guilty.

Other sources of bias in the investigation were evident.

(A) The focus on individuals within the bureaucratic system (the hospital) rather than the malfunctioning of the system itself.

(B) The media was accused of biased and sensationalized reporting of unfounded allegations and suspicions. As criminal lawyer Clayton Ruby stated, the media, when reporting on the inquiry, ignored their usual rules of fairness and thus held some responsibility for the damage done to reputations.

(C) Justice Grange tended to assume that a nurse was responsible for the murders and to disregard other evidence. He also disregarded the evidence that tended to raise questions about whether or not the "excess" of digoxin could have resulted from measurement error or some alternative explanation.

(D) The television coverage overemphasized the putative guilt of the nurses. The cameras tended to zoom in on the nurses' faces or hands as they were giving evidence, but rarely seemed to zoom in on those of the police or the lawyers. Such camera work emphasized the discomfort of the nurses and encouraged a picture of them as probably guilty.

Among the issues that were raised for nurses by the events at Sick Children's and the Grange Inquiry are the following.

Box 13.7 continued

(1) Nurses have little status or authority within the hospital. Yet they are held responsible or accountable for their work.

(2) In contrast to the continuing low status of nurses, their clinical roles have grown for three reasons: (a) the increased number of critical care patients in hospitals, (b) the increased number of specialists involved with each individual patient, and (c) the increased use of technology, all of which the nurse must co-ordinate.

(3) The Associate Administrator: Nursing – the highest level in the nursing echelon – was three administrative levels below that of the senior management of the hospital. Nurses thus had no access to the most senior levels of hospital management.

(4) Nurses are expected to be generalists and to move easily from one part of the hospital to another, from one type of care to another. They are expected to perform duties at night that they are not permitted to perform in the day, because during the day only the doctor is thought to have the ability to perform them.

(5) The bureaucratic structure, the assumption that nurses are generalists, and the low level in the hierarchy held by top nursing administrators limit the opportunities for nurses' advancement. In addition, nurses suffer from burnout, job stress, low job satisfaction, and the like.

(6) The events surrounding the inquiry also reinforced the notion that the "feminine" skills of nurturing and caring are much less important than the masculine skills of curing and "analysis."

(7) It became clear that the image of nursing was infused with negative stereotypes and myths, similar to the negative stereotypes of women in society.

By articulating the issues, publishing the book, lobbying various levels of government, and dealing with the hospital bureaucracy, nurses are beginning to make some changes. Destructive as the events at Sick Children's were, the outcome for the nursing profession in the long run may be hopeful.

The RNAO Responds: A Nursing Perspective on the Events at the Hospital for Sick Children and the Grange Inquiry (Toronto: Registered Nurses Association of Ontario, April, 1987).

other workers. Thus, although there are more than 4.5 times as many nurses as physicians, nurses and their colleagues received only one-half of the expenditures of physicians. According to Dickinson and Hay (1988: 65), the average salary of physicians was nearly 3.5 times the average salary of nurses in 1978: $52,499 (physician) as compared to $15,307 (nurse).

Two explanations are given for the division of labour by sex in health care (Navarro, 1975). The first is the socio-economic roles of the family in society, and of men and women within the family. The main function of the male in the family is to be the breadwinner, and therefore he is an active member of the labour force. The female's function is largely the maintenance of the labour force through her husband and the reproduction of the labour force through her children. Because of this division of labour within the family, the employer pays for the work of one and often receives the benefits of the work of two. The male worker receives financial remuneration, which he can then share as he sees fit. The reward for the female is considered to be the "emotional satisfaction" she derives from caring for her family, as well as the financial support provided by her husband.

The second reason for the division of labour by sex is the economic utility for owners of capital of having a "reserve army," a temporary and readily available cadre, of workers. In periods of high employment and production, or during war or other periods when the labour force requires additional workers temporarily and at short notice, women can be brought out of their homes and given work, for wages, in the paid labour force. Because women are still seen as primarily responsible for and rewarded by the family and domestic sphere, their jobs tend to be marginal to the paid labour force and poorly rewarded in terms of income, status, and working conditions.

Why is there a new focus on women's health and on research into women's health? According to the *Women's Health Office Newsletter* of McMaster University, men's bodies have been considered normative for major research trials such as myocardial infarction. Drug use and reactions have been tested on the bodies of young men. Women's bodies have seldom been studied in major investigations of significant causes of morbidity and mortality because women's bodies function, in part, cyclically, complicating testing and analysis. Nevertheless, as the U.S. National Institute of Health has stated, "the under-representation of women in such studies

has resulted in significant gaps in knowledge." To redress this lack, new research efforts are being addressed to women's health issues, and women's bodies are being studied along with those of men.

The following are some of the most important reasons cited for the new focus on women's health: women comprise 52 per cent of the population; women use disproportionately more medical care services than men; women have ongoing health problems associated with their reproductive systems; some diseases affect women and men differently and at different rates. Men and women have different risk factors and require different interventions. Women's multiple roles may affect their health. Women are usually gatekeepers and custodians of health care for the entire family; a woman's health affects the health of her family. Women live longer than men and have more chronic illnesses than men, thus it is hoped that addressing women's health needs alongside those of men may contribute to new paradigms of understanding (*Women's Health Office Newsletter*, 1994: 17).

This new focus on women's health has the potential of either contributing to an increased or a decreased medicalization of women's bodies. As the number of women scientists and medical practitioners increases, and as long as women become more vocal and more powerful – as they have with respect to two major women's health issues of the late twentieth century, breast cancer and midwifery – women's health issues and female-defined medical concerns may be increasingly addressed.

Women as Hidden Healers

Recently the analysis of the division by sex of the medical labour force has been expanded from the formal medical care system to the informal, from the discussion of health care in the paid labour force to health care in the unpaid labour force. Just as earlier feminists focused attention on the essential place of domestic labour in the economy, feminist scholars today are re-examining the contribution made in the home to the health field. They are finding that women have always been carers and healers: they have cared for their children and other family members during times of sickness; they have provided information about health and healing remedies. However, the pivotal role of women as invisible healers, negotiators, and mediators has long been overlooked.

Female health caring is constrained by both gender and privacy.

It is shaped, at least, by two convergent sets of social relations. First, by a sexual division of labor in which men make money and women keep the family going, and second, by a spatial division of labor where the community becomes the setting for routine care and maintenance and the institutions of medicine are the location for the acquisition and application of specialist skill. (Graham, 1985: 25-26)

Women's health work in the home involves most of domestic life. It includes cooking, cleaning, and providing a secure, warm, and stable home environment. It involves feeding others with adequate nutrients, clothing the family in seasonally appropriate clothes, and providing soap and water and the other necessary accoutrements to good hygiene. Women also provide a social/emotional environment conducive to good mental health. Women are usually responsible for teaching other family members about the ways and means of maintaining and promoting health and preventing illness both by example and by giving information. Women also usually decide when medical intervention is required, make relevant appointments, and see that family members are able to keep the appointments.

Women's health work takes place when all family members are healthy and also when they are not. Because fewer people are kept in institutions and because increasingly many diseases are chronic rather than acute, health care in the home is increasingly demanding. It may prevent women from maintaining or seeking outside employment. It may be a serious threat to the woman's own mental or physical health. As with other domestic work, women's health-care work in the home has been overlooked and underestimated. Studies are only now being published documenting the nature, experience, and extensiveness of domestic health labour.

The Medicalization of Women's Lives

Every stage of a woman's life has been subject to medical scrutiny. Pregnancy and childbirth, which many would argue are natural events, are seen as occasions for medical intervention. From the "diagnosis" of pregnancy to the delivery itself, medical care dominates women's experience of childbirth. Births take place (usually) in hospitals, in high-tech delivery rooms, amid glaring lights in sterile conditions, and in an equally sterile emotional atmosphere, with anesthetics to dull pain, surgery to speed delivery (whether Caesarean section or merely episiotomy), forceps to change the position of the baby, and fetal monitoring to check on the progress of the baby. Women at puberty may experience menstrual cramps or irregularity; doctors give them pills to help. Some women have noticed that they feel differently at different times during their monthly cycles. Mood and physiological changes such as depression and breast-swelling have been diagnosed as PMS or premenstrual syndrome. Hormones are used to smooth out the emotional ups and downs and the physical discomforts. Menopause has been defined as a deficiency disease. Again, hormones are prescribed. Many types of contraception are available only from physicians – the "pill," the IUD, the cervical cap, and the diaphragm. Indeed, most types of contraception are for women. It is women who are to take responsibility, with the help of their doctors, for preventing pregnancy. Wherever abortions are legal, they are performed by doctors.

Even old age, a condition experienced by women more than men, is generating new medical specialties – geriatrics and geriatric psychiatry.

Deborah Findlay's work would suggest that it is impossible to separate socio-cultural norms regarding the good and bad woman from the conceptualizations made by obstetricians and gynecologists regarding female physiology, pregnancy, and labour. Rather than being characterized by objectivity, as medical science claims to be, Findlay demonstrates how "objectivity" in medical science is better characterized as a resource that enables physicians to define and regulate the social world of women by surveillance and by labelling some body behaviours as normal and others as pathological (Findlay, 1993). It is not that medical science sometimes errs and bias creeps in. On the contrary, the argument sustained by Findlay is that medical science is integrally rooted and bound up in socio-cultural categories of gendered normality and deviance.

Women's emotions, too, are frequently medicalized. The oppressive experience of living in a sexist society characterized by unequal access to opportunities and rewards in the family, the society, the economy, and the political sphere is bound to result in some costs, emotional and otherwise, to women. Yet they are likely told that "it is all in their head," and are prescribed counselling, psychiatric treatment, and mood-altering or other psychotropic drugs. As well, there are numerous historical examples of sexism in the social construction of medical diagnoses and treatment. In the nineteenth century, middle- and upper middle-class women were described as weak and as vulnerable to their reproductive systems. The belief was that each individual human organism had only a fixed and limited amount of energy. Because childbearing and the provision of an heir for her husband was thought to be her most important role in life, the woman was admonished not to seek education or to work, either inside or outside the home. Sickly, weak, and hysterical, they were expected to rest – to lead a life of invalidism.

Ill health became a virtue, a sign of the refinement of middle-class women. Good health, by contrast, came to indicate a coarser makeup. Art and popular literature presented the female heroine as frail and pale, and thus beautiful. To achieve this desired look, women drank vinegar and took arsenic (*Ideas*, 1983). Not only were doctors telling women they were basically sick, but a whole patent medicine industry developed to deal with the problem. The following advertisement illustrates the ways in which women's lives were defined as inherently diseased.

> Lydia E. Pinkham's vegetable compound – the only positive care and legitimate remedy for the peculiar weaknesses and ailments of women. It cures the worst forms of female complaints – that bearing down feeling, weak back, falling and displacement of the womb – inflammation, ovarian troubles. And it is invaluable to the change of life. Dissolves and expels tumours from the uterus at an early stage and checks any tendency to cancerous tumour. Subdues faintness, excitability, nervous prostration, exhaustion – and envigorates [*sic*] the whole system! (*Ibid.*: 4)

Women were taught that they were controlled by their reproductive systems. A gynecology textbook used by Canadian doctors in 1890 states, "Women exist for the sake of their wombs" (Mitchinson, 1987). All women's natural physical processes were seen as potential diseases. Menstruation required rest. Pregnancy and childbirth required medical treatment. Menopause, as the symbolic end to the meaningful life of women, was expected to leave women depressed. Masturbation was evil, so heinous that the only appropriate cure was a clitorectomy. This was a very popular medical procedure and was used to treat almost anything that was considered "unfeminine" (Barker-Benfield, 1976). Because of the central role of women's reproductive systems, women were advised to forgo education. It was thought that it would deplete the body of the vital energy required for pregnancy, childbearing, and lactation, which were, after all, the most important functions of women.

The medical indications for "ovariotomies" (the removal of the ovaries) included a wide variety of behaviours and attitudes, including:

> troublesomeness, eating like a ploughman, masturbation, attempted suicide, erotic tendencies, persecution mania, simple cussedness, and anything untoward in female behaviour. (*Ideas*, 1983: 6)

If they were not subjected to gynecological surgery, women who were suffering were given the "rest cure." This involved isolation in a dark room, eating a bland, boring diet of soft foods, and receiving no company except the doctor and nurse. This sensory deprivation was for the purpose of resting the brain or inducing the cessation of thought – thinking, after all, was the cause of the problems in the first place.

In the nineteenth century Charlotte Perkins Gilman was prescribed this rest cure and became famous when she complained about the misery of her domestic life with its myriad restrictions. In *The Yellow Wallpaper* (reissued in 1973), Gilman describes how she just about "went crazy" under the prescribed regime, until she realized that to heal herself she had to leave her husband and her upper-class domestic life and pursue a career in writing and feminist leadership.

In contrast, working-class women were expected to work constantly and hard even while bearing many children. Employers did not give them time off for pregnancy, childbirth, or for any illness, "feminine or not." A day's absence, in fact, could cost a woman her job. The lives of working-class women were exceedingly hard. They commonly worked every day of the week, even until midnight. Industrial accidents and occupational hazards were frequent. Contagious diseases always attacked the crowded tenement flats of the poor first. There was no cult of invalidism here.

A number of feminist social commentators have noted that the social/sexual control effected through gynecological surgery and the rest cures prescribed by male medical doctors in the nineteenth century has been extended today by other means: women today receive psychotherapy and tranquilizers to keep them in their places and to help them cope with the tribulations of life in a sexist and patriarchal society.

Cooperstock and Lennard (1987) have documented the ways in which tranquilizers, one tool used by psychiatrists and general practitioners, are often used by women as aids in helping them cope with their "assigned" and "expected" social roles. In this study, women themselves talked of taking tranquilizers: (1) to help manage their feelings of exhaustion, busyness, and stress resulting from childbearing responsibilities; (2) as an aid in minimizing their feelings of anxiety and upset because they lacked time to be themselves and to do what they wanted as independent human beings (e.g., write); and (3) to enable them to continue to live with their husbands and families under situations of grave unhappiness, but without options. Two illustrations follow:

I take it to protect the family from my irritability because the kids are kids . . . I'm biding my time. One of these days I'm going to leave kit and kaboodle and walk out on him. Then maybe I won't need any more Valium.

I would like to be off in Australia somewhere, writing. You know, do my own work. But having to stop the writing to get the supper on, it irritates me. And there are so many irritations during the day. But I cannot change the situation because of my family. (Cooperstock and Lennard, 1987: 319)

A number of types of gynecological surgery are still used today. In fact, obstetricians and gynecologists claim to be "the spokesmen for women's health care" (Scully, 1980: 15). Many women consult obstetricians or gynecologists for any woman's "problem" and for routine pre-natal and post-natal care, childbirth, abortions, and birth-control prescriptions, insertions, and fittings, as well as for care during the menopause.

Diana Scully (1980) examined, through a participant observation study, the work of these specialists and published the results in *Men Who Control Women's Health: The Miseducation of Obstetrician-Gynecologists*. She showed how the male-dominated profession is infused with sexist ideas, dating back to the struggle engaged in by obstetrics and gynecology to legitimate its place in the world of medicine in the nineteenth century. Among residents and interns women are seen as objects whose fundamental purpose is to provide teaching and learning material for the future obstetricians and gynecologists. Female medical students are under-represented in the specialty as compared with other medical specialties.

Michelle Harrison, in her autobiography, *A Woman in Residence* (1982), documented this finding through her personal experience as a resident in obstetrics and gynecology. Harrison described case after case of the prevalence of sexism; of the desire to get things done quickly, according to medically defined bureaucratic protocol; of the frequent use of surgery; and of the preference for working with unusual cases. All this frequently takes place while ignoring the experience of the individual woman, her questions, her desires, and her particular situation. As evidence of sexism, Harrison recorded some of the verbal attacks made on women in labour by doctors. "You should have thought of that nine months ago"; "Why didn't you say 'no'?" She records one doctor angrily shouting to a woman who was being examined and whose vagina was tight and was resisting the doctor's pushing fingers, "You had no trouble separating your legs nine months ago, did you?" No matter whether the issue was birth control, childbirth, a D&C, a hysterectomy, or some other medical procedure, Harrison was able to document from personal experience the aggressive practices of the obstetricians/gynecologists.

A number of researchers have shown that rates of hysterectomies are excessive (see Fisher, 1986, for a review of the literature). The American Medical Association, through research it sponsored, has placed the hysterectomy and the D&C (dilation of the cervix and curettage of the uterus) as coming second – after surgery on the knee – in unnecessary surgical procedures (Scully, 1980: 17). Some have noted that this increase in women's surgery follows a decline in birth rates; consequently, physicians need to find new work opportunities. One of the best predictors of the rate of hysterectomies in a population is the number of surgeons who are paid on a fee-for-service basis.

The definition of a diseased womb is

indeed a very broad one. What is an acceptable level of symptomatology (irregular bleeding, fibroid tumours, menopausal flush or flash) in one area is unacceptable to some doctors and thus to their patients in another area. As John Bunker (1970, 1976), for instance, has noted, there are twice as many surgeons in the U.S. in proportion to the population as there are in England and Wales. There are also more than twice as many hysterectomies.

A hysterectomy terminates the production of some necessary hormones. Pharmaceutical companies have busily synthesized artificial hormones that can be administered to women after they have had the uterus removed. This raises another possible explanation for the increased rates of hysterectomy – the pharmaceutical industry's need for profit. Through drug salesmen and expensive advertising (see Chapter Fifteen), doctors and medical students are led to believe that a cure for the "side effects" of the hysterectomy caused by lack of hormone production can be remedied simply through the prescription of a synthetic product.

Explaining the Medicalization of Women's Lives

Scholars have described the many ways that the contemporary medical system tends to medicalize women's lives and experiences. Women are a population at particular risk from medicalization for birth control, pregnancy, childbirth, the new reproductive technologies, menstruation, pre-menstruation, menopause, their physical appearance, and the whole realm of women's emotional well-being and stability. Except for the very limited specialty of urology, there is no medical specialty of doctors trained to work on men as there is for women – obstetricians and gynecologists. Men's life stages and life cycles have, relatively speaking, not even been noticed or studied yet.

Doctors and women patients of certain classes have together "generated" and "marketed" certain types of diseases associated with the very facts of women's lives, such as pregnancy and childbirth, contraception, menstrual cycles, menopause, weight, and various types of unhappiness (Reissman, 1983). In so doing, they have served the short-term needs of one another. Physicians have been able to grow in power and authority, and their incomes have risen. In return, women have been provided with relief from pain and acknowledgement of their special needs for rest, mood-altering chemicals, freedom from unwanted burdens, and the like. The cost to women has been their lack of knowledge about their own bodies and power to determine what actions they should take. The solution is not to repudiate medicine entirely. Rather, claims Reissman (1983), the solution is to "alter the ownership, production and use of scientific knowledge."

For certain problems in our lives, real demedicalization is necessary. Experiences such as routine childbirth, menopause, or weight in excess of cultural norms should not be defined solely or primarily in medical terms, and medical-technical treatments should not be seen as automatically

Box 13.8 Healthy People

The U.S. Department of Health and Human Services has published a book entitled *Healthy People 2000: National Health Promotion and Disease Prevention Objectives*. This document outlines specific health/disease rate targets and the means for reaching these targets by the year 2000. To increase the overall health of the population the document suggests changes in all of the following health promotion areas: physical activity and fitness, nutrition, tobacco, alcohol, other drugs, family planning, mental health and mental disorders, violent and abusive behaviour, educational and community-based programs. The recommended changes with regard to health protection include unintentional injuries, occupational health and safety, environmental health, food and drug safety, and oral health.

---

appropriate solutions to these problems. For other conditions where medicine may be of assistance, the challenge will be to differentiate the beneficial treatments from those that are harmful and useless (Reissman, 1983: 118).

Sexism Today

Doctors in Canada and the U.S. are responding to charges of sexism by a commitment to address the concerns that women, both inside and outside of the profession, have expressed. In 1987 the Ad Hoc Committee on Women's Health Issues of the Ontario Medical Association (OMA) presented a report to the OMA acknowledging health issues unique to women that require special care and thought. The report made a number of recommendations that would require the OMA to take an active stance in defining women's health issues and in changing medical practice, medical education, and research investment so as to better address the concerns of women physicians as well as women patients (Report to the OMA Board of Directors, 1987).

Summary

(1) Two main critiques of the contemporary medical care system are that it is dominated by the medical model of health and illness and that there is a great deal of sexism in medicine and in the experiences of health and illness. Implications of the medical model are: advances against disease only happen through scientific investigation, the focus being on the causally tied pathological alterations within the cells of the body. Disease is considered an undesirable and abnormal functioning within the individual.

(2) Two models that compete with the medical model are the environmental social-structural model, which focuses on the causes of disease that lie outside the individual organism, and the lifestyle model, which focuses on understanding disease as the result of actions based on personal decisions regarding lifestyle.

(3) A complete medical care system would include something of all three models. The Canadian medical system almost exclusively relies on the medical model, which has

limited relevance to cures for the most significant modern diseases: those that are chronic. Medical intervention has enabled many people to live who would have previously died, thus increasing the burdens of chronic illness and chronic disability.

(4) The Canadian government has begun to be directed to a broader vision of disease causation and treatment. Canadian health ministries currently emphasize disease prevention and health promotion strategies through media propaganda, health education, and community development.

(5) There are large differences between males and females in income and power in the medical labour force today. Women can be seen as hidden healers, through their role in the family. Deinstitutionalization and the increasing chronic nature of many diseases will intensify the burden of home-based caregivers.

(6) New trends in Canadian society include the women's movement, which has demanded knowledge and information about women's bodies and has discovered the negative effects that medicine has had on women's physical and mental health. Another trend is the changing disease profile. The main concerns of health-care services in the future will be caring for an older population and for individuals with mental health problems.

(7) Sexism and patriarchy are prevalent in the organization and provision of medical care and in the construction of categories of illness and relevant treatments. Women's lives and experiences have been medicalized by the contemporary medical system. The OMA is beginning to change this through changing medical practice and medical education.

CHAPTER FOURTEEN

Nurses, Chiropractors, Naturopaths, and Midwives

WHAT IS THE RELATIONSHIP of the work of the allopathic doctor to other health-care providers? Do doctors compete with nurses or do they work in different spheres? Can the work of the nurse be seen as "separate but equal" to the work of the doctor? Why have chiro-practors and naturopaths been called quacks? What is their role in the provision of health care? How do they relate to allo-pathic doctors? Do alternate health-care providers receive financial support from the state? Are midwives' rates of mother and infant mortality higher than those of obstetricians and gynecologists? These are among the questions that will be addressed in this chapter.

Nursing: The Historical Context

Some form of nursing has always been avail-able for sick people, even if it was only un-skilled assistance such as help with feeding and being made comfortable. Usually nurs-ing was women's work, an extension of domestic responsibilities (O'Brien, 1989). Today, nursing is a complex paraprofessional occupation mostly taking place in a hospital run by a rigidly hierarchical bureaucracy segregated "according to sex and race, power and pay, specialty and education" (Armstrong et al., 1993: 11), and equipped with complex technology (about 80 per cent of all nurses work in hospitals). The duties of the hospital nurse are extremely varied, specialized, and clearly limited within a complex division of labour. Some nursing roles, such as those in the coronary care unit (CCU) or the intensive care unit (ICU), involve highly skilled medical management and quick decision-making in operating complex machinery. Thus, one nurse in an ICU or CCU may spend her/his working hours monitor-ing and recording the readouts on a series of machines maintaining minimal bodily func-tioning to keep a patient alive. Another, working in a psychiatric clinic, may spend her/his days in group and individual therapy, where the tasks involve listening and communicating on an emotional level with patients. But wherever nursing is taking place, it is almost always under the jurisdic-tion of a physician.

The rise of Christianity led to distinctive roles for nurses. Healing and caring for the sick came to be thought of as acts of Christian charity, and nursing became a full-time occupation for the sisters of the church: they worked with the sick, founded hospitals, and provided bedside care. Commitment to the sick was an acceptable role for devout Christian women because it was an occupation controlled by the Church. Nursing sisters were known to refuse the orders of doctors, and even to refuse to work for certain types of patients when such work violated their Christian convictions.

Military nursing, too, has a long history. As early as the thirteenth century there is evidence that the Knights of St. John of Jerusalem admitted women into their order so that they could nurse the wounded. In the seventeenth century the Knights of Malta maintained a type of nursing service in their hospital at Valetta. Two knights were assigned to each of the towns surrounding the harbour. Each pair of knights had four nurses "to assist them in their rounds," their duties being to carry supplies to the sick and the poor, to see that the physicians appointed to visit them attended to their duties and that patients received the proper care and medicine (Nicholson, 1967: 16).

After the Reformation, nursing disappeared as a respectable service for devout Christian women in countries where the Catholic Church and its organizations were destroyed. Those who could afford it were doctored and nursed at home. Hospitals were built, but they were primarily for the poor and indigent, and were filthy, malodorous, and overcrowded. They had no adequate drainage. The beds were seldom changed. Patients shared the same dirty sheets. It was not realized that fresh air was healthy and so windows were frequently boarded up. Germs spread and multiplied in such a filthy and airless atmosphere. Often even the workmen who came to whitewash the walls became ill.

The only women working in these hospitals were those who had no choice – the poor, infirm, and old, sometimes patients who had recovered. At this time a stereotype emerged of the nurse as a drunken, poverty-stricken old woman who lived and ate with the sick. Hospitals were considered places to go to die, or to go when there was no other choice. They were infested with rats, had contaminated water supplies, and were frequently inhabited by patients with contagious diseases such as cholera, typhoid, and smallpox; it is no wonder that the mortality rate, even among the nurses, was often very high (frequently as high as 20 or 30 per cent).

Marie Rollet Hébert is thought to have been the first person to provide nursing care in what is now Canada. She and her husband, a surgeon-apothecary, worked together to help the sick from the time of their arrival in Quebec in 1617. Later, other nursing sisters, members of various religious orders, immigrated to what are now Quebec City and Montreal, established hospitals, and treated those wounded in the wars. These nurses were more like doctors than modern nurses; they made and administered medicines and performed surgery. They administered the hospitals and the missions they founded.

Box 14.1 The Nightingale Pledge

I solemnly pledge myself before God, and in the presence of this assembly, to pass my life in purity and to practise my profession faithfully; I will abstain from whatever is deleterious and mischievous and will not take or knowingly administer any harmful drug. I will do all in my power to maintain and elevate the standard of my profession, and will hold in confidence all personal matters committed to my keeping and all family affairs coming to my knowledge in the practice of my profession. With loyalty will I endeavour to aid the physician in his work and devote myself to the welfare of those committed to my care. (The Nightingale Pledge was formulated in 1893 by a committee chaired by Mrs. Lystra E. Gretter, R.N. First Administered to the 1893 graduating class of Farrand Training School, Harper Hospital, Detroit.)

Gloria M. Grippando, *Nursing Perspectives and Issues*, 3rd Edition (Albany, N.Y.: Delmar, 1986).

By the eighteenth and nineteenth centuries, epidemics of smallpox, influenza, measles, scarlet fever, typhoid, typhus, and tuberculosis threatened the health of the people. In response, nursing sisters from several orders established hospitals as places where the sick could be segregated and cared for. (Much of the population, at the time, was migrant and homeless.)

The Story of Florence Nightingale

The story of the transition of hospital nursing from a duty performed out of charity and for the love of God, or by poor women who had no option, to an occupation requiring proper training must begin with Florence Nightingale. She revolutionized nursing work and laid the foundations for the modern, full-time occupation of nursing. "On February the 7th, 1837, God spoke to me and called me to his service" (Bull, 1985: 15). Another time she wrote, "I craved for some regular occupation, for something worth doing instead of frittering away my time with useless trifles." This statement has been taken to be the origin of her motivation to serve the sick despite years of opposition by her wealthy mother and sister and the pleas of several ardent suitors.

Florence was born into a wealthy English family in 1820. The upper-class Victorian woman was expected to marry, provide heirs for her husband, run his household, and be a decorative companion at social events. Ideally, her days would have been taken up with organizing the servants and governesses in the household, perhaps engaging in some fancywork, and meeting with other women concerning some charitable cause. As has been said, nursing at that time was for the most part done by the indigent. It most certainly was not a suitable occupation for someone of Florence's social standing.

Nevertheless, she was committed to making something special of her life in the service of God. Her first experience of the kind of service God might be calling her to occurred one summer when the family was holidaying at their summer place, LeaHurst, in Derbyshire. Florence met and helped a number of poor cottagers, taking them food, medicine, and clothing. Later she nursed her sick grandmother and an orphaned baby. When she learned of a school for the training of nurses located in Germany, the Kaiserwerth Hospital, she visited it. Her sister Parthe was so furious at this "inappropriate" behaviour that she screamed and threw her bracelets at her sister. Nor was her mother supportive.

Her father, having seen Florence refuse suitors, read and study mathematics late into the night, and maintain her fervent commitment to God's call, finally weakened. Two rich, aristocratic friends, Sidney and Elizabeth Herbert, supported Florence and spread the word that she was England's leading expert in matters of health. When a director was required for a nursing home, the Institution for the Care of Sick Gentlewomen in Distressed Circumstances, the Herberts recommended Florence. She accepted the position and turned the nursing home into a very good and well-run hospital.

Slightly more than one year later, after she had gained invaluable experience in running the hospital, Sidney Herbert, who was then Secretary of War, asked Florence to go to Turkey to nurse British soldiers injured in the Crimean War. The war, to that point, had been disastrous for the British. They had been unprepared. There were widespread shortages of equipment, food, bedding, and medical supplies. The soldiers were expected to live on mouldy biscuits and salt pork. They slept in the mud in clothes and blankets that were stiff with blood and crawling with lice. There was no water for washing, and all the drains were blocked. Almost every man had diarrhea, but there were neither diets nor special medicines to relieve it (Bull, 1985: 35). There were no hot drinks, because the necessities for lighting a fire were not available.

Florence was asked to raise and take a group of 40 nurses into this chaotic and squalid situation. She advertised in London and beyond, but was only able to find 38 women, some of whom were religious sisters, with the qualities she required. Nightingale and her staff set out for the Crimea. She laid down strict orders to be followed: all were to be considered equal; all were to obey her. They were to share food and accommodation. All wore uniforms comprised of grey dresses and white caps.

Nightingale and her nurses arrived at the Barroch Hospital in Scutari ready to work and armed with financial resources. They were met with hostility by the doctors, who refused to let them see patients and welcomed them by providing all 38 women with only six dirty, small rooms (one of which had a corpse in it). They were given no furniture, lighting, or food. Florence had experienced much opposition in her life; she was prepared to wait. She told her nurses to make bandages, and offered the doctors milk puddings for the patients. These tasks seemed "suitable" for women, and so the doctors accepted the milk puddings. Less than one week later there was another battle and a huge number of casualties. The doctors, overwhelmed by the enormity of the disaster and the tasks that lay ahead of them, and reassured that the nurses were willing to obey (they had waited) and that they could provide "feminine services" (making milk puddings), asked the nurses to help out.

Nightingale ordered food, cutlery, china, soap, bedpans, and operating tables. Her nurses sewed clean cotton bags for straw mattresses. Two hundred men were hired to clean the lavatories and basins. The floors were scrubbed for the first time in anyone's memory. Soldiers' wives were recruited to wash clothes and bedding. Still the mortality rate did not decline significantly. Nightingale, who knew far more than the doctors about the importance of fresh air, cleanliness, and clean water, had the plumbing inspected. The body of a dead horse was found blocking the pipes and contaminating the water supply of the entire hospital. Once this was cleaned out the mortality rate dropped dramatically.

Nightingale was viewed as a heroine both within the hospital and without. Within the hospital she was known as the lady with the lamp. She offered all kinds of services, even banking and letterwriting, to her patients. At the end of the war the grateful soldiers dedicated one day's pay to her for the establishment of the Nightingale School of Nursing and the Nightingale Fund. Outside the hospital, at home in Britain, she was heralded as the most important woman of her time. There were at this time no female judges, MPs, or civil servants. No woman had ever taken charge of

an institution. Florence Nightingale was a heroine to Britain and to the Western world.

Her greatest achievements were in public health and its reforms. Her writing and lobbying in this area continued long after she returned from the Crimea even though she spent the rest of her life, essentially, bedridden. It has been suggested, in fact, that Nightingale's work was the beginning of modern epidemiology. Her work is also thought to have provided the model for the early training of nurses. Eventually, after several unsuccessful attempts, the first nursing school in Canada, on the Nightingale model, was established at the General and Marine Hospital in St. Catharines, Ontario, in 1874. A few years later the Toronto General (1881) and the Montreal General (1880) hospitals were established. Nursing students often comprised almost the whole staff of these hospitals (Jensen, 1988).

The long-term effects of Nightingale's work on the practice of nursing are often considered to have been equivocal, however (Reverby, 1987). On the one hand, she herself exhibited enormous strength and commitment and was able to garner extensive personal power as a leader in epidemiological research models, in health-care policy, in the management of hospitals, in military nursing, and as the most important role model for the secular occupation of nursing in her society. On the other hand, she ensured that the women who worked as nurses for her were taught to be handmaidens to men (doctors) and to be "mother-surrogates" to patients. Even though Nightingale studied hard to become

Box 14.2 Things You Never Learned in Nursing School

Submitted by Dorothy Fulford, Employment Relations Officer, Ottawa Regional Office. The following job description was given to floor nurses by a hospital in 1887.

In addition to caring for your patients, each nurse will follow these regulations:

1. Daily sweep and mop the floors of your ward. Dust the patient's furniture and window sills.
2. Maintain an even temperature in your ward by bringing in a scuttle of coal for the day's business.
3. Light is important to observe the patient's condition. Therefore, each day fill kerosene lamps, clean chimneys and trim wicks. Wash the windows once a week.
4. The nurse's notes are important in aiding the physician's work. Make your pens carefully; you may whittle nibs to your individual taste.
5. Each nurse on day duty will report every day at 7 a.m. and leave at 8 p.m. except on the Sabbath, on which day you will be off from 12 noon to 2 p.m.

6. Graduate nurses in good standing with the director of nurses will be given an evening off each week for courting purposes or two evenings a week if you go regularly to church.
7. Each nurse should lay aside from each pay day a goodly sum of her earnings for her benefits during her declining years so that she will not become a burden. For example, if you earn $30 a month you should set aside $15.
8. Any nurse who smokes, uses liquor in any form, gets her hair done at a beauty shop, or frequents dance halls will give the director of nurses good reason to suspect her worth, intentions and integrity.
9. The nurse who performs her labours and serves her patients and doctors without fault for five years will be given an increase of five cents a day, providing there are no hospital debts outstanding.

Gloria M. Grippando, *Nursing Perspectives and Issues*, 3rd Edition (Albany, N.Y.: Delmar, 1986), p. 59.

educated and rejected her own assigned gender role, she expected her nurses to be subservient, obedient, and docile in their relationships with medical doctors. Nursing, under Nightingale, was to be a woman's job. Only women had the necessary character and qualities. While carving out a respectable occupation for women, she also reinforced a ghettoized and subordinated female labour force that is still in place today. This model of nursing supported the traditional stereotype of the physician as father figure and the nurse as mother.

Nursing Today: Issues of Sexism, Managerial Ideology, and Hospital Organization

Nurses comprise about two-thirds of all medical care providers. As Table 14.1 demonstrates, the proportion of nurses in the population has increased substantially since 1965. The number and proportion of registered nursing assistants has also grown significantly over the same period of time. More than 80 per cent of nurses work in health-care institutions – hospitals and

TABLE 14.1 Number of Physicians and Nurses, and Population per Physician and Nurse in Canada, 1901-1989

Year	PHYSICIANS Number	PHYSICIANS Population per physician*	NURSES Number**	NURSES Population per nurse*
1901	5,442	978	280	19,014
1911	7,411	970	5,600	1,284
1921	8,706	1,008	21,385	410
1931	10,020	1,034	20,462	506
1941	11,873	968	25,826	441
1951	14,325	976	41,088	325
1961	21,290	857	70,647	258
1971	32,942	659	148,767	146
1981	45,542	538	206,184	119
1982	47,384	521	214,989	115
1983	48,860	510	218,344	114
1984	49,916	503	222,960	113
1985	51,948	487	229,650	110
1986	53,207	479	237,181	107
1987	55,275	467	241,955	107
1988	57,405	455	249,827	104
1989	58,942	449	252,189	105

*Based on census data.

**Registered nurses for 1941 to 1975; census figures for 1931 (graduate nurses) and earlier years (nurses). Excludes Newfoundland prior to 1961; excludes Yukon and Northwest Territories prior to 1941. The 1921 figure includes nurses-in-training. Figures from 1981 to 1989 include only nurses registered during the first four months (three in Quebec) of the registration period and registered in the same province in which they work or reside.

Sources: For 1901 to 1971, Statistics Canada, *Historical Statistics of Canada* (2nd ed.). Ottawa: Minister of Supply and Services, Series B82-92, 1983. For 1981 to 1989, Health and Welfare Canada, *Health Personnel in Canada.* Ottawa: 1991.

long-term care institutions (Abelson *et al.*, 1983: 13). Table 14.2 shows that nursing was the sixth largest category in the female work force in 1981.

Critical analysis of the work of the contemporary nurse tends to follow one of four lines: (1) the patriarchal/sexist nature of the content of the work and of the position of the nurse in the medical labour force, (2) the managerial revolution in nursing practice today, (3) the impact of the bureaucracy of the hospital on the working life of the nurse, and (4) the impact of cutbacks on the quality and safety of nursing work.

Sexism

Sexism is ubiquitous in the medical labour force. Doctors have usually been men and nurses have usually been women. The role of the nurse, in fact the very name for that role, refers to female functions. "Nursing"

TABLE 14.2 Largest Occupations in the Female Labour Force, 1971 and 1981

GROWTH IN FEMALE EMPLOYMENT
1971-1981

Occupation	1981 Total Female Labour Force (000s)	1971 Total Female Labour Force (000s)	Rank
Secretaries and stenographers	368	128	2
Bookkeepers and accounting clerks	332	196	1
Sales persons	292	117	4
Tellers and cashiers	229	117	4
Waitresses	201	96	5
Nurses	168	67	6
Elementary and kindergarten teachers	140	19	18
General office clerks	115	36	13
Typists and clerk-typists	103	18	19
Cleaners	97	42	10
Sewing machine operators	93	43	8

Source: *Census of Canada*, 1981, Catalogue 92-920. Reproduced with permission of the Minister of Supply and Services, Canada, 1989.

comes from the word "nutrire," and means to nourish and suckle. Both historically and today, nursing refers both to a mother's action in suckling or breastfeeding her baby and to the act of caring for the sick. Because the earliest nurses were nuns working for the glory of God, nurses have long been called sisters. In England head nurses in hospitals are still called sisters. The very concepts of caring, nurturing, feeding, and tending to the sick are inextricably tied up with ideas about women (Reverby, 1987; Growe, 1991). To be a woman is to be a nurse, and to be a nurse is to be a woman (the few men who train or are educated to be nurses usually end up moving up the administrative ladder).

Florence Nightingale's views of nurses and their training reinforced the image of the nurse. It was her view that while nurses could be trained in some of the detailed duties of bedside care, the most fundamental aspects of the nursing occupation could not be learned. Just as it was impossible to train a person to be a mother, it was impossible to train a woman to nurse. The components of a woman's character, her selfless devotion to others and her obedience to those in authority, could not be learned or taught. Entrance requirements to nursing schools thus always included an interview in which these nebulous characteristics were evaluated. Fine tuning was still necessary, but it could only be managed if the woman was first of all "successful as a woman." One of the first hospitals to train

Box 14.3 The Doctor-Nurse Game

Leonard I. Stein's paper "The Doctor-Nurse Game" (1987) describes one strategy used by doctors and nurses to manage the contradictory position of the doctor who has more power and authority and the nurse who frequently has more information and knowledge, both about particular patients and their health and well-being on a day-to-day basis while in the hospital, and about hospital routines and common medical practices. This contradiction is especially acute when the doctor in question is a resident, intern, or new graduate. Stein's point is that nurses frequently make suggestions to doctors about how to treat certain situations and cases but that such suggestions must be handled with great subtlety and caution and even disguised. The object of the game is for the nurse to make recommendations to the doctor all the while pretending to be passive. On the other hand, the doctor must ask for advice without appearing to do so. A typical scene would proceed as follows:

Nurse A: Mr. Brown has been complaining of pains in his legs for more than six hours today. He appears to be quite uncomfortable.

Doctor G: Is this a new symptom for Mr. Brown or is it recurring?

Nurse A: Mr. Brown complained of a similar pain last week when he was admitted. He was given xxxx and it seemed to diminish.

Doctor B: OK. Let's try xxxx. What dosage did he require to get relief?

Nurse A: 3 m/hour.

Doctor G: OK. xxxx, 3 m/hour, nurse, please.

Nurse A: Thank you, doctor.

As Stein indicates, the game plan is taught to the nursing students at the same time as they learn the other aspects of nursing care. Doctors usually start to learn once they actually begin to practise in a system with nurses.

---

nurses according to the Nightingale model, the New England Hospital for Women and Children in Boston, included among the requirements for admission "that an applicant be between 21 and 31 years old, be well and strong, and have a good reputation as to character and disposition, with a good knowledge of general housework desirable" (Punnett, 1976: 5).

This sexism in ideology is reflected in differences in pay, authority, responsibility, prestige, and working conditions between men and women. (See Chapter Thirteen for a more complete discussion of sexism.)

Physicians have the advantage over nurses in all of these respects. The nurse has responsibility for carrying out the doctor's orders, but no authority to change them when they are incorrect.

Most nurses in a hospital work shifts. Their working conditions are inferior to those of physicians in a number of ways: (1) their working hours are rigid; (2) night and evening shift work is regularly required; (3) they are restricted in their work to the location of the hospital; and (4) they are at risk because of the occupational health and safety conditions in hospitals (Walters, 1994;

Walters and Haines, 1989); (5) their work-loads are excessive (Walters and Haines, 1989); (6) they are, because of underfunding, often forced to do housekeeping and other non-nursing work (Stelling, 1994). Thus, sexism restricts women to an occupational ghetto, and that ghetto provides them with few rewards and no or only limited author-ity (Warburton and Carroll, 1988).

Managerial Ideology in Hospitals

The second critical issue is the dominance of a managerial ideology in hospital manage-ment systems. This ideology is the outcome of an historical trend that reflects the devel-opment of the money economy, bureaucracy, capitalism, and rationalization. Managerial ideology assumes that it is the job of man-agers to run organizations as efficiently as possible so as to provide adequate service at minimal cost. In Canada hospitals are still, by and large, funded through the public sector – federal and provincial funding pro-vides operating grants, and local municipal-ities also provide some funds. Hospital managers are accountable to boards who are faced always with limited financial resources and the necessity of setting priori-ties. Payments for nursing services must compete with needs for cleaning, equipment upkeep, purchase of new and increasingly expensive technology, and payments to the myriad other hospital personnel, including occupational and physical therapists, social workers, staff physicians, and others.

Several rationalized management sys-tems have been developed for use in hospi-tals. One is case mix groupings (CMGs) – a technique for detailing the specific duties and the time they take in order to deal with the average patient with a particular diag-nosis. Productivity and cost-effectiveness are the goals. The nursing service must attempt to work within the specified time limits and still provide adequate nursing care to the patients. Case mix groupings are a set of mutually exclusive categories that can be used for describing patients' clinical attributes. Patients within a particular CMG are believed to require roughly equivalent regimens of care and hence are believed to consume similar amounts of hospital re-sources (May and Wasserman, 1984: 548).

Case mix grouping assumes that all patients with a particular medical condi-tion will require similar medical treatment (e.g., childbirth by Caesarean section, or chemotherapy with stage-one lung cancer). A cost is then assigned to the nursing care required by the typical patient with that specific diagnosis. Each of the patients in the hospital is accorded a particular time/ cost value. Nurses, nursing assistants, orderlies, and other medical care personnel can then be assigned to various wards for specific lengths of time to provide prede-fined services.

According to Campbell (1988), such ration-alization has numerous deleterious effects on the working lives of nurses. In many ways such a system demeans the authority of the nurse and diminishes her powers of decision-making. Individual patient needs are assessed, not in a holistic way, as a result of experience gained during the practice of the art and the science of nursing, but by a predetermined, quantified, and remote system. Nurses know that individual patients always vary from the norm in some way or another. Yet time for each patient has been allocated by the classification system. The nurse, then, is constrained, by virtue of the time available to her, to behave to all patients, regardless of their individual differences, in a certain predefined way. Not only is such a requirement destructive to the morale of the nurse, but it may be dangerous or harmful to the patient.

Staffing assignments based on the information provided by the CMGs do not allow for the fact that, just as patients differ from the norm, so, too, do nurses. For instance, a small nurse may need help in turning a large patient. Turning may be required hourly. Yet, if the assignment has not considered such characteristics of staff/patient interaction as relative size, it cannot predict costs accurately. In this case the smaller nurse will have to get the assistance of a larger nurse, a nursing assistant, or an orderly. Finding the person to provide such assistance will take additional time; the person who provided the assistance will have to deduct the time taken from his/her total time available for care. A generous allotment of flexible time could enable the nursing staff to manage such situations. However, such flexibility does not exist in the climate of cost-containment – even profit orientation – that typifies the contemporary hospital system. For nurses, the outcomes of such management systems include decreased job satisfaction, burnout, and stress (Conley and Maukasch, 1988). For patients, the outcomes can include poor quality of care, slower recuperation,

and, ultimately, greater long-term susceptibility to ill health.

Bureaucratic Hospital Organization

The third area for the critical analysis of the nursing profession today is related to the effect of the hospital's bureaucratic structure on nursing. In *Men and Women of the Corporation* (1977), Kanter argues that power structures and opportunity structures within an organization have significant impacts on the work lives of the people in the organization. Opportunity consists of the available career expectations and the probability of reaching these expectations. The structure of opportunity within the organization is determined by rates of promotion, locations for promotion, jobs that lead to promotion, and the like. People who have little opportunity tend to: (1) have lower self-esteem and sense of self-determination, or lack confidence in their ability to change the system; (2) seek satisfaction outside of work; (3) compare themselves with others on the same organizational level rather than with people on a higher level; (4) limit their aspirations; (5) be critical of managers and those in powerful positions; (6) be less likely to expect change; and (7) be more likely to complain.

Opportunities for nurses are severely lacking in modern hospital organizations. The structure of promotion in the hospital limits vertical mobility for nurses. The major option open to a nurse with aspirations is to leave nursing and become an administrator. The first step would be to become a head nurse on a ward, with day-to-day

Box 14.4 Mistakes and Complaints

"Vocabulary of complaint" is a term used by Turner (1987) in his analysis of the ways that nurses express the stress and frustration they experience in their working lives because of rigid bureaucratic organization, the prevalent managerial ideology, sexism, and relative powerlessness in the medical labour force. According to several pieces of research, such resistance to working conditions tends to be expressed indirectly through addiction to drugs and alcohol (it is estimated that approximately 10 per cent of all nurses are affected), burnout (which is evidenced in the high rate of turnover among nurses and the high rate leaving nursing altogether), and over-compliance (Milgram, 1974). There is some evidence that nurses have been willing to follow doctors' orders even when they could be detrimental to the health and well-being of the patient.

Turner (1987) coined the term "vocabulary of complaint" because he noted that there is a split in nursing discourse between complaint and compliance. As nurses learn the formal public vocabulary of nursing and its ideology, they also learn a less public and defensive discourse of complaint whereby they are able to manage some of their frustrations.

While the official occupational ideology specifies how in principle tasks are to be accomplished, the vocabulary of complaint outlines methods of survival on the job which have the consequence of delegitimizing the authority of the formal structure of operations within the bureaucracy. (Turner, 1987)

The vocabulary of complaint serves to provide nurses with some sense of shared culture and solidarity to set against their lack of autonomy and authority, coupled with excessive responsibility. Such vocabulary includes the notion that not only do nurses have a unique and special knowledge base and skill with respect to health care, but also that doctors and their medicines can be downright destructive to patients. The second theme of the vocabulary is one that delegitimizes the hierarchy and authority system in the hospital. The third serves to diminish the idealism of the new nurse through the description of nursing as poorly rewarded, emotionally and physically burdensome drudgery. Unfortunately, from the perspective of change, such a vocabulary serves as a safety valve for the frustrations and this wards off any attempts at real institutional or collective change.

---

responsibility for the running of the ward and for managing a team of nurses, nurses' aides, and orderlies. Further movement up the system takes the nurse further and further from the actual practice of nursing.

Power is the capacity to mobilize resources (Kanter, 1977). Power includes the discretionary ability to make decisions that affect the organization; the visibility of the job; the relevance of the job to current organizational problems; and opportunities for promotion. Nurses, by Kanter's criteria, have very little power. Two examples of powerlessness experienced as responsibility without authority captured the imagination and the political agenda of Canadian nurses. The first was the case of three intensive care nurses at Mount Sinai Hospital in Toronto.

These three nurses refused to admit a critically ill patient one night because they knew that they did not have the staff available to provide the one-to-one care the patient required. They reported their decision and the reasons for it to the admitting doctor and the nursing supervisor. The result was that they were suspended without pay. They took their case to court but lost. "Nurses are to obey first and grieve [submit their grievances] later," according to the courts (Wilson, 1987: 22).

The arrest of Susan Nelles for the murder of four infants at the Hospital for Sick Children in Toronto was the second event. Even though physicians and others were routinely on the ward and were also frequently involved with the patients, it was a nurse who was first blamed. Furthermore, many nurses (and others) were incensed by the fact that the televised hearing showed that the Grange Inquiry treated doctors differently from nurses – dramatically so. Doctors, on the assumption that they were innocent, were questioned with deference and respect. Nurses were questioned under the assumption of guilt and suspicion. "It was as if the police assumed, if it wasn't this nurse who committed the crime, which nurse was it?" (*ibid.*: 27).

Through these two examples, nurses learned that they were the first to be blamed (had the most responsibility) but had little respect or authority.

Cutbacks

Today, in the mid-1990s, there is a grave threat to the financial future of the health-care sector. Already, cutbacks are having significant effects on hospitals. The overall number of beds in hospitals has already declined but hospitals are actually treating more patients (Rachlis and Kushner, 1994), hospital patients are sicker (Stelling, 1994), and medications and treatments are more frequent and more complex. Thus, nurses report that they are increasingly overworked and frequently have to work overtime, without pay, just to get their work done and the patients and charts ready for the next shift. Jobs are being eliminated or made insecure; nurses are being laid off and those who remain must work harder, under worsening conditions, with decreasing opportunity to provide the care they are trained to provide (Armstrong *et al.*, 1993).

Nursing as a Profession

Nurses have been striving in many ways to reach professional status. They have done this through: (1) increasing educational requirements (to include, by the year 2000, graduation from university); (2) forming their own "college" to handle questions of practice and the discipline of members; (3) attempting to carve out a body of knowledge that would be separate from that used by other medical care workers; and (4) emphasizing the special qualities and skills that nurses have and that physicians do not have.

Freidson (1970: 49) argues that nursing is still, however, a paramedical occupation, as are the occupations of laboratory technicians and physical therapists, because of four characteristics they all share: (1) the technical knowledge used by the paramedical occupations is usually developed and legitimated by physicians; (2) the tasks of paramedicals are usually designed to help physicians fulfil their more "important" duties; (3) paramedicals usually work at the request of the physician; and (4) they are accorded less prestige than the medical profession. The medical profession is unique in the sense that no other profession has such a bevy of supportive occupational groups enabling it to do its work. While lawyers use members of other occupational groups regularly, there is no way that these groups (e.g., accountants, real estate agents, court clerks, bailiffs, and so on) are considered paralegals.

The modern occupation of nursing has developed out of the context of an historical subservience to the medical profession. Nursing tasks, roles, rights, and duties have arisen to satisfy and please physicians. From the day that Florence Nightingale and her nurses in the Crimea first waited to nurse until the doctors gave the orders, nurses have waited on doctors. Contemporary nurses, in an effort to enhance their position in the medical labour force and/or to achieve the status of a profession, have taken a number of actions. Krause lists these: (1) the shift to university training, (2) the takeover of physicians' dirty work, (3) the use of managerial ideology, (4) taking control of the technology, and (5) unionizing (1978: 52). Each of these will be examined in turn.

The Shift to University Training

There is no question that the Canadian Nursing Association wants all nurses to have a baccalaureate in nursing in order to practise after the year 2000. The B.Sc.N. degree program is designed to increase the credibility of the nurse by providing a theoretical background to nursing practice. It is expected:

> to provide a broader base of preparation with a stronger theoretical perspective that enables nurses to care for patients in a wide variety of settings. Program content provides the links between the societal needs and the health care system today, and also prepares the graduates to work, now and in the future, beyond acute care settings and into the community. With a focus on critical thinking, independent decision-making, research, leadership, and management skills, the professional nurse is prepared to apply a holistic approach to the patient and family-centred care that is reflective of her/his advanced clinical and theoretical preparation. (Kirkby, 1986: 14)

A number of phrases in the preceding statement illustrate something of the unique focus that the Canadian Association of University Schools of Nursing and the Registered Nurses' Association of Ontario are attempting to put on nursing. If nurses do indeed engage in critical thinking, independent decision-making, research, and leadership, it is only in the context of the definitions of appropriate care legitimated and instituted by management through such techniques as the case mix groupings discussed earlier, and under the authority of physicians. Thus, while there is a definite move toward requiring a university degree in order to practise, there is still no significant change in the hospital authority structure (Armstrong et al., 1993). Even nurses with M.Sc.N. or Ph.D. degrees are not able to claim professional autonomy vis-à-vis the physician.

Taking Over the Work of the Physician

Nurses have taken over some of the "dirty work" of physicians, including such routine tasks as monitoring blood pressure, setting up intravenous infusions, and giving medications. At the same time, they have passed some of their own "dirty work" down the line to registered nursing assistants, e.g., bed-making, bathing, feeding, and other actions for the personal care of patients. Both of these shifts have served to shore up the relative importance of nurses in the medical labour force: they have accepted some of the higher-status tasks of physicians and rejected some of their own, less desirable tasks.

Managerial Ideology and Nursing

Managerial ideology applied to nursing has led to a multitude of divisions within the occupational group. These include divisions among nurses by specialty (e.g., intensive care nurses, pediatric nurses), divisions by type of training (hospital- or university-based training), and divisions by level of responsibility (head nurse, staff nurse). Furthermore, registered nurses have divided the occupations below them in the medical labour force into various classes of subordinates, even while trying to maintain control as the governing body over registered nursing assistants and midwives (Wysong, 1986: 9). Below these groups are nursing aides or assistants. The impact of managerial ideology is noticeable both in the increase in the introduction of such efficiency measures as CMGs and other means of speeding up the delivery of nursing care, and in the "ideological subordination" of nurses to the interests of the organization or hospital, rather than to the patient or the nursing community (Warburton and Carroll, 1988).

Rejecting High-Tech Medicine

Nurses in Canada are seeking advancement for their occupation through an anti-technology ideology. Rather than emphasizing their skill and expertise in handling new technologies, nurses are rejecting the value of the medical model and arguing that medical care should be based much more broadly on holistic care, health promotion, and disease prevention. They emphasize the importance of their caring approach versus a curing approach. This argument is particularly persuasive given the current demography of illness, i.e., the increasing rates of chronic illness and the aging population. "Expanded rehabilitation-nursing services will be required, due to an increase in chronic illness and stress-related health problems, and will necessitate expanded community health teaching and counselling services" (Kirkby, 1986: 14). The following statement of nursing ideology illustrates something of this trend: "Nurses need to function in the roles of caregiver, teacher, counsellor and patient advocate. Nurses also coordinate the health care provided by many health care professionals" (*ibid.*).

Unionization

Unionization has had a positive effect on at least one particular aspect of the status of nursing – income. Paradoxically, however, unionization has also served to proletarianize nurses and to establish their position as members of the skilled working class who work for wages. More research needs to be undertaken on the short- and long-term consequences of the unionization process.

In 1939 a group of Quebec City nurses negotiated the first employment contract. By 1945, the British Columbia nurses' professional association assisted in unionizing the nurses of the province. Almost 20 years later nurses in the rest of the country took the course of unionization. By 1980 nurses in each province were represented by two provincial organizations – a professional body and a nurses' union. In 1981 both of these – the union and professional association – became national organizations (Jensen, 1988).

Alternatives to Allopathic Medical Practitioners

Allopathic medicine refers to the type of healing based on a theory of opposites, or on the assumption that opposites cure. Health is natural, not just the preferred state of the body. Disease is an unwanted aberration that attacks the equilibrium of the body. The germ theory of disease provides the basic assumptions upon which the growing edge of allopathic medicine has developed. It is the root from which the monumental scientific and biomedical research industry has grown. Insofar as disease is fundamentally an unnatural, abnormal invasion into an otherwise healthy body, treatment involves an attack on the enemy – the disease. Such treatments as surgery, medication, and radiotherapy are the outcomes of this type of thinking.

A variety of criticisms of this medical model have been offered (see Chapter Thirteen). Advocates of holistic health criticize the medical model, not because of its focus on the individual but rather because of what they claim is its reductionistic focus on and the mechanical and biomedical treatment of the physical body. Instead, they advocate treating the whole person – body, mind, and spirit – through a combination of methods that are best suited to a particular individual. Among the methods chosen are massage, acupuncture, visualization (imaging a healthy body), meditation, prayer, psychic and faith healing, chiropractic, and naturopathy.

"Healers" who offer various services beside or in competition with the allopathic physician stand in differing relationships to the dominant medical profession (Wardwell, 1988). The five types of relationships are: (1) ancillary workers whose work is controlled by the medical profession, e.g., nurses; (2) limited medical practitioners who offer non-competing, limited, and parallel services such as dentists; (3) marginal practitioners who offer "complete" or near "complete" alternatives to the medical profession but do not have widespread state support, e.g., naturopaths or chiropractors; (4) quasi-practitioners who operate from a fundamentally different set of epistemological assumptions, e.g., psychic healers; and (5) parallel professions who offer services that are considered almost equal to those offered by the allopathic practitioner, such as osteopaths.

Willis (1983) gives a more critical analysis of the division of labour among health-care workers (see Chapter Thirteen). He sees the varying relationships to the medical profession as the result of the three different strategies used by allopathic practitioners to control the medical marketplace. He argues that in Australia, midwifery was subordinated to medicare, optometry was limited in the sorts of work that could be done, and chiropractic was excluded from the medical labour force (although over the years chiropractic has gained in numbers of practitioners and in perceived legitimacy).

The purpose here is to describe the philosophies and the occupational status of each of three alternatives to the physician: chiropractors, because they are the largest competing alternative health-care occupation in Canada and in the world; naturopaths, because of their growing importance in Canadian society; and midwives, because of their centrality to issues concerning women's health and because of their increasing legitimacy as an alternative to obstetrician-based delivery.

Each of the three alternatives can be seen as a different type of holistic health care. In each case the focus is on the individual as a "whole," body, mind, and soul. Whereas allopaths focus on the biological system within the human body and tend to look for a specific mechanical cause of a bodily disturbance, holistic practitioners tend to focus on the individual as a whole, as he or she relates to the social world. In addition, while allopathic medicine tends to focus on the treatment of disease, holistic practitioners are concerned with early prevention of disease. As alternatives to physicians, holistic practitioners fit within the lifestyle model of illness described in Chapter Thirteen. Thus, they tend to individualize medical/social problems, while ignoring the social-structural and environmental causes of illness.

Chiropractic

The field of chiropractic has been a frequently misunderstood occupation. It is passionately supported by some and passionately repudiated by others. Spinal manipulation is an ancient technique legitimated even by Hippocrates: "Look well to the spine, for many diseases have their origins in dislocations of the vertebral column" (cited in Caplan, 1984). Yet one chiropractor commented that when he first started to practise 40 years ago, he was refused admittance to service clubs, and people would make a point of stopping him on the street to call him a quack (personal communication). From the beginning, American allopathic doctors vehemently opposed chiropractic. The American Medical Association, as late as the mid-1980s, was saying that there is no scientific evidence for chiropractic and warned the public against the untold dangers of submitting to such treatment (Caplan, 1984).

The founder of chiropractic, Daniel David Palmer, was born in Port Perry, Ontario, in 1845, and moved to Davenport, Iowa, when he was twenty. His work as a healer was based at first on magnetic currents through the laying on of hands. His first success in spinal manipulation is said to have occurred in September, 1895, when a deaf janitor in Palmer's apartment building dropped by to be examined. After discovering that the man had been deaf for seventeen years and that his deafness had begun when he had exerted himself and felt something give way in his back, Palmer manipulated the man's spine. This immediately restored his hearing (Langone, 1982). Encouraged by this remarkable healing, Palmer investigated the impact of vertebral displacements on human disease. He called the newly discovered technique chiropractic, from a combination of two Greek words: cheir and practikas, meaning "done by hand" (Salmon, 1984).

Palmer campaigned for the legitimation and popularization of this new method of healing. He founded the Palmer School of Chiropractic, whose only admission requirement was the $450 admission fee. In 1906 Palmer was charged with practising without a license and put in jail. It was not until 1913 that the first state, Kansas, passed licensing laws to allow the practice of chiropractic.

Palmer's most important pupil was his own son, Bartlett Joshua, or B.J. For 50 years or so, B.J. was able to popularize chiropractic to such an extent that he died a multi-millionaire. He was a gifted salesman who developed mail-order diplomas and advertising strategies that spread chiropractic around the world. Apparently B.J. advertised extensively for students, emphasizing the lack of exams or other requirements, lectured on business psychology, and wrote books with titles such as *Radio Salesmanship*. He was fond of making up slogans and having them engraved on the school's walls. One such slogan was, "Early to bed, early to rise; work like hell and advertise" (Weil, 1983: 130). B.J.'s sense of humour was also exhibited in the following question and answer, included in his book, *Questions and Answers About Chiropractic*, published in 1952. "Q. What are the principal functions of the spine? A. (1) To support the head; (2) to support the ribs; (3) to support the chiropractor" (quoted *ibid.*).

B.J. Palmer believed that vertebral subluxation (misalignment) was the cause of all disease, and thus that chiropractic was a complete system that could cure all disease. His belief that chiropractic was adequate to deal with all problems led to a major schism in 1924, when B.J. Palmer introduced an expensive new piece of equipment – the neurocalometer – and insisted that all chiropractic offices rent one from the Palmer School at $2,500 per annum. The major dissenter was an Oklahoma City lawyer who had become a chiropractor. In response to the unilateral dictate to buy this equipment, he established his own school of chiropractic. Within a few years he developed the theory that chiropractors should use other methods as well as spinal adjustment, methods such as nutrition and physical therapy. Such a combination is called "mix," and this chiropractic philosophy is called Mixer; B.J. Palmer's philosophy, in contrast, is called Straight. The schism continues today. However, Mixers are in the majority.

Chiropractic Theory and the Possible Future of Chiropractic

There are two patterns of practice based on spinal manipulation: osteopathy, founded in 1894 by Andrew Taylor Still, and chiropractic, founded by Daniel David Palmer in 1895. Osteopaths, who now include surgery and chemotherapy among their treatments, have achieved considerable legitimation, particularly in the United States. Chiropractic has now achieved a good deal of legitimation in Britain, Europe, and Canada. But as recently as 1971 the AMA established a Committee on Quackery whose purposes included the containment and eventual elimination of chiropractic (Wardwell, 1988: 174-84). It had become a significant competitor to allopathic doctors. By the 1960s, 3 million

Americans were visiting over 20,000 chiropractors and spending $300 million. In spite of the efforts of the AMA, today chiropractic has undisputed although limited legitimacy in the U.S. and Canada. For instance, chiropractors are included (up to a limit) under medicare in Canada and medicaid in the U.S. Chiropractic education receives federal funds. Chiropractic is now the major healing occupation in competition with allopathic medicine throughout the world.

The theory of chiropractic is based on the idea that vertebral misalignment can, by interfering with the patterns of the nervous system, cause a wide variety of disorders, including peptic ulcers, diabetes, and high blood pressure. In fact, anything that can be said to result from or to develop out of the context of a depressed immune system can be treated by chiropractic.

Chiropractic theory has been distinguished from allopathic theory along three lines (Caplan, 1984). In the first place, allopathic medicine views the symptoms as evidence of the disease and as a result of a simultaneously occurring disease process. The removal of symptoms is tantamount to the removal of disease for the allopath. Chiropractic, on the other hand, theorizes that symptoms are the result of a long-term pathological functioning of the organism. Disease precedes symptoms for a long period of time, perhaps years. The removal of the spinal subluxation allows the body to heal itself, in chiropractic.

The second distinction lies in the competing views of the role of pathogens in disease. The allopathic understanding is that pathogens of a specific type and frequency invade the body and begin any of a number of different disease processes. Killing the pathogens thus becomes a goal of treatment and the basis of scientific medicine. All practitioners not subscribing to this view are by definition "unscientific" and have been derogatorily referred to as cultists and quacks (*ibid.*: 84). Chiropractic says that pathogens are a necessary but not a sufficient condition for the initiation of a disease process. Before pathogens can take effect, the body must be vulnerable. While a number of factors can enervate the body, including genetic defects, poor nutrition, and stress, vertebral subluxation is also important.

The third distinction rests in the fact that allopathic doctors, even when they grant chiropractic some legitimacy, limit it to specific musculoskeletal conditions. Chiropractors, on the other hand, view their work as holistic, preventive care.

There are several alternative future scenarios. First, chiropractic could remain, as it is now, a marginal occupation only partially financed through medicare. Second, allopathic physicians could adopt chiropractic techniques to use themselves in addition to the traditional surgical and chemotherapeutic techniques. Third, chiropractors could be subjugated, as nurses and pharmacists have been, to work only under the jurisdiction of allopathic doctors. Fourth, chiropractic could increase in status and legitimacy and could become an equal competitor with allopathic medicine for funding. Or chiropractic could increase in status and legitimacy to be truly supported by medicare as an alternative to allopathy of equal status.

Current Status of Chiropractic in Canada

In Canada, the Canadian Chiropractic Association is an association of Mixers only. Statements drawn from the Ontario Chiropractic Association brochure *Facts About Chiropractic* describe the current philosophy of chiropractic:

> Chiropractic is the science which concerns itself with the relationship between structure, primarily the spine, and function, primarily the nervous system, of the human body as the relationship may affect the restoration and preservation of health.
>
> Historically, the profession's central concern has been the effect of improper function of the spinal vertebrae on the function and expression of the nervous system and, to a lesser extent, vascular system, vital elements of which are intimately linked to the spine.
>
> The major therapy is manual adjustment or manipulation of the spine and extremities. Adjunctive therapies include the use of massage, heat, light, ultrasound, electrotherapy, specialized exercise programs, and forms of muscle testing and technique.
>
> Chiropractic is not only interested in specific ailments; it has pioneered the holistic approach to health care now widely espoused in North America, and this is seen in chiropractic education and practice in the areas of nutrition, lifestyle,

exercise habits, sleeping habits, lifting techniques, etc.

The number of chiropractors has grown from about 100 in 1906 to approximately 30,000 around the world today (Coburn and Biggs, 1986). There are approximately 20,000 chiropractors in the United States and 2,600 in Canada, of whom 1,000 are in Ontario. Chiropractors constitute the third largest group of primary medical care practitioners, after physicians and dentists. The number of chiropractors in Canada increased rapidly after World War Two, when the Canadian Memorial Chiropractic College was established in 1945. The Department of Veterans Affairs gave an early impetus to the development of chiropractic in Canada when it funded the education of 250 veterans who desired training in chiropractic. Today all the provinces in Canada except Newfoundland license chiropractors.

X-ray examinations, general examinations, and treatment by chiropractors have been funded under national medical insurance since 1970. However, only a limited number of visits per year are covered in every province where they are licensed, except in Saskatchewan, where the full services are covered by medicare, medicaid, and most health insurance plans. Chiropractic services are covered under Workers' Compensation in seven provinces (Biggs, 1988). However, the relative cost of chiropractic services as a proportion of the total expenditures on health is negligible compared to the approximately 14.5 per cent of the health-care budget currently expended on allopathic practitioners.

The standards for admission to the Canadian Memorial Chiropractic College are comparable to those for other health occupations such as dentistry, pharmacy, optometry, and medicine. To enter, a student must have a minimum of two years of university in the sciences, and preference is given to applicants with the B.Sc. The four-year program is based on studies of human anatomy and related basic sciences, including x-rays, diagnostic skills, and clinical studies. Graduates see themselves as a part of a health-care team. They neither claim nor want to promote the view that their type of health care is the only useful model. In fact, according to Kelner *et al.* (1980: 80-81), they expect to refer at least a third of their patients to allopathic physicians.

The practice of chiropractic in Canada today is largely limited to musculoskeletal disorders such as headaches, neck and back pain, and soreness in the limbs. In addition, a wide range of functional and internal disorders are caused fully or in part by spinal dysfunction. Numerous recent studies have demonstrated but not explained the superiority of chiropractic in treating neck and back injuries. There is also evidence to suggest the efficacy of chiropractic in treating a broader spectrum of disorders, including epilepsy, asthma, and diabetes (Caplan, 1984). The future prospects for the development and spread of chiropractic look very promising.

Box 14.5 Alternative Medicine Is Growing in Popularity

Recent estimates show that at least as many as one-third of the population of Americans use alternative health care. According to a recent article published in the *New England Journal of Medicine* (1993) based on a telephone survey of 1,539 adults, 34 per cent used at least one unconventional therapy in 1990. In order of use, these alternative or complementary treatments include: relaxation techniques, chiropractic, massage, imagery, spiritual healing, commercial weight loss programs, macrobiotics and other lifestyle diets, herbal medicine, mega-vitamins, self-help groups, energy healing, biofeedback, hypnosis, homeopathy, acupuncture, and folk remedies. In all, in 1990,

Americans made 425 million visits to the practitioners of these alternatives, as compared to 388 million visits to allopathic doctors. The cost of these alternatives totalled $13.7 billion and people paid for about three-quarters of it themselves. By contrast, Americans spent $839 billion, about 60 times more, on allopathic care.

The most prevalent reason for seeking alternative health care is backache, followed by anxiety, headaches, chronic pain, and cancer. Almost everyone with a serious disease also visited an allopathic practitioner. However, three-quarters never tell their allopathic physician about their visits.

Naturopathy

Naturopathy is a form of holistic health care considered by its practitioners to be relevant to all the disabilities and diseases that might bring a patient to a "doctor's" office. It is based on the assumption that health and illness are both natural components of a total human being – spirit, body, and mind. Just as individuals are unique, this philosophy proposes, so, too, is each individual's sickness unique to him or her. Healing depends on the activation of the normal healing processes of the human body. Naturopathy, based on these principles, has been defined by the Canadian Naturopathic Association as follows:

The philosophy, science and art of healing which assists the self-recuperative processes of the body by the use of natural, biochemical and psychological forces, in an integrated system for optimum psychological effect, through the application of the natural healing laws of life to the whole man for the restoration and maintenance of health, and the prevention of disease. (Canadian Naturopathic Association, 1966: 7)

Naturopathy has its philosophic roots in Greek medicine and in the work of Hippocrates, with his emphasis on the importance of the body's own healing powers. Its modern forefather was a German physician named Samuel Hahnemann (1755-1843), who was the founder of homeopathy, the most important part of naturopathic medical practice. Hahnemann based his later research and practice on the fact that when he took quinine, which was the treatment for malaria, it caused the symptoms of malaria. Homeopathy, from the Greek words "homoios pathos," which mean "similar sickness," is based on a number of principles that are in opposition to allopathic medicine. These include the following (Coulter, 1984).

(1) The key to the cure of illness is embodied in the principle of similars, i.e., minute dosages of a natural substance known to cause similar symptoms to those indicative of the disease are administered as treatment.

(2) Different people react differently to the same illness because each person is unique.

(3) The body should only receive one remedy at a time, otherwise the body's healing powers will be divided. The physician looks for the "most similar" remedy, not those which just seem superficially to be similar.

(4) The physician should administer the minimum dosage required by the patient. This minimum dose will provide the same curative powers as a larger dose.

Sickness is thought of as a message rather than a biological pathology. Sickness provides a crisis through which the person can re-evaluate his or her own life (body, mind, and soul). Thus, symptoms are not viewed as signs of disease, nor is the goal to eradicate symptoms through heroic measures such as surgical removal or chemical destruction. Rather, symptoms are indications of a healing crisis to be enhanced by the administration of minimum doses of a substance that causes the same symptoms. Naturopathic diagnosis is based on knowledge, not of the disease's normal or standard

course, effects, and treatment, but rather on the unique and idiosyncratic characteristics of the interaction of the individual person and the particular patterning of symptoms. Thus, naturopathic medicine does not treat diseases but stimulates the individual's vital healing force.

Although there are numerous and complex methodological difficulties in comparing the efficiency of allopathic and homeopathic remedies, research has begun to confirm the value of homeopathy in, for instance, the treatment of rheumatoid arthritis (Gibson *et al.*, 1980).

Current Status of Naturopathy in Canada

Homeopathy arrived in the United States in 1825. By 1844, some of the most prominent allopathic physicians had adopted its principles and established the American Institute of Homeopathy, the country's first medical association (Coulter, 1984). The allopaths were threatened by this move and in 1846 founded a competing organization, the American Medical Association. Professional contact between the allopaths and the homeopaths was prohibited by the AMA's newly established Code of Ethics (Coulter, 1984).

The formal system of naturopathic treatment was established by Benedict Lust (1872-1945), who founded the first naturopathic college in 1900 in New York City. Lust's primary method of treatment was hydrotherapy, but this was enhanced by a variety of other techniques, including homeopathy, botanical remedies, nutritional therapy, psychology, massage, and manipulation (Weil, 1983).

There was hostility between allopaths and naturopaths until around the turn of the century, when the number of naturopaths began to decline dramatically, so that they no longer posed a threat. Since then naturopathy has never constituted a well-organized occupational group but rather a loose assortment of holistic practitioners providing a variety of healing modalities. Until recently most naturopaths were also chiropractors.

In Ontario, the Drugless Practitioners Act of 1925 can be seen as the starting point for the formal recognition by the province of naturopathy as a distinct healing occupation. By this Act a Board of Regents, composed of five appointees of the Lieutenant-Governor and made up of allopathic and chiropractic practitioners, was set up to regulate chiropractors and naturopaths. While naturopaths were allowed to practise, they were not formally included under this Act until 1944. In 1948, chiropractors were limited to "hands only, spine only" (Gort, 1986: 87). In the following year, 1949, the Ontario Naturopathic Association was formed by chiropractors who had been deprived of the right to the full range of practice (i.e., including naturopathic medicines) by the 1948 decision. By 1952 each form of drugless therapy was allowed to govern itself because it had become increasingly difficult for one board to regulate practitioners with diverse education, training, and practice.

Naturopathy in Ontario

There are more than 140 naturopaths practising in Ontario today. They are not covered by medicare. The justification for the exclusion of naturopaths from financial support was threefold: (1) the occupation was not growing, (2) they were not "scientifically oriented" (Mills, 1964, II: 80), and (3) there was no Canadian College of Naturopathy (nor was there any support for its establishment). While one dedicated person, Larry Schnell, began a lobbying campaign in 1964 to get naturopathy financial support under state insurance, most naturopaths feared that they would have to see too many patients and thus would be unable to practise in the way they believed was appropriate.

In 1978 the Ontario College of Naturopathic Medicine was opened; it offered a post-graduate program to allopaths, chiropractors, and naturopaths to help them upgrade their qualifications. In 1982 the Ontario Ministry of Health appointed a Health Professions Legislative Review Committee to determine which occupations needed regulation for the protection of the public interest. It was decided by this committee that naturopathy should not be regulated by the state but by its own members. The argument was that since (1) naturopathy was not scientific and (2) there was no standardization of practice, it posed a risk of harm to the patients and should not be regulated and thereby given official sanction.

The naturopaths responded in a brief called *Naturopathic Medicine and Health Care in Ontario*, delivered in July, 1983, to Keith

Norton, the Minister of Health at the time. The brief justified the value and efficacy of naturopaths as well-educated and trained healers, and made it clear that the Ontario naturopaths felt that it was allopaths who were trying to exclude them from practice. It levied a series of sharp criticisms at the costliness and narrowness of allopathic medicine, its mechanistic fallacy, and negative social consequences. The final statement in the ONA document highlighted and summarized their critique of allopathic medicine: "We find little cause to applaud allopathy's near monopoly of the medical services of this province" (ONA, 1983: 21).

Among the recommendations made by the brief were: the establishment of a separate Act to regulate naturopathic medicine, the right to use the title "doctor," the right to use the terms "naturopathic medicine" and "naturopathic physician," the elimination of the term "drugless practitioner" in favour of "naturopathic physician," the right to hospital privileges, the right to refer to various laboratory and x-ray services, and the right to authorize medical exemptions and to sign medical documents.

In the same year, 1983, the Ontario College of Naturopathic Medicine became the first Canadian institution to offer naturopaths an undergraduate education. By 1986 the planned deregulation, in spite of the organized opposition of the ONA, was passed by the Ontario government. In spite of signs of the resurgence of naturopathy – rising numbers of practitioners and the presence of a college to provide training – it seems that naturopathy in Ontario is at a standstill because of deregulation.

One of the most difficult issues faced by naturopathic medicine in its struggle for legitimacy lies in the nature of its science. Allopathic medicine claims to be scientific because it is modelled on the prototypical science of nineteenth-century physics. The goal of this model of research is the establishment of universally true causal laws. Individual differences are ignored in favour of generalizations. Homeopathy, on the other hand, is based on a different model of science. This assumes a universally true set of principles that are applied differentially to each individual. In contrast, allopathic scientific doctrine changes considerably over time. With each new discovery some part of allopathic medicine is often nullified.

If the trend in the United States is any indication of the future for naturopathy, then it looks bright. From its heyday in the nineteenth century, when there were 15,000 practitioners, fourteen medical schools, dozens of periodicals, and organized groups in every state and large city (Coulter, 1984: 72), naturopathy declined to a low of about 100 practitioners in 1950. In the 1960s, however, this trend was reversed, and today there are thousands of mainstream practitioners (allopaths, chiropractors, nurses, and clinical psychologists) who use at least some naturopathic methods. There are four or five large firms engaged in the manufacture and export of homeopathic remedies. And in the 1980s two states, at least, have adopted licensing laws for the practice of naturopathic medicine. As the holistic health movement grows, naturopathy will likely become increasingly legitimate.

Midwifery

Have you thought about your own birth experience? Have your parents told you about it? Was your father in the delivery room with your mother? Did your mother have any surgical intervention when she gave birth to you? Do you know anyone who has given birth at home or in an institution other than a hospital? Are you aware of the sorts of services that midwives offer to the mother and father before, during, and after the birth? What is the legal status of the midwife in Canada?

The final section of this chapter describes the history of midwives, their work, and their contemporary situation. The term "midwife" comes from old English and means "with the women." The French translation is *sage-femme*, or "wise-woman." Essentially, the work of the midwife is to assist women as they prepare for and give birth and as they learn to care for their new offspring. Midwives have usually practised their work in the home, where the woman is surrounded by her friends and family. Birth in this environment is seen as a natural event, much like sleeping, eating, and relating sexually.

Based on their assumption that birth was fundamentally a natural event, midwives tended not to use artificial or mechanical means to interfere with birth. For instance, a breech birth (upside down, feet first presentation) would be predicted by a midwife through a manual examination. Massage of the fetus in the uterus would be used to turn the baby for a normal delivery. Manual manipulation rather than forceps would be used to move the baby down the birth canal. Pain would be handled by rubbing the back, shoulders, and neck of the woman in labour, and with encouraging words. The woman in labour would be relaxed and comforted by the familiar atmosphere and the presence of a person or persons with experience.

In order to understand the position of midwifery in Canadian society today, a brief overview of its history is necessary. Birthing assistance has almost always and everywhere been the responsibility of women. The practice of midwifery can be traced to ancient history. The Bible includes a number of references. Exodus 1:15-22 is one example: two midwives, Shiprah and Puah, refused to obey the orders of the King of Egypt that all male infants be put to death. The written work of classical Greek and Roman physicians such as Hippocrates and Galen documents the prevalence of midwives. Male physicians were summoned only when special difficulties developed (Litoff, 1978). Until the Middle Ages, women who had themselves given birth were acceptable attendants at other births.

Up to the fourteenth century, midwifery flourished in Europe. Then came the witch hunts in which women healers of all sorts, and particularly midwives, were put to death. Midwives were especially vulnerable because of their close association with the placenta, which was believed to be an essential ingredient in witchcraft. By the end of the fifteenth century, witch-hunting had declined. In Britain, and soon elsewhere in Europe, midwives gained formal legitimacy under the Medical Act of 1512.

This Act was to be administered and enforced by local churches. Licensing depended on the good character of the aspiring midwife and required that she be hardworking, faithful, and prepared to provide service whenever it was needed, both to the rich and to the poor. She was also forbidden to use witchcraft, charms, or prayers that were not dictated by the Church. Men were not allowed to be present at birth (*ibid.*). Then in 1642, the College of Physicians gained authority to license midwives. For the next 300 years the legitimacy, recognition, and power of the male midwives, now obstetricians, grew in England and Europe while those of female midwives declined.

In the last quarter of the nineteenth century, British midwives made numerous attempts to gain legitimacy and recognition as an autonomous professional group. In 1893, a committee of the British House of Commons reported on the significant rate of maternal and infant mortality. They attributed these deaths to poorly trained, unregulated midwives and recommended that midwives be registered (Eberts, 1987). Most doctors rejected this suggestion because they argued that midwives did not have the necessary training. The Royal British Nursing Association supported the proposal to create a new occupational category: obstetric midwives or nurses.

Public opinion in favour of midwifery seemed to be growing. By the early part of the twentieth century, midwives in Britain were legitimized through regulation. With the establishment of the National Health Service in 1948, midwives were employed as a form of public health nurse. In 1968, legislation in Britain expanded the role of the midwife.

The earliest mention of midwifery among non-Native Canadian women appears in a deed in the Montreal archives, which reveals that the women of Ville-Marie, in a meeting held on February 12, 1713, elected a community midwife named Catherine Guertin. Also, in the English settlement of Lunenburg, Nova Scotia, Col. Sutherland, the Commander, wrote to the British government in 1755 asking that two pounds per year be paid to the two practising midwives. There are apparently no other regions in Canada where midwives were on government salary (Abbott, 1931).

Until the middle of the nineteenth century, most births in Canada took place in the presence of midwives, and until 1938 most births took place in the home (Oppenheimer, 1983), although from the mid-nineteenth century home births were frequently attended by physicians. From 1809 until 1895 the legal position of midwives was uncertain and changeable. Whenever midwives did not pose a threat to the work of allopaths they were allowed to practise; but where they did compete with allopaths, primarily in the cities and among the middle and upper classes, there were attempts to restrict them (Wertz and Wertz, 1977; Barrington, 1985; Ehrenreich and English, 1973a; Biggs, 1983).

Statistics indicate that in 1899 about 3 per cent of all Ontario births were attended by midwives and 16 per cent were attended by doctors (Biggs, 1983). However, as the legal status of the midwife was questionable, the number given for midwife-attended births is probably an underestimate. Over the duration of most of the nineteenth century, apparently, midwives garnered a significant amount of community and media support. The *Globe* newspaper opposed a medical monopoly of childbirth until 1895, when a bill to reinstate licensing of midwives was vehemently defeated in the legislature. At this juncture the *Globe* reversed its position.

In the last part of the nineteenth century and on into the twentieth century, the importance of the doctors' exclusive right to attend births grew. Today the midwife is in a changing and somewhat vulnerable legal position all across Canada with the exception of Ontario, where midwives are now being included in medicare on a salaried basis and trained for practice (Burtch, 1994); on the other hand, the position of the allopathic physician, and particularly the specialist obstetrician, has grown and is now secure. Two provinces, Alberta and British Columbia, are currently taking steps to formally recognize midwifery. In the meantime, midwives risk prosecution if anything goes wrong. How did such a reversal of trends occur? What are the social circumstances that forced the demise of the midwife and turned the position of the allopath into a monopoly? Just as social conditions contributed to medicalization in general (see Chapter Ten), social conditions affected the changing status of midwives; but elements of sexism and patriarchy were additional factors in the conflict between the male-dominated and female-dominated occupations.

Issues in the Practice of Midwifery

In addition to sexism and patriarchy, other social forces responsible for the contemporary pattern of high-tech, hospital-based, interventionist-oriented, physician-attended births include (1) bureaucratization and hospitalization, (2) the profit motive, (3) the public health movement, (4) the emphasis on safety and pain relief in childbirth, and (5) the campaign for ascendancy waged by physicians.

BUREAUCRATIZATION AND HOSPITALIZATION

Before 1880 hospital care was almost entirely for the poor and those suffering the wounds of war. The York Hospital, later to become Toronto General Hospital, opened in 1829 as the first real hospital in Ontario. It was established to treat the veterans of the War of 1812 and destitute immigrants. The first institution for women who needed care during childbirth, The Society for the Relief of Women During Their Confinement, was established as a charity in 1820. It provided the services of midwife-nurses and doctors, and also clothing for the mother and child, and food. In 1848 the first hospital, the Toronto General Dispensary and Lying-In Hospital, was established to provide care for destitute women and a place for training medical midwives. Hospitals at this time were still clearly for the poor and working classes. About the turn of the century there was a concerted effort made to centralize medical care in the hospital. With this move came the further consolidation of hospital births by male midwives or obstetricians. By 1950, Canadian midwifery had all but disappeared.

Concern with modesty spread the belief that decent people did not have their babies at home. Childbirth literature of the time emphasized the complicated, scientifically managed birth event. The new techniques, including the induction of labour, anesthesia, and the use of mechanical and surgical tools, further entrenched the notions that (1) the hospital was the only place for this potentially complex and dangerous birth procedure, and (2) the skilled hands of the equipped and trained obstetrician were the only ones appropriate for delivery. This belief was entrenched further by the fact that the majority of new immigrants who began to populate the cities used midwives. Middle- and upper-class women tried to distance themselves from the poor and immigrant women and from their typical childbirth practices. This, too, reinforced the growing belief in the appropriateness of the hospital as the place of choice for childbirth.

The First World War had an impact on the growth of scientifically managed hospital births. Childbirth came to be described as a dangerous process, and women, with war casualties fresh in their minds, were increasingly concerned to deliver their babies in a safe environment. Many were impressed with the physician's argument that infant and maternal mortality rates could only be substantially decreased when childbirth was recognized as a complicated medical condition (Litoff, 1978). The

growth of hospitals provided additional beds. Automobiles and roads cut down the time it took to get to the hospital after labour contractions had started.

PROFITS FOR DOCTORS

Physicians themselves had a lot to lose if midwives were given free reign to practise. According to Biggs (1983), midwives (in about 1873) charged approximately two dollars per birth, whereas male doctors charged five dollars. Biggs quotes from a Canadian letter in the *Lancet* in which the writer expresses the view that the doctors should have a monopoly over birth, at least in part because they have to invest so many years and so much money in their education. As she states, physicians felt that they should be "protected most stringently against the meddlesome interference on the part of old women," and that the amount of money lost through the competition with midwives would constitute a "decent living for [his] small family" (1983: 28).

THE PUBLIC HEALTH MOVEMENT

During the last half of the nineteenth century and into this century, Canada, along with the rest of the Western industrialized world, has witnessed a dramatic decrease in mortality rates. Research has emphasized the important roles played by improvements in nutrition, birth control, and sanitation in this decrease. The decline in maternal mortality rates corresponded to the general decrease in mortality, and coincided with the increasing prestige of the physician and the growing belief that medicine could cure all ills. All these developments furthered the move to obstetrician-centred, hospital-based births.

THE EMPHASIS ON SAFETY AND PAIN RELIEF IN CHILDBIRTH

One strong element in the move to birth taking place in hospitals was women's search for relief from the pains of childbirth. The desire to obliterate these pains can best be seen against the backdrop of the gender roles and fashions of the nineteenth century. The major causes of such pain were rooted in the cultural constraints on middle-class women, which demanded a certain delicate beauty. This could best be achieved by wearing boned corsets that constricted their waists, rib cages, and pelvises. Sensitivity to pain was considered feminine, and thus women's pain threshold tended to be low. Women were encouraged to see themselves as fragile, sickly, and weak. For middle-class women, comfort during birth came, at the turn of the century, to mean the obliteration of consciousness through twilight sleep. Since twilight sleep (a combination of morphine and scopolamine) could be monitored more effectively in the hospital, women sought to give birth in hospital.

Precautionary measures, including the enema and shave, were developed to preserve septic conditions and prevent puerperal fever. Forceps were developed in the nineteenth century; their use became standard practice because they could speed the birth process if it were slow. Caesarean section, a potentially live-saving procedure to be used when the labour was ineffective or the pelvis too small, became a frequent and even dangerously overused procedure. By the 1930s there were real safety advantages, including blood transfusions and antibiotics, in treating problematic births at the hospital. Yet as Barrington (1985) ruefully points out, much of the necessity for blood transfusions and antibiotics resulted from hospital-caused infections and doctor-caused hemorrhages.

THE CAMPAIGN FOR ASCENDENCY WAGED BY THE PHYSICIAN

Doctors described midwives as "dirty, ignorant and dangerous" (Biggs, 1983: 31). Devitt (1977: 89) quotes a midwife's statement published in 1906 that the obstetricians thought of her as the "typical old, gin-fingering, guzzling midwife with her pockets full of forcing drops, her mouth full of snuff, her fingers full of dirt, and her brains full of arrogance and superstition." Biggs (1983: 23) quotes the following doctor's view of midwives from the *British American Journal of Medical and Physical Science*:

And when we consider the enormous error which they (midwives) are continually perpetuating and the valuable lives which are frequently sacrificed to their ignorance, the more speedily some legislative interference is taken with respect to them, the better the community at large.

Physicians gained ideological superiority over midwives by portraying their own work as scientific and buttressed by safe and efficient tools such as forceps, all of which would improve the likelihood of safe birth for mother and infant. Given that physicians were educated men from good families and were coming to have high status, social power, and also often much political influence, it is no wonder that they were successful in promulgating the view that their births were the best births.

The Present Status of Midwives

The women's movement has been critical of the medical profession for denying the autonomy of women. It has also spearheaded a movement for women's rights over childbirth. Romelis (1985) discussed five specific components of this movement, including groups advocating (1) natural childbirth, (2) the Leboyer method, involving gentle birthing and immersion in water, (3) alternative in-hospital births, (4) home births, and (5) non-hospital birth centres. The role of midwifery fits within this movement, and is also part of a movement that abhors modern hospital practices such as "invasive diagnostic procedures, induction and acceleration of labour, reliance on drugs for pain, routine electronic fetal monitoring, dramatically increasing Caesarean-section rates, and separation of mother and baby after the birth" (Romelis, 1985: 185).

While the legal status of midwives in Canada (except in Ontario) appears to be ambiguous, i.e., in most provinces they are neither legal nor illegal, there are probably somewhat more than 100 midwives practising today in Canada (Barrington, 1985: 38). Midwifery training for nurse-midwives who work in isolated areas is offered at three Canadian universities: the University of Alberta, Dalhousie University, and Memorial University.

While the Canadian Medical Association has rejected the need for midwifery, a task force sponsored by the Ontario government in 1986 advocated the licensing and training of midwives (Fry, 1987). In response to the task force, the Ontario Minister of Health announced that midwifery would be established as a recognized part of Ontario's health-care system (Brown, 1988: 875).

In the fall of 1991 Ontario passed Bill 56: The Midwifery Act, which established midwifery as a distinct, self-regulating profession. Ontario was the first province to recognize and license midwives. Along with this legislation the College of Midwives is to oversee standards of professional ethics, grant licences, and address complaints. The cost, which will involve salary for primary responsibility for 40 deliveries and secondary responsibility for 40 other deliveries per year of midwifery services, will be covered by medicare. Education and training for midwifery is being made available at several institutions of higher education in Ontario, including McMaster and Ryerson universities (a four-year post-secondary degree). Midwives in Ontario are now allowed to conduct low-risk deliveries at home or hospital, to undertake certain surgical procedures (e.g., episiotomy), and to prescribe. Sixty-two midwives had already graduated from upgrading courses for practising midwives, as of October, 1993 (*Kitchener-Waterloo Record*, October 2, 1993).

Summary

(1) Some sort of nursing function has always been associated with illness and treatment. Florence Nightingale was the founder of the contemporary system of nursing care. She led a handpicked team of 38 nurses to serve in the Crimean War. Nightingale pioneered sanitary practices and caused the mortality rate to drop. She also provided many other services for the patients.

(2) Although Nightingale established nursing as an important profession in society, she did so by reinforcing a ghettoized and subordinated female labour force. The view of nursing as a woman's job is still in place today.

(3) Critical analysis of the work of the contemporary nurse tends to follow one of three lines: the patriarchal/sexist nature of the content of the work and the low position of nursing in the hierarchy of the medical labour force, the managerial revolution in nursing practice, and the impact of the hospital bureaucracy on the working life of the nurse.

(4) Nursing has been striving to reach the status of a profession by: increasing educational requirements, forming a licensing body, attempting to carve out a body of knowledge separate from that of physicians, and emphasizing nurses' special qualities and skills. However, some argue that nursing is a paramedical occupation and not a profession, because: the technical knowledge that nurses use is created, developed, and legitimated by physicians; the tasks nurses perform are less "important" than those of doctors; and nursing has less prestige than the medical profession.

(5) Allopathic medicine is the type of medical practice engaged in by medical doctors. It is based on the theory that opposites cure. Treatment involves an attack on the disease. Conversely, the holistic paradigm of health emphasizes treating the whole person. Three alternatives to allopathic medicine are chiropractic, naturopathy, and midwifery.

(6) Daniel David Palmer founded chiropractic in 1895. Attempts to legitimate and popularize this method of healing were stunted by allopathic practitioners. Distinguishing features of chiropractic are: through the manipulation and correction of spinal subluxation the body heals itself; the goal is not to kill germs but to make the body less vulnerable to germs; the work is viewed as holistic, preventive care. Although chiropractic in Canada today is limited to musculoskeletal disorders, its legitimacy continues to grow.

(7) Naturopathy is based on the assumption that health and illness are both natural components of a total but unique human being – spirit, body, and mind. Healing depends on the activation of the normal healing processes of the human body. Naturopathy, begun by a German physician, reached the U.S. in 1825. It was met with opposition by most allopaths. More and

more mainstream practitioners use at least some naturopathic methods today.

(8) Midwifery is based on the belief that birth is a natural process, therefore the use of artificial or mechanical means to interfere with birth is avoided. Midwives have almost always been women. Attempts by midwives to gain legitimacy have been met by rejection because doctors argue they do not have the necessary training, and because their presence threatened the work of allopaths.

(9) The social conditions that allowed allopathic practitioners to achieve a monopoly of the childbirth process also contributed to the contemporary pattern of high-tech, hospital-based, interventionist-oriented, physician-attended births.

Some such social conditions are: patriarchy and sexism, bureaucratization and hospitalization, the profit motive, the association of the medical model with the success of the public health movement, the growth in measures for safety and pain relief, and the campaign for ascendancy waged by the physician.

(10) While the legal status of midwives is still ambiguous in most provinces, the Ontario Midwifery Act passed in 1991 provides legislation to license and regulate midwives. With this legal recognition a salary and workload level has been established and institutionalized as part of the health-care delivery system of Ontario. It is likely that this action will be the beginning of a trend across Canada.

CHAPTER FIFTEEN

The Medical-Industrial Complex

THE MEDICAL-INDUSTRIAL complex is a large and growing network of private and public corporations engaged in the business of supplying medical care for a profit. Included in the medical-industrial complex are proprietary hospitals and nursing homes, home-care services, diagnostic services, including the tremendously expensive CAT scanners and MRIs, hemodialysis, the pharmaceutical companies, and the medical tools and technology supply companies. The pharmaceutical industry, an important component of the medical-industrial complex, will be discussed in detail in this chapter.

When do you decide to go to the pharmacy for over-the-counter medication from the shelves or to consult with the pharmacist? Do you try to become aware of the side effects of various over-the-counter medications? When you have been prescribed a drug, do you generally ask the doctor about side effects? Do you take your medication exactly as directed – over the length of time suggested and at the prescribed intervals? Do you believe that all the drugs available in Canada have been adequately tested and are safe? What safety precautions are available

in the Third World? Have you heard of thalidomide or DES? Do you know some of the devastating results of their use? To what extent is the pharmaceutical industry driven by an interest in profit and to what extent is the industry motivated by the desire to serve those suffering sickness and pain? These are among the questions that will be addressed in this chapter.

Drug Use

The discussion of the use of drugs and the pharmaceutical industry in contemporary Canadian society will be fairly wide-ranging because a discussion of drugs involves a large number of sociological issues. The drug industry is also a major actor in Canadian medical care. Canadians spent $9.2 billion, almost 14 per cent of the total 1991 health budget, on pharmaceuticals (Rachlis and Kushner, 1994: 125). It may be helpful to begin with a pictorial representation of the various levels of analysis that will be considered (see Figure 15.1).

At the first level the discussion will focus on the ways in which the socio-demographic

FIGURE 15.1 Factors Affecting Drug Use in Canada

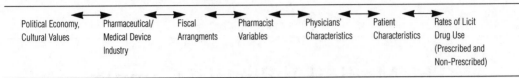

characteristics of patients – age, gender, class – correlate with their drug-taking habits. The social characteristics of physicians, too, influence drug prescription – characteristics such as the form of medical practice (e.g., solo as compared with group practice), the amount of continuing education, and the size of the practice.

Pharmacists and the pharmaceutical industry, too, are important forces in drug use. Cost, availability, advertising, and special pricing arrangements are among those known to influence physicians' decisions on what to prescribe and pharmacists' decisions on what to stock. Drug payment schemes – whether state controlled, those offered by private insurance, or individual payment alternatives – affect drug use. Much of the recent sociological research on drug use in Canada has focused on the pharmaceutical industry, its organization and structure, and especially on the fact that under capitalism it must be guided by the profit motive, which at times conflicts with the goal of promoting health.

Rates of Drug Use and Patient Variables: Age, Gender, and Class

There is considerable evidence that drugs are frequently over-prescribed in Canada.

For example, it has been estimated that between one-third and two-thirds of the antibiotics used are unnecessary or inappropriate (*ibid.*). Two groups are particularly vulnerable to misprescribing – the elderly and women. One expert has estimated that one of the side effects of over- or misprescribing among the elderly is at least 200,000 illnesses due to bad reactions to drugs (many of which may not have been needed in the first place) (*ibid.*). Moreover, there is a long history of evidence that women are especially likely to be over- or misprescribed psychotropic drugs or sedatives (Harding, 1994b).

There are strong relationships between gender, age, and drug use. The drugs cited in Figure 15.2 were prescription and included vitamins, pain relievers, heart and blood pressure medicine, cold remedies, skin ointment, tranquilizers, stomach medicine, antibiotics, and laxatives.

The rate of use varies by sex and age. Moreover, the incidence of multiple drug use is well substantiated in the literature (Lesage, 1991; Smith and Buckwalter, 1992). Although people over 65 comprise about 12 per cent of the population they use approximately 25-30 per cent of all licit drugs (Smith and Buckwalter, 1992). An elderly person uses an average of 13 prescriptions a

FIGURE 15.2 Per Cent of Population Using a Prescription Drug

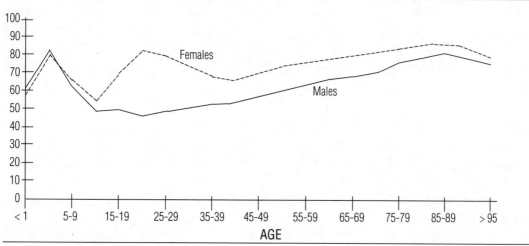

Source: Quinn *et al.*, 1992.

year. The average elderly woman takes five to seven prescriptions every day and two to three over-the-counter medications (Rachlis and Kushner, 1989: 101). Drug interactions can affect the absorption rate, distribution throughout the body, metabolism, and the elimination of a drug, and they can diminish or exacerbate the pharmacological effects of another drug. Drug effects are known to be somewhat different among the elderly. They are tested and prescribed on the basis that the average ingester is a male in his thirties. For those reasons, the elderly are twice as likely to experience drug reactions as younger adults (Pulliam *et al.*, 1988). Some evidence suggests that as many as 77 per cent of the admissions of the elderly to hospital may result from drug overdoses or side effects (*ibid.*).

Over-the-counter drug use may exacerbate the problems noted above. One study indicated that as many as 70 per cent of the elderly take over-the-counter medication without discussing this with their physicians. The incidence of medication errors by the elderly has been well documented. These include: forgetting to take medication, taking smaller or larger dosages than prescribed, taking medication for the wrong reasons, inability to read labels, difficulty of opening containers, and impaired memory (*ibid.*). Various consequences of drug misuse have been documented, including falls, dizziness, varieties of illnesses, and even death.

If current trends continue, drug use will increase for the following reasons: (1) the rate of drug use for those over 65 is higher than for any other age group, and this is the part of the population that is growing most quickly; (2) the supply of physicians per population in Canada has been steadily

TABLE 15.1 Percentage of Population 65 Years and Over, Canada and Selected Countries, 1970 and 1980 and Projections for 1990, 2000, and 2010

COUNTRY		1970	1980	1990	2000	2010
Canada	%	7.9	8.9	9.8	10.4	11.4
United States	%	9.8	11.3	12.7	13.1	n.a.
Japan	%	7.1	8.9	10.9	14.9	17.4
France	%	12.9	13.7	13.2	14.6	14.4
W. Germany	%	13.2	15.0	14.1	15.4	18.7
Sweden	%	13.7	16.2	17.5	16.7	18.0
U.K.	%	12.9	14.9	15.5	15.3	15.6
Europe	%	11.4	13.0	13.2	14.5	15.3
World	%	5.5	5.09	6.0	6.6	7.0

Source: United Nations, *Demographic Indicators of Countries: Estimates and Projections as Assessed in 1980* (New York: United Nations, 1982).

growing since 1967; and (3) the ratio of pharmacists to the population is also increasing (*Canada Year Book*, 1988). As Table 15.1 indicates, this aging trend is typical throughout the Western industrialized world and in the world population in total.

Females are consistently heavier prescription drug users than males (Figure 15.2). This pattern of greater use of prescription drugs among females is true of a wide variety of drugs. There are a few exceptions, chiefly in diagnostic and treatment drugs relevant to heart disease. Researchers have recently noted that there has been an overemphasis on diagnosing and treating heart disease in men, with the result that it has often been overlooked in women. At times this has been costly to women, who have gone to the hospital with signs and symptoms of heart disease only to be sent home – to a fatal heart attack.

Recently, policy-makers and researchers have turned their attention to some of the unique characteristics of heart disease in women in an attempt to reverse this growing rate.

Those in the lower income groups spend a greater percentage of their incomes on prescription drugs (Statistics Canada, 1981). This may, however, be the result of more expensive drugs rather than more drugs, as there is some evidence that prices for the same drug can vary up to 300 per cent, depending on the location. In Toronto, one study noted, the highest prices for drugs were charged in the areas of the city where the poorest people lived (*Globe and Mail*, June 12, 1970: 10). Indeed, recent research has shown that those with less education and lower family income are more likely prescribed mood-altering drugs (Rawson and D'Arcy, 1991). As Harding

Box 15.1 The Case of Valium

Valium, a trade name for diazepam, is one of the top-selling drugs. It is a tranquilizer, introduced into the market in 1963 by Roche. The company advertised Valium as useful for the treatment of a wide range of problems, including "psychic support for the tense insomniac," for the "always weary," and for the housewife "with too little time to pursue a vocation for which she has spent many years in training," thereby creating a rather broad market for the drug. In 1980, Roche was convicted by the Supreme Court of Ontario and fined $50,000 for violation of a section of the Combines Act, which makes it an offence to sell a product at an unreasonably low price if the effect is to lessen or eliminate competition (Lexchin, 1984: 20). Apparently, in order to secure a place in the market, and in close competition with a similar tranquilizer, Horner's Vivol, Roche, between 1970 and 1974, had given away Valium worth an estimated $5 million at market price. The reasons for this generosity became clear during testimony before the Supreme Court of Ontario. Charles Nowotry, Roche's secretary-treasurer, explained that the drug was given away in order to get the drug into hospital pharmacies, where it would be picked up by the leading doctors, who would then influence their residents and interns to prescribe it on a continuing basis (Lexchin, 1984).

says, "a process of medicalization of the symptoms of poverty and other problems of lower socio-economic positions may be occurring" (1994b: 171).

Psychoactive Drugs

Mood-modifying drugs are among the most heavily prescribed and often misprescribed drugs in Canada (Lexchin, 1984: 154). While data on total Canadian consumption are incomplete, a recent federal government publication, *The Effects of Tranquillization: Benzodiazepine Use in Canada*, provides useful estimates based on data gathered from a variety of small studies. The following findings are based on these sample surveys: (1) in 1971, 13 per cent of the residents of Metropolitan Toronto said they had used tranquilizers during the previous year; (2) in 1976, 14 per cent of Ontario residents said they had used tranquilizers during the year; and (3) in 1977, a cross-Canada survey indicated that 6 per cent of Canadians had taken a tranquilizer during the previous fourteen days.

Saskatchewan has a provincial drug plan and maintains a computerized listing of usage rates for various medical and pharmaceutical services. Thus, complete data on prescription rates (which do not necessarily correspond exactly to usage rates) are available in Saskatchewan (Harding, 1994b). In the last reported year (1989), 23 per cent of the provincial population received a prescription for central nervous system disorder.

A report in the *Globe and Mail* (Lexchin, 1988b: A7) described some of the reasons that doctors give for over-prescribing tranquilizers to women. The views expressed by the doctors are exemplified by the fol-

TABLE 15.2 Percentage of Population Receiving a CNS Drug Prescription, September, 1975-March, 1977 and 1989

	September, 1975-March, 1977	1989
Men	23.9%	20%
Women	31.9	26
Senior Men*	45.9	41
Senior Women*	59.7	53
Total Population	27.9	

*This includes all men and women 60 years or older in the earlier study and those 65 and older in the later study.
Sources: Harding and Curran, 1978: 5, 13-16; Quinn *et al.*, 1991: 80; Harding, 1994b: 157-81.

lowing quotations: "It's constitutional. The female's nervous system is more sensitive." "They're affected by problems and emotional upsets more." "That's the way the Lord made them . . . females have more time to indulge in neuroses than men." "They're bored often, and frustrated." "As they get older, there's the menopause which we men do not indulge in." The elderly, too, receive significantly more tranquilizers than people from the younger age ranges of the population (Cooperstock and Hill, 1982).

Table 15.2 indicates the proportions of men, women, and the elderly who are prescribed just one type of drug, a CNS (central nervous system) prescription drug. CNS drugs are thought to be among the most over- and misprescribed of drugs because the decision to use them is generally based on subjective indicators, given to the doctor by the patient, of various sorts of personal suffering or unhappiness. In other words, CNS drugs often represent the medicalization of social problems (see Waitzkin, 1989, for a study of how this happens in the doctor's office).

As Table 15.2 shows, women more than men, the elderly more than the non-elderly, and elderly women more than elderly men are likely to be prescribed CNS drugs. There is some evidence that the elderly are seen by the medical profession as less treatable and more helpless and depressed (Clark *et al.*, 1991). They may then be more likely to receive CNS drugs in aid of social rather than medical problems.

Canada's Health Promotion Survey (1987), using a method of random digit dialling, contacted over 11,000 Canadians over 15 years of age. The respondents were asked whether they had taken any of five psychoactive drugs in the previous twelve months. Approximately 8 per cent reported using sleeping pills, 6 per cent tranquilizers, 2 per cent stimulants, 1 per cent cocaine, and 6 per cent marijuana or hashish. These percentages represent 1.6 million sleeping pill users, 1.2 million tranquilizer and cannabis users, about 400,000 users of stimulants, and 200,000 cocaine users.

The use of these psychoactive drugs varies consistently by age, sex, marital status, and income. Here again, tranquilizers and sleeping pills are more likely to be consumed by women than men (10 per cent of women and 6 per cent of men used sleeping pills; 8 per cent of women and 5 per cent of men used tranquilizers) and are more likely consumed by the elderly than the young: 21 per cent of those 65 and over used sleeping pills, and only 5 per cent of those between 25-34. Those of the lowest income and educational groups are more likely to use tranquilizers. Sleeping pills are, by contrast, as likely to be used by those at the top as those at the bottom of the income and education hierarchies. Divorced, separated, and widowed people are more likely than single and married people to use both sleeping pills and tranquilizers.

Aside from women and the aged, two other groups, available data suggest, use high levels of mood modifiers: those who are unemployed and poor and those who are chronically ill. Findings suggest that those who are unemployed, retired, or not in the labour force for other reasons are higher users of psychoactive or mood-altering drugs than those who are active in the labour force. This is true of both sexes and all age groups. One study reported a significant correlation between the amount of time spent working outside the home and drug consumption. "Eleven percent of those in full-time jobs, 19 percent of those with part-time jobs, and 25 percent of those who were home full-time, reported use in the previous two weeks" (Cooperstock and Hill, 1982: 11).

Patterns in the prescription of mood-altering drugs are consistent with an hypothesis that would tie their use to social and economic conditions. As mentioned, Rawson and D'Arcy (1991) report higher use among the less educated and among families of a lower income level. They reported a doubling of the rate of use by lower as opposed to higher family income groups. Such findings add credence to the possibility that poverty and associated problems may tend to be medicalized.

A large Ontario study found that one-quarter of another high-risk group, the chronically ill, took prescribed tranquilizers, and 27 per cent of these were taking two or more psychoactive drugs simultaneously (Cooperstock and Hill, 1982: 15). Finally, almost one in two of the people in institutions are on tranquilizers regardless of their diagnosis. As Cooperstock and Hill note, because the usage rates of the institutionalized do not depend on diagnosis, drugs may be used primarily to manage patients within institutions.

It is clear that a good proportion of these mood-modifying drugs are given for social and personal problems, not for medical ones (Cooperstock and Lennard, 1988). Pharmaceutical companies often advertise their drugs as solutions to social problems. For instance, methylphenidate, a stimulant, is promoted as a help for coping with the stressful pace of modern urban living (Vavasour and Mennie, 1984). Any number of personal and social problems, including the empty-nest syndrome, marital conflict, family stress, delinquency, and loneliness are being treated with mood-modifying or psychotropic drugs (see Cooperstock and Lennard, 1987). Drug advertising promotes the idea that women are emotional and suffer more from psychosomatic illnesses. Reliance on psychotropic drugs can have numerous negative consequences for the individual. Such drugs may also mask the need for social change that could prevent the troublesome situation.

Box 15.2 Halcion

The stories often begin with the taking of a pill and end with a murder. A woman, Ilo Grundberg, shot her 83-year-old mother eight times in the face while being under the influence of the sleeping pill, Halcion. While taking Halcion, another woman stabbed her two sons. Despite her claims that Halcion was the cause, she was convicted twice of murder and eventually committed suicide.

Examining psychiatrists testified that Ilo Grundberg was involuntarily intoxicated from the effects of Halcion when she killed her mother. Halcion is intended only for short-term use, yet Grundberg's doctor had prescribed the drug to her for much of the preceding year. Grundberg had grown increasingly paranoid and agitated while on the drug. After being acquitted of murder, as there was no clear motive for the murder and Grundberg had little memory of it, she became involved in a $21 million civil suit against Upjohn, the Michigan-based manufacturer of Halcion. She charged that Halcion is a defective drug and Upjohn failed to warn the public about its sometimes severe reactions. Upjohn retaliated by stating that in no way was the company negligent and there was absolutely no connection between the murder and the drug.

Halcion, used by more than seven million Americans and sold in more than 90 countries, is Upjohn's second largest moneymaker. Its annual world-wide sales are $250 million – $100 million in the United States. It is quite obviously liked by patients and doctors alike.

Since Halcion's arrival on the market, critics have debated whether or not Halcion is more dangerous than other benzodiazepine drugs. The critics claim that it is more likely to cause such nervous-system disturbances as amnesia, anxiety, delusions, and hostility. They state that Upjohn and the FDA have done very little to protect the pill-taking public. Upjohn responds by saying that Halcion is no more likely to cause adverse reactions than any other sleeping pill.

Since its entrance onto the market, Halcion has led a controversial life. When it first became available in the early 1970s, it appeared to be the answer to many problems having to do with side effects resulting from the use of benzodiazepines. Clinical studies showed that while people fell asleep quickly, users experienced virtually no grogginess the next day as it cleared the system quickly. By 1979, however, there were reports of peculiar psychiatric changes in those using Halcion. These side effects included amnesia, hallucinations, paranoia, and verbal and physical aggression. Records show a long history of concern over the use of Halcion. For example, a series of evaluations were written recommending against its approval by the FDA. The drug was eventually approved despite protests.

New drugs have a tendency to reveal their true colours over time. The FDA maintains a system of post-marketing surveillance. Doctors and drug companies file reports describing any adverse reactions to the drugs they prescribe or sell. Unfortunately, many adverse reactions never get reported. A drug's record can also be skewed by factors such as its manufacturer's reporting practices, the kinds of patients who happen to take it, and the amount of publicity it receives. In 1987, it was noted that during its first three years on the market, Halcion had collected approximately eight to thirty times as many adverse reaction reports as other more popular drugs.

Upjohn may believe that adverse reactions to Halcion may be no different from reactions to other benzodiazepines, but the company has never convincingly explained why Halcion has a record for generating strange stories. For instance, a 1987 doctor's report described episodes of delirium, sleepwalking,

Box 15.2 continued

---

and amnesia in five elderly hospital patients who were receiving as little as an eighth of a milligram dose of Halcion. One man was found trying to turn somersaults in his room while others were trying to flee the hospital in their pyjamas. Not one of them remembered their adventures in the morning.

Halcion's critics feel that the drug poses risks other drugs may not have and therefore it should be taken off the market. Many people, however, prefer the periods of uneasiness to the heavy hangovers other sedatives can cause.

"Sweet Dreams or Nightmare?" *Newsweek*, August 19, 1991.
"Horror Stories Start with Pill, End in Murder," *Winnipeg Free Press*, December 20, 1992, p. A11.

Physicians and Prescribing

There is a high correlation between the number of visits to a physician and the number of prescriptions. In Chapter Ten it was shown that the number of physicians in a society affects the degree of medicalization. Since physicians in general have a tendency toward medical intervention – prescribe drugs, for example – when confronted with a patient exhibiting a problem, their everyday work will affect the rate of drug use. But social characteristics and conditions affect the rates at which doctors prescribe drugs. The Birmingham Research Unit of the Royal College of General Practitioners in Britain (1977) found a tenfold difference in the rate at which family practitioners prescribed psychotropic drugs, ranging from 40 per 1,000 patients to 415 per 1,000 patients. The rates at which physicians prescribe drugs are tremendously variable (Hemminki, 1975).

Research has documented significant deficiencies in doctors' knowledge about drugs. Lexchin (1984) reports that fully 27 per cent of the practising physicians in Ontario and 25 per cent of those in Nova Scotia had inadequate knowledge about the uses of antibiotics and sulphonamides; only 41 per cent in Ontario and 12 per cent in Nova Scotia were skilled in their use; the remainder were in an intermediate position.

Pharmaceutical firms are a major source of information for doctors. Many doctors (28 per cent) say that they learn much of what they know about medicines and their "appropriate" use from drug firms (Wilson *et al.*, 1963). There is no lack of evidence to document that the drug company representatives who call on physicians regularly with brochures, samples, and gifts are a major source of information about drugs for doctors (Hemminki, 1975). If expenditures by drug companies on promotion and advertising are any indication (twice as much as on research and development), then promotion and advertising have an important impact on prescribing (Rachlis and Kushner, 1994). In the United States, pharmaceutical companies spend more on advertising than do either alcohol or tobacco companies. Drug advertisements have repeatedly been criticized as misleading and incomplete, and as portraying people in stereotypical ways. For instance, the elderly may be seen engaging in passive activities wearing depressed faces or acting childishly, playing with childish toys (Foster and Huffman, 1995). Advertisements recommend drugs for such a variety of everyday, "normal" concerns that they seem to be suggested as useful to everyone at least some of the time. Drug prescribing, then, is often a symptom of the tendency toward the medicalization of social issues.

Another major source of information for most Canadian doctors is the *Compendium of Pharmaceuticals and Specialties (CPS)*, published by the Canadian Pharmaceutical Association. Although it has been recommended by the *Canadian Medical Association Journal* (Lexchin, 1984), there is no question that it is inadequate in a number of ways. It is not comprehensive, and it is known to have continued recommending certain drugs long after research documenting destructive side effects had been published in medical journals. The most thorough study of its value found that 46.3 per cent of the drugs recommended by the *Compendium* were "probably useless, obsolete, or irrational mixtures" (Bell and Osterman, 1983). Well-known risks and negative effects were ignored in over 60 per cent of the drugs listed. There were scientific errors regarding the biochemical effects of nearly 40 per cent of the entries. Bell and Osterman were led to conclude that the present version of the *CPS* is basically a tool to promote the interests of drug companies.

What other factors influence doctors' decisions about prescribing drugs? The few studies available indicate that the level of medical education has an effect. For example, Becker *et al.* (1982) studied the rate at which physicians prescribed chloramphenicol (because this drug has potentially fatal complications, lower rates of prescription were seen to indicate better prescribing): younger physicians who had more years of post-graduate education had lower and more appropriate rates for prescribing this drug.

Freidson (1975) distinguished between client-dependent and colleague-dependent forms of medical practice. His argument is that regulation is more effective in colleague-dependent than client-dependent practices. When doctors have to account to other doctors for the diagnoses they make and the treatments they choose, they are likely to exhibit higher medical standards than when the doctor is primarily seeking to satisfy the patient. Following this line of reasoning, it can be predicted that doctors who are involved in medical networks or in some form of group practice are more likely to have appropriate prescribing habits.

Surprisingly, perhaps, especially as it contradicts the declarations made by many practising doctors, physicians who see fewer patients may spend more time with them but do prescribe more medications. A Canadian study compared a group of doctors who worked on the basis of fee-for-service with a group of doctors who worked on salary (Lexchin, 1984). Approximately one-half of the doctors in private clinics, as compared to one-quarter in community health centres, prescribed drugs inappropriately. Salaried physicians were also more likely to warn patients of side effects and other potential problems. The researchers explained that because salaried physicians had more time per patient, they were able to take the time to prescribe appropriately and to explain when and how to use the drug, as well as what side effects to look out for.

Several studies have shown that both drug advertisements and drug detail men and women (pharmaceutical company representatives who visit doctors with samples and information about drugs) (Lexchin, 1994a) significantly affect the drug-prescribing habits of doctors. Lexchin (1994b) indicates that much of the over-prescription of antibiotics, ulcer medications, and anti-hypertensives results from drug advertising. The Canadian Medical Association and its journal historically defend the pharmaceutical industry and support its viewpoints (Lexchin, 1994a).

Pharmacists

There is very little sociological analysis of the role of pharmacists with respect to prescription and non-prescription drugs. There is no question, however, but that they do have considerable discretionary influence by making recommendations both to doctors and to individuals who shop for over-the-counter medical assistance. Consumers frequently ask the pharmacist to recommend "something" – a non-prescription drug – for a cough, sleeplessness, pain,

or anxiety. The pharmacist may suggest a particular brand-name drug or a range of suitable products of different brand names. A number of factors will affect the pharmacist's recommendation.

The cost of a particular drug varies greatly from pharmacy to pharmacy, and even within the same pharmacy. One survey found that the prices of fifteen of the most common drugs varied by as much as 89 per cent from one outlet to another. Even within the same outlet, price differences as great as 130 per cent over a two-week period have been noted (Allentuk, 1978: 68). Drug salesmen, advertising, and gifts given by pharmaceutical companies are all likely to have an effect on recommendations made by pharmacists. Pharmacists are frequently owners of the drug stores or pharmacies in which they work. This being the case, they are most likely influenced in their decisions by the necessity of turning a profit. Pharmaceutical companies vary their drug costs from store to store and situation to situation at times. It is in the interest of the pharmacist's economic position that he or she recommends the drug that provides the greatest profit margin.

Pharmacists do have some discretionary power, too, when presented with a drug prescription. Unless the physician has written "no substitution" on it, pharmacists are free to dispense any company's brand of a particular drug. To encourage pharmacists to dispense their own brand of drug, pharmaceutical companies may discount the price on the given product. Discount pricing, in that it provides the pharmacists with the most room in which to make a profit, often encourages the pharmacist to use a particular drug. Discount pricing also affects provincial revenues in provinces with drug plans. Under discount pricing, a particular drug company will charge the pharmacist one price for the drug, e.g., $20 for 50 pills, and the pharmacist may charge the purchaser, who may be reimbursed through a government or other drug plan, a substantially different price, e.g., $100 for 50 pills. The pharmacist thereby realizes a profit of $80 on that particular drug company's product. The provincial governments, the consumers, and the drug insurance plans are the losers. That this is not an insignificant cost to the provincial government is noted by Lexchin (1988a), who quotes a news article estimating that discount pricing was costing provincial drug plans $40-$60 million annually across the country.

The Pharmaceutical Industry in Canada

The Canadian drug industry has always been divided into domestically owned companies, the first one founded by E.B. Schuttleworth in Toronto in 1879, and foreign-owned subsidiaries, the first one established in Windsor by Parke, Davis and Company (Lexchin, 1984: 331). The industry grew slowly (the foreign-owned companies stayed in Canada because they could obtain tariff and tax advantages) until the 1940s. The antibiotic revolution and the development of medications to control patients in mental hospitals spurred the growth of the industry. Economies of scale became

possible in the manufacture of these drugs, and production was centralized. This, plus the increasing openness of world trade, meant that the small Canadian companies could not compete with the larger foreign-owned companies. After World War Two only one Canadian company of any consequence was left – Connaught Laboratories. Today, subsidiaries of multinationals control 90 per cent of the Canadian market (*ibid*: 33). These large multinationals belong to the Pharmaceutical Manufacturers Association of Canada, a very effective lobby/pressure group that has its headquarters in Ottawa (Rachlis and Kushner, 1994).

The pharmaceutical industry is one of the more profitable manufacturing industries in Canada (see Table 15.3). High profitability and predictions of consistent growth in view of long-term demographics (e.g., the aging of the population) provide reason for continued optimism. Furthermore, this appears to be a low-risk industry. A recent Canadian study indicated that the drug industry was ranked 67th in terms of risk, almost the lowest in Canada. What strategies do the drug companies use to maintain their profitability? What is the impact of the great financial success of the multinational drug companies on the health of Canadians and on the health of the people of the Third World? In what way do the profit-making strategies of the drug companies affect health negatively? These three questions will be discussed here.

The pharmaceutical industry is successful in maintaining its position as one of the most profitable industries through a variety of strategies, including: (1) the absence of price

TABLE 15.3 Rate of Return on Equity Before Taxes, 1978-1987

Year	Pharmaceutical Industry	All Manufacturing	Ranking of Pharmaceutical Industry out of 87 Manufacturing Industries
1978	22.7	17.4	20
1979	28.3	21.9	17
1980	30.1	20.1	10
1981	31.0	17.4	6
1982	30.0	5.4	7
1983	33.9	9.9	3
1984	40.3	15.7	2
1985	41.1	12.7	3
1986	45.5	14.9	1
1987	42.2	16.2	1

Source: Statistics Canada, *Corporation Financial Statistics: Detailed income and retained earning statistics for 182 industries* (Ottawa: Supply and Services Canada, various years).

competition; (2) patent protection; (3) competition and drug development focused on drugs with potential for widespread use (and thus profit) rather than drugs for the rare condition; (4) production of brand-name rather than generic products; (5) drug distribution (dumping) in the Third World; (6) advertising and providing select information to physicians and consumers.

First, one of the major reasons for the high profits in the drug industry is that the selling price of drugs is not necessarily related to drug production costs (Lexchin, 1984: 41-48). In the absence of price competition the manufacturer is free to determine prices in the interest of maximizing benefits to the company (i.e., profits). Variations in cost that bear little or no relationship to

manufacturing costs have been noted. For example, Allentuk (1978: 66) points out that "Carroll Labs simultaneously sold 400-milligram Meprobamate tranquillizer tablets at $3.15 per hundred and 200-milligram tablets of Meprobamate at $3.35 per hundred."

Patent protection is the second technique used to maintain the high level of profits. Patents serve to limit competition. Once the company has invented a new drug, patent protection gives the company an exclusive right to manufacture and distribute the drug for a period of years. This period can be extended if the company takes out additional patents on associated drugs. Of 1,335 drugs listed in the Ontario Drug Benefit Formulary on July 1, 1982, almost 75 per cent were available from only one manufacturer. Drug companies claim that patent protection allows them to pay for the research necessary for the invention of new drugs. However, this argument can be seriously challenged because much of the research done in any drug company is directed toward developing imitative medicines that can compete with products already successfully developed and marketed. The pharmaceutical industry spends heavily on advertising, drug promotion, and lobbying. Apparently, 20 per cent of its total sales revenues were invested in promotion. This is approximately double the amount that is spent on research (Rachlis and Kushner, 1992).

Government reports circulated in the 1960s indicated that patent protection was a major barrier to competition and lower prices in the pharmaceutical industry. Bill C-102, which stipulated an amendment to the Patent Act that would allow drugs still under patent to be licensed for import, was passed in 1969. Imported chemicals are the ingredients for generic drugs, which are less expensive. The multinationals objected to Bill C-102 and lobbied against it. As a result of their lobbying efforts and the ensuing conflict, the federal government commissioned Harry Eastman, an economist, to lead an inquiry into the pharmaceutical industry. Eastman's 1985 report demonstrated that the result of Bill C-102 has been of benefit to Canadian people because it stimulated competition and saved Canadians $211 million on pharmaceutical costs in one year – 1983.

The government, responding to the pharmaceutical industry and to the Eastman report, passed a compromise bill, C-22. This legislation gives companies up to ten years of patent protection, but in return the Pharmaceutical Manufacturers Association of Canada promised to invest $1.4 billion in research and development by 1995 and to create 3,000 new jobs. Many citizens' organizations, such as the Consumer Association of Canada, opposed the bill, arguing that drug costs could rise by over $300 million by 1995 (Lexchin, 1988c).

The third strategy described by Lexchin (1988b) is that the drug companies rarely do research or attempt to develop medicines in areas where there is unlikely to be a large market or with a view to addressing the needs of people with less-than-common diseases. Rather, they tend to produce new drugs based on similar existing drugs, thus circumventing patent protection for large markets that have already been developed. As an example, there are numerous anti-inflammatory (anti-arthritis) and benzodiazepine (minor tranquilizer) drugs currently on the market in Canada. Such a wide choice is of virtually no therapeutic or medical value. However, because there is a huge market, every pharmaceutical company enters the market with a similar product. As a result there are 20,000 pharmaceutical products based on only 700 active ingredients.

Fourth, the widespread use of brand-name rather than generic products contributes significantly to the profits of the pharmaceutical companies. (The generic name is the scientific name for a particular drug, while the brand name is the name given to the drug by the pharmaceutical company that produces the drug in question.) On average, generics cost 67.5 per cent less than the most expensive brand-name equivalents. It is clearly to the advantage of the pharmaceutical industry to promote brand-name products. Although these are usually prescribed, the pharmacist may substitute a generic if it is available (unless otherwise directed by a physician). Unfortunately, few are available. Of all the products listed in the Ontario Drug Benefit Formulary in 1982, generic equivalents were available for only 91. In 1980, only 8.7 million of 175 million prescriptions in Canada were filled by generics (Lexchin, 1984: 62).

Fifth, the history of the pharmaceutical industry is replete with stories of drug-related illness and death. The "dumping" of out-of-date drugs, which have been removed from sale in the industrialized world, in the developing countries is one cause. Health-destroying side effects of many drugs, whether they are taken alone or in combination with other drugs, is another problem. The industry gives rise to another risk by marketing for a wide variety of symptoms drugs that are only appropriate for a limited number of purposes.

The case of the Dalkon Shield IUD (an intra-uterine device for birth control) is one example of the great health costs of a profit-driven industry (Vavasour and Mennie, 1984). The Dalkon Shield went on the market in 1971 in the United States. By early 1972 there were numerous reports of adverse reactions such as pelvic inflammatory disease, blood poisoning, and tubal pregnancies. By 1974, seventeen people had died from its use. Because the U.S. market began to look very poor, the manufacturer offered the Dalkon Shield to developing nations at a discounted price of 48 per cent of the original price (*ibid.*). The shields were distributed to the developing nations even though they were unsterilized, and nine out of ten lacked the necessary inserter. Furthermore, only one out of 1,000 was distributed with any instructions for insertion (and their insertion is a delicate and potentially dangerous procedure). In 1975 the United States banned the Dalkon Shield, but by 1979 they were still being sold in the Third World (Ehrenreich *et al.*, 1979).

Clioquinol has been long promoted for the prevention and treatment of a wide variety of non-specific travellers' diarrheas. However, there is no adequate evidence that it is effective (Vavasour and Mennie, 1984). On the other hand, the dangers of clioquinol have been proven. Clioquinol has caused many thousands of cases of SMON – subacute myelo optic neuropathy – a condition that involves continuous pain, paralysis, blindness, and, in some cases, death. In 1977 a Japanese court determined that clioquinol had caused at least 10,000 cases of SMON. The manufacturer admitted responsibility for the tragedy and apologized to the victims and families. However, its concern did not prevent the company from marketing the drug in Malaysia, Thailand, and Kenya as late as 1980 (Muller, 1982).

Chloramphenicol has been severely restricted in the U.S. since 1961 because of its association with a severe and often fatal blood disease – aplastic anemia. As late as

Box 15.3 Drug Promotion in the Developing World

At least 20 per cent of the sales revenue of the pharmaceutical industry is spent on drug promotion. Promotion includes traditional advertising in both medical journals and the mass media and outright gifts to physicians, ranging from a pen-and-pencil set to a Caribbean or European vacation for two. Sponsorship of medical and associated conferences and drug-related research are two other drug promotion strategies. According to an article in *The Times* (London), bribes by drug companies are sometimes used to help sell drugs: "A doctor has claimed that he was offered ten pounds a time by a drug company salesman for prescribing a certain drug and for filling in a form on the drug's effects." Drug promotion is excessive, and puts pressure on physicians; in addition, it is also frequently false or partly false. "Promotional materials exaggerate effectiveness, gloss over dangers and try to downgrade the competition."

In the developing world, the problems associated with drug promotion are even more acute, and the pressure greater. Drugs are sometimes promoted for diseases for which they are inappropriate, and no information is given about how they are to be used or what their hazards or side effects are. Anabolic steroids can, for example, stunt growth, cause liver tumours, and bring about irreversible masculinization in females. In Britain, anabolic steroids are cautiously recommended for osteoporosis, renal failure, terminal malignancies, and anaplastic anaemia. Yet, in the *African Monthly Index of Medical Specialties*, anabolic steroids have been suggested for malnutrition, weight loss, exhaustion, lack of appetite, and excessive fatigue in children.

1973 the manufacturer was still recommending chloramphenicol in Latin America for a wide variety of conditions, such as sore throats, ear infections, and pneumonia (Muller, 1982).

The sixth strategy of the profit-driven pharmaceutical industry is to provide only select information to doctors and consumers about the efficiency and safety of various drugs. Drug salesmen provide doctors with information about the usefulness and value of various drugs. Doctors are influenced by this information to prescribe one medication over another. However, the pharmaceutical companies sometimes give biased or insufficient information about their drugs. As a consequence, the more doctors rely on such sources for information about drugs, the more likely they are to prescribe incorrectly (Lexchin, 1988a: 505).

The Case of Thalidomide

Just as the pharmaceutical companies have shown that their marketing strategies in the Third World take health less seriously than profits, so, too, have profits come first in Canada at times, and with deleterious consequences. Probably the incident with the most visibly tragic consequences was the thalidomide disaster, which resulted in the birth of some 115 babies in Canada with phocomelia (the absence of limbs and the presence of seal-like flippers instead).

Thalidomide was developed by a West German company in 1954. It was called

GRIPPEX, and was initially recommended for the treatment of respiratory infections, colds, coughs, flu, nervousness, and neuralgic and migraine headaches. It was widely available without prescription, quite cheap, and therefore very accessible. It was manufactured in West Germany, Canada, Great Britain, Italy, Sweden, and Switzerland under thirty-seven different brand names (Klass, 1975: 92). Later it was marketed in Germany as the "safest" sleeping pill available because it was impossible to take enough at any time to commit suicide. It was advertised in Great Britain (where it was called Distavel) as so safe that the picture accompanying an advertisement was of a little child in front of a medicine chest. The caption read, "This child's life may depend on the safety of Distavel."

By the summer of 1959 there were a number of reports in Germany, Australia, and Britain of serious side effects. These indicated that the drug caused nerve damage, affected balance, and caused tingling in the hands and feet. This should have been a warning about the potency of the drug and its effects on the central nervous system (Winsor, 1973). But the manufacturer continued to market the drug in Germany and licensed another company to produce and market the drug in Canada and the U.S. The drug was tested briefly in the U.S. and then samples were distributed. It was manufactured, beginning April 1, 1961, under the name KEVADON in Canada. A warning was included in the package about peripheral neuritis. It was distributed in Canada under a number of different names.

By December 1, 1961, two representatives of the German companies reported to Ottawa that a number of babies with congenital malformations had been born in Germany and that the mothers of these babies had taken thalidomide. Rather than contacting the research centres in Germany, England, and Australia directly, the Canadian government relied on the ambiguous and evasive reports presented by the companies involved. It was not until three months later, on March 2, 1962, that the Food and Drug Directorate of National Health and Welfare decided to withdraw the drug, claiming that, until then, the evidence for its removal was "only statistical" (*ibid.*). Removing the drug was complicated. Unlike France, Belgium, the U.S., and Britain, Canada did not require the drug manufacturers to label the drug with its international name, thalidomide, under which its side effects were being publicized (*Kitchener-Waterloo Record*, September 27, 1972). As a result, a number of pharmacists were not aware that their shelves contained the drug in question. By the time it was removed the damage had been done.

Approximately 115 babies were born in Canada with phocomelia. Other external defects included small ears, eye defects, depressed noses, and facial tumours. Internal problems were found in the cardiovascular system and the intestinal tract. There were several cases of missing organs, such as gall bladder or liver. These physical defects meant emotional traumas for the mothers, fathers, siblings, and other family members, as well as for anyone who was involved with the "thalidomide babies." In

Box 15.4 Thalidomide's After-Effects Today

A task force established in 1987 asked the federal government for more than $6 million in damages for the victims of thalidomide, as well as $50,000 in tax-exempt living expenses for each person affected. The thalidomide "victims" organized to ask for justice for those who have suffered as a result of the drug. Many of the affected Canadians have additional medical problems, including back pains and circulatory difficulties. They are now entering their thirties. Their parents are dying or becoming too old to care for them now or in the future. In its official report the task force said:

Thalidomide victims are currently facing enormous difficulties in many aspects of their lives: education, employment, careers, housing, transportation, insurance, daily living, socialization, sexuality and recreation. In short, their entire lives have been affected.

In 1991 the federal government awarded them $8.5 million (Priest, 1993).

Globe and Mail, February 14, 1989.

some communities the birth of the deformed children made local newspaper headlines, and townspeople "flocked" to the hospital to see for themselves. Some people blamed the mothers for having taken the drug. Whole families were stigmatized. There were approximately 3,000 disabled babies in West Germany and 500 in Great Britain. When Belgium, Sweden, Portugal, and other European countries are included, the number of deformed babies reached 8,000 (Steacy, 1989).

Very few cases ever occurred in the U.S. Dr. Frances Kelsey, the medical officer who reviewed safety data for the Food and Drug Administration, was sceptical and critical of the drug. She had, by chance, read a letter to the editor in a medical journal, which presented negative information about the drug (*Kitchener-Waterloo Record*, August 17, 1962). She was dissatisfied with the available information on the safety of

the product and did not allow it to be marketed. In particular, she was concerned that the drug could cross the placenta. Only the few samples of the drug given to doctors were ever used. Apparently Dr. Kelsey resisted extraordinary pressure from the drug company, which made "no less than 50 approaches or submissions to the FDA" (Winsor, 1973).

The drug companies had used a number of tactics to increase the sales of thalidomide. One involved planting an article in the June, 1961, issue of the *American Journal of Obstetrics and Gynecology*, allegedly written by Dr. Ray Neilson of Cincinnati. The article said that the drug was safe for pregnant women (Winsor, 1973). Later, Dr. Neilson admitted that the article had been written by the medical research director for the manufacturer and was based on incomplete evidence, i.e., on evidence only that the drug was harmless when taken in the last few months of pregnancy, when, of course, the limbs had already developed. Nevertheless, it was advertised as safe for pregnant women when it was clearly known to be unsafe when ingested in the early months.

Today Canadians are being reminded of this tragedy and of the outcome for the people who suffered physical deformities and emotional and social scars as a result. Recently the Canadian Broadcasting Company aired a documentary on the events and their aftermath, called "Broken Promises." This documentary revealed the culpability of the Canadian government in failing to keep the drug out of Canada. The *Globe and Mail* (February 15, 1989) reported on the program as follows:

> the most striking impression left by "Broken Promises" is that a number of pharmaceutical companies, druggists, doctors and prosthesis manufacturers have callously exploited the victims of Thalidomide with the crudest and most obvious motive – profit.

In addition, the Canadian government was criticized for failing to provide compensation.

The Case of DES

From the 1940s through the 1960s, many physicians prescribed the synthetic estrogen hormone DES (diethylstilbestrol, or simply stilbestrol) to pregnant women who had histories of miscarriage, diabetes, or toxemia of pregnancy. More than 4 million women took DES over this period. Approximately 2 million male and 2 million female children of these women had been exposed to DES in the U.S.; there are approximately 400,000 children of "DES mothers" in Canada. These children have developed a number of abnormalities, including a rare vaginal cancer, adenocarcinoma, and a variety of apparently benign structural changes of the uterus, cervix, and vagina. As many as 97 per cent of the DES daughters have cervical abnormalities. Adenosis, the most common problem, is estimated to occur in 43 to 95 per cent of the women. About one-half of DES daughters have had or may have problems with pregnancy, including primary infertility (difficulty in becoming pregnant), premature births, stillbirths, and ectopic pregnancies (gestation outside the uterus). Ectopic pregnancies, which may be dangerous to the mother-to-be as well as the fetus, appear to occur in five times as many DES daughters as in other women. Problems have been seen in DES sons as well. About 30 per cent have genital tract and semen abnormalities, including cysts and extremely small and undescended testicles. The impact of DES has become known over the last decade or so.

Other Negative Effects of the Pharmaceutical Industry

Finally, the pharmaceutical industry must also be criticized for causing ill health in another way. As Harding (1987: 552) suggests, "the pharmaceutical industry is an outgrowth of the interlocking petrochemical

Box 15.5 The Discovery of the DES Problem in Canada

In 1982, Harriet Simand was a healthy, twenty-one-year-old philosophy student at McGill University in Montreal. Suddenly, after what was supposed to be a routine medical examination, Harriet was diagnosed with a rare cancer: clear cell adenocarcinoma of the cervix and vagina. Harriet had to have a hysterectomy. She was given an 85 per cent chance of survival after five years. Later she discovered that the cancer was linked to DES, a wonder drug marketed between 1941 and 1971 to prevent miscarriage. Harriet's mother had tried unsuccessfully to carry a baby to term eight years before becoming pregnant with Harriet. To prevent another miscarriage, the doctors prescribed DES.

At the time DES was prescribed for a variety of "feminine conditions," including irregular bleeding and spotting, menopause, and as a "morning-after pill" to prevent successful implantation after conception. In fact, it was prescribed as a "morning-after pill" to numerous college and university students who had engaged in unprotected sexual intercourse and did not want to get pregnant.

Approximately one out of every thousand daughters of women prescribed DES will develop cancer. DES daughters also have an increased risk of contracting a precancerous condition called cervical dyplasia, of giving birth prematurely, of miscarriage during the second trimester, of ectopic pregnancy, of genital organ malformations, and perhaps of breast cancer. DES sons have an increased risk of undescended testicles, a condition which is sometimes related to testicular cancer. They may also have low sperm counts and abnormal sperm formation.

When Harriet Simand learned that she had this rare form of cancer, she began to read up on the subject and to ask questions. She discovered that American medical journals had already published a number of papers on some of the results of DES. Harriet was shocked to discover that the Canadian government was not taking action to inform people of the potential long-term intergenerational effects of DES, even though early screening is known to have beneficial effects on the outcome of the disease. Partly because of the lobbying efforts of women's groups in the U.S., the American federal government had sponsored screening clinics for DES daughters and sons and a widespread information campaign. Doctors were encouraged to inform them of patients who had ever been prescribed DES.

None of this had happened in Canada. Yet Harriet and her mother Shirley, working together, found that there were at least six known DES-related cancers in Quebec, and the same number in Ontario, and that approximately 100,000 people had been exposed to the drug.

In response, the Simands founded DES Action Canada. They set up an office and received a $50,000 annual grant from the federal government. Numerous newspaper and magazine stories have publicized their concerns. Many doctors have volunteered to help in screening DES daughters and in contacting patients who had taken the drug. The National Film Board has produced a film on the subject. DES Action Canada hosted an international conference in 1986. Nine chapters of the organization have been established across the country. About 30,000 Canadians have called DES Action Canada for advice and assistance.

Harriet and Shirley Simand have shown what people can do to lobby the government.

industry, which also produces pesticides, herbicides and fertilizers." Toxins from the petrochemical industry have been responsible for environmental health calamities. The public health dangers of this industry are therefore not confined to those that arise from the adverse effects of the drugs themselves.

Issues in Drug Regulation

Governments can have an important role in the regulation of the drug industry and ultimately in the drug-related health of its citizens. However, most government regulations are inadequate. As a result: (1) half the drugs now on the Canadian market have never passed modern tests regarding safety or effectiveness; (2) even where regulations are in place in the industrialized world, substandard drugs are being marketed and distributed overseas; (3) drug companies seem to have a monopoly on the information available to doctors as well as on the side effects of various drugs.

However stringent the laws, the government cannot guarantee that any drug is safe for all the uses to which it may be put. It is not difficult to imagine a situation in which a drug is prescribed for one use or for one person, and is then used again by the same person on another occasion when the symptoms seem to be similar, or is passed on to a friend or a family member who seems to have the same problem. People often regulate their drug use, ignoring the specific directions given by the physician and/or the pharmacist. *People*

have been known to develop drug allergies very suddenly. And many people stop taking antibiotics once they start to feel better, even though the doctor advised that they finish the complete supply.

While the government can insist that patients be told about drug interactions, it cannot regulate the actual mixture of drugs taken by anyone individual. Some drugs react negatively when taken in conjunction with alcohol: 82 per cent of all Canadians drink (Canada's Health Promotion Survey, *Technical Report*, 1988). One characteristic of many alcoholics is that they try to keep their drinking habit secret – even from their doctor. Untold problems result from drug-alcohol interactions. In addition, a number of drug-related problems or side effects are only discovered after long-term use. For these and other reasons, the drug regulations established by the government can only be considered as partial protection.

The Canadian government is also in a weak position because of the Canadian branch-plant economy. Most drugs are developed and tested elsewhere. The Canadian government frequently relies on tests done abroad by other governmental bodies or by the drug firms' research departments. This raises complex problems of biased information from pharmaceutical companies and political problems of intergovernmental relations. Moreover, as Lexchin documents, in spite of the fact that Canadian drug laws are among the strictest in the world, there are still major gaps that may jeopardize people's health. These gaps, Lexchin argues, are not accidents, nor are they idiosyncratic; rather, they are

Box 15.6 The Purpose of the Health Protection Branch

The following is a statement from the *Canada Year Book* (1994: 48) on health protection. This statement reflects the formal views of the responsibilities of the Health Protection Branch of the federal government. It may be interesting for you to analyse this statement as ideology and to evaluate it against the research reported in this book.

The federal and provincial governments protect the public from unsafe food, drugs and cosmetics, and from unsafe radiation emitting devices and medical devices. Federal programs also protect Canadians from harmful micro-biological agents, environmental pollutants and contaminants and fraudulent drugs and devices.

Standards for food safety, cleanliness and nutrition quality are determined by laboratory research and data from Canadian and international sources. The standards cover the use of food additives and maximum levels of agricultural chemicals permitted in foods. They are maintained by regular inspections of domestic and imported foods. In addition, all food additives and agricultural chemicals are thoroughly evaluated before they can be used in food sold in Canada.

Manufacturers of new drugs must give government regulators information about the drug, including the results of pre-clinical and clinical studies, and its therapeutic properties and possible side effects. Market approval is governed by assessments of the drug's risks and benefits, and by evaluations of its safety and effectiveness. Once a new drug has reached the market, manufacturers are required to report adverse reactions. Further assessments are conducted to determine whether any reactions are serious enough to signal the need for additional evaluation or action.

To combat disease in Canada, the national laboratory of Health and Welfare Canada monitors seasonal maladies (such as influenza), sexually transmitted diseases, non-communicable diseases and potential disease threats. The surveillance helps identify risk factors and changes in disease patterns. By investigating outbreaks of communicable diseases and by providing laboratory diagnostic services, the department is able to rapidly detect and control infections. In recent years, provincial laboratories have contributed to control of infectious diseases through unique testing agents and through improved procedures for training and for diagnostic quality assurance.

In the environmental health field, government agencies study the effects of potentially hazardous chemical and physical environments. Recent studies have investigated the effects of tobacco smoke, pesticides and household products.

---

the result of the structure of the drug regulation body of the federal government – the Health Protection Branch – and its close interlocking ties with the international pharmaceutical industry organization and the Pharmaceutical Manufacturing Association of Canada. These two groups interact through an extensive system of liaison committees that allow the PMAC to participate at all stages in drug regulation, policy development, and implementation (Lexchin, 1990). It is not hard to imagine that there could be situations in which there is a conflict of interest.

Medical Devices and Bioengineering

Companies that produce and sell various medical devices such as artificial heart valves, artificial limbs, kidney dialysis machines, anesthesiology equipment, surgical equipment, and heart pacemakers are among the largest growth industries in the world. Sales increased world-wide from $17.5 billion in 1976 to $48.1 billion in 1986. Canadian sales of more than 300,000 types of medical devices topped $1.4 billion in 1986 (Regush, 1987). There are uneven regulations controlling the industry. Das Gupta, who was head of the Bureau of Radiation and Medical Devices, has said that he could document more than 200 deaths and 800 injuries simply from voluntarily submitted reports from hospitals, doctors, and coroners. In fact, he estimated that if all the cases were known the numbers would be much higher. For instance, he suggested that anesthesia equipment alone probably accounted for 200 deaths per year in Canada in the early to mid-eighties. Anesthesia gas lines, he said, were simply patched together with masking tape in a number of major hospitals (Regush, 1987).

> At present a manufacturer who wants to market a medical device in Canada has only to provide the bureau [Bureau of Radiation and Medical Devices of the federal Health Protection Branch] with a device notification in most cases – ten days after it is available for sale – detailing the directions for its use, its purpose, and its model number. (*Ibid.*: 254)

The government does not generally require evidence concerning the potential for harm or benefit, or the safety of the various devices. The only exceptions are for tampons, condoms, contact (intra-ocular) lenses, and devices that are implanted more than thirty days in the body. These are only accepted for marketing after the government has examined the evidence, provided by the manufacturing firm itself, as to the safety of the device.

The government, through the Bureau of Radiation and Medical Devices of the Health Protection Branch of National Health and Welfare, has not developed its own safety standards in any systematic way. In fact, it has relied on the industry to police itself. Unfortunately, such policing has not always been adequate. There is evidence that, for instance, up to 50 per cent of all the medical devices delivered to hospitals have failed to meet the minimum standards for safety established by the independent Canadian Standards Association. Some of the inadequacies were minor, but some were potentially life-threatening (*ibid.*). The case of the Même breast implants was perhaps the most widely publicized of recent health and safety hazards in the Canadian medical device marketplace.

Among the responsibilities of biomedical engineers are the evaluation and testing of equipment, the investigation and explanation of the causes of accidents, and the supervision of the repair of biomedical equipment. There is, however, a shortage of such personnel in Canada, owing in part to the lack of training programs. From what limited information is available, it

Box 15.7 The Debate About the New Reproductive Technologies

Reproductive technologies have created a scientific revolution with repercussions as wide-ranging and as ethically questionable as the hydrogen bomb several decades ago. The development of reproductive technology forces us to make decisions about what kind of world we want to live in and how much control we want over that world.

The new reproduction technologies include sex selection techniques, artificial insemination, and test tube techniques including in-vitro fertilization, embryo transplants, and surrogate motherhood. Such techniques are no longer rare or unusual. Without much public debate, they have become institutionalized as legitimate options for infertile couples or couples who want to select the characteristics (gender, IQ, colouring) of their offspring. The Ontario government has made a significant investment of more than $7 million in the province's five infertility clinics. One hundred thirty couples have been successfully "treated," resulting in the birth of 207 babies at an average cost of $34,000 each. This is not an insignificant figure when compared to the 240,000 Canadian couples who are childless.

The problem here is that the reproduction technologies ought not to be considered morally neutral medical interventions. Rather, they must be seen as bearing, as reproduction always does bear, significant social, political, economic, and moral ramifications. A concern with eugenics has, for instance, been the cause of the most significant attempt at genocide in this century – the Holocaust, the attempted extermination of the Jews by the Nazis in Germany. The root of the difficulty with all of the new reproductive technologies is that power over their development is primarily under the control of the medical and legal professions. It is, then, the values of these groups, and not the values of all the citizens of the country, that will determine the goals of reproduction, the "types" of people who are fit to be parents, and the "types" of offspring who are most "desirable" for the society of the future.

---

is clear that the whole issue of medical devices and bioengineering needs a great deal of research.

Summary

(1) There is a correlation between people's socio-demographic characteristics and their drug-taking habits. The heaviest users of drugs are those people under five and over sixty-five years of age. Females tend to use prescription drugs and visit doctors more frequently than males. Those in lower income groups spend a greater proportion of their income on prescription drugs than those in higher income groups.

(2) Psychoactive drugs are one of the most heavily prescribed and often misprescribed drugs in Canada. Females, the elderly, and the unemployed are high users. Chronically ill patients are often prescribed two or more psychoactive drugs simultaneously. A good proportion of these mood-modifying drugs is given for social and personal reasons and not for medical problems.

Box 15.8 Premature Adoption of New Technology

There are a number of reasons why modern capitalist societies are prone to adopt technology prematurely. Butler (1993) documents three major forces: key societal values (e.g., placing a huge value on science and technology), federal government policies (e.g., policies that place the burden of responsibility for determining safety, quality, and efficacy of new technologies on their manufacturers), and reimbursement policies and economic incentives (policies that enable the costs of new technologies to be reimbursed adequately so as to ensure their profitability). These forces make it possible and even likely that new technologies are adopted prematurely even when this is at variance with physician opinion and of unproven medical benefit. Technological favouritism is characteristic of a variety of technologies (overuse of cardiac pacemakers, gastro-intestinal endoscopy, and coronary bypass surgery have been estimated at 20 per cent, 17 per cent, and 15 per cent, respectively). It is especially prevalent in regard to childbearing women (particularly with respect to electronic ultrasound, electronic fetal monitoring, and Caesarean section). It has been estimated, for instance, that the overuse Caesarean sections is as high as 50 per cent. Moreover, there are cases where women have been given Caesarean sections even against their will after obstetricians have asked for and received a court intervention (*ibid.*).

(3) There are large differences in rates of prescription from doctor to doctor. Doctors receive much of their information about drugs from pharmaceutical companies. The drug promotion and advertising strategies used by these companies have an important impact on prescribing. Other factors in the rate at which doctors prescribe drugs include education, type of practice, and method of remuneration.

(4) Pharmacists tend to recommend non-prescription drugs that will maximize their profit. Pharmacists may choose between a brand name (more expensive) and a generic drug for a customer when filling a prescription. Pharmaceutical companies offer incentives to ensure that the pharmacist will choose their brand.

(5) Multinationals control 90 per cent of the Canadian prescription drug market. Pharmaceutical manufacturing is one of the more profitable manufacturing activities in Canada. Some of the reasons for this are: the absence of price competition, the presence of patent protection, price-fixing, discount pricing, advertising, and drug distribution (dumping) in the Third World.

(6) There are several instances where profits have come before health in Canada. One is the case of thalidomide, which resulted in the birth of 115 babies in Canada with phocomelia. DES is another drug that was used by pregnant women with disastrous consequences. It is now known that it has caused many abnormalities in the reproductive systems of the offspring of these mothers.

(7) The government is not able to regulate the use of drugs in Canada adequately. This means that half the drugs now on the Canadian market have never passed modern tests regarding safety or

Box 15.9 Sue Rodriguez

Contemporary issues that are a direct result of the bur-geoning availability of medical care and technological and pharmaceutical solutions are the uses of euthana-sia, the right to die, dying with dignity, and living wills. Sue Rodriguez, a woman with amyotrophic lateral sclerosis (ALS) unsuccessfully fought, up to the Supreme Court, for her right to choose her own time of death with the help of a physician. Rodriguez chose her time of death in any case, and was helped, albeit illegally, by a physician who has remained anonymous to prevent charges being laid.

Public opinion polls over the last forty years docu-ment a trend toward increasingly liberal attitudes to euthanasia. The Institute for Behavioural Research at York University recently completed a study (1992) of the attitudes of a randomly selected 600 people living in Toronto. Almost 61 per cent indicated they would approve of euthanasia, always, most of the time, or sometimes; 18 per cent agreed that they would approve euthanasia under some conditions (but rarely); and 21 per cent said they would never approve euthanasia (Pollard, 1994: 3).

Sue Rodriguez was diagnosed with ALS, often called Lou Gehrig's Disease after the famous American baseball player who died from it. ALS is a degenerative and chronic condition, a motor neuron disease that at the time of her diagnosis was essentially untreatable. Death was the inevitable result after a period of time of increasing disability and increasing dependence on others for the basics of life. Rodriguez decided that at some point in the illness trajectory she would decide that her life was not worth living any longer. She also felt that by the time she felt that way she would be unable to commit suicide – she would lack the minimal muscle power for such an act. She decided that she wanted to die at a time of her own choosing and with the assistance of a physician who would ensure a rel-atively painless death.

To Rodriguez the phrase used in *Final Exit* by Derek Humphrey, "self-deliverance," rang true. In her own book, *Uncommon Will* (pp. 36-37), Rodriguez's commit-ment is described as follows:

> That night she decided to commit suicide. Sue made the decision without fuss, hesitation or moral argu-ment. The idea came and she accepted it. It felt right. The dominant nature of the disease was now clear. It would rob her of everything that gave value to her life and then slowly, at its own leisure, kill her. . . . Sue was moving toward the position that she would steadfastly and publicly maintain for the next two years – that the quality of life is the issue of life, and that a life deprived of quality was not worth living.

The story of her fight to have someone help her commit suicide, first through the British Columbia courts and finally before the Supreme Court of Canada, became public record and garnered considerable media attention in the years 1992-94. The moral and legal questions were hotly debated in magazines, on televi-sion, and on the radio. In the end Sue lost her legal battle: she did not have the right to assisted suicide.

However, she educated people across the country about ALS and about the issues surrounding "the right to die." At the end an anonymous doctor assisted her with her suicide and she was accompanied as she died by Svend Robinson, the NDP MP who had become a close friend and supporter through her battle for assistance to die with dignity and at a time of her own choosing.

Lisa Hobbs Birnie and Sue Rodriguez, *Uncommon Will: The Death and Life of Sue Rodriguez* (Toronto: Macmillan, 1994).

effectiveness. Drug companies seem to have a monopoly on the information available to doctors as well as on the side effects of various drugs.

(8) The medical devices industry is a profitable and growing industry. Except for devices to be used within the body, the government does not require evidence as to the harm or benefit or safety of medical devices.

Bibliography

Abelson, J. Paddon, and C. Strohmenger. 1983. *Perspectives on Health*. Ottawa: Statistics Canada.

Abbott, Maude. 1931. *The History of Medicine in the Province of Quebec*. Montreal: McGill University Press.

Academic American Encyclopedia. 1980. Princeton, N.J.: Arete Publishing Company.

"Accessibility to Higher Education – New Trend Data." 1988. *CAUT Bulletin* (Ottawa), June: 15.

Achilles, Rona. 1990. *Desperately Seeking Babies: New Technologies of Hope and Despair*. London: Routledge.

Achterberg, Jeanne. 1985. *Imagery in Healing*. Boston: Shambhala.

Ackerknect, E.H. 1968. *A Short History of Psychiatry*. New York: Hafner Press.

Ackerman-Ross, F.S., and N. Sochat. 1980. "Close Encounters of the Medical Kind: Attitudes toward Male and Female Physicians," *Social Science and Medicine*, 14A: 61-64.

Active Health Report – Perspective on Canada's Health Promotion Survey – 1985. 1987. Ottawa: Minister of National Health and Welfare.

Alford, R. 1971. "The Political Economy of Health Care: Dynamics without Change," *Politics and Society*, 2: 127-64.

Allentuck, Allen. 1978. *Who Speaks for the Patient?* Toronto: Burns & MacEachern.

Alonzo, Angelo A. 1992. "Health Behavior: Issues, Dilemmas and Explorations Toward a Paradigm," paper presented at the annual meeting of the American Sociological Association, Pittsburgh, August.

Altman, D. 1986. *AIDS in the Mind of America*. New York: Anchor Press/Doubleday.

Aneshensel, Carol, S. Leonard, and I. Pearlin. 1987. "Structural Contexts of Sex Differences in Stress," in R.C. Barneth, L. Biener, and G.K. Baruch, eds., *Stress*. New York: Free Press: 75-95.

Angus, Douglas E. 1984. "Health Care Costs: Past, Present, and (Can We Forecast?) the Future," working paper. Ottawa: University of Ottawa.

Angus, Douglas E. 1987. "Health Care Costs," in D. Coburn *et al.*, eds., *Health and Canadian Society: Sociological Perspectives*. Second Edition. Toronto: Fitzhenry & Whiteside: 57-72.

Angus, D.E., and P. Manga. 1985. *National Health Strategies: time for a new perspective*. Ottawa: University of Ottawa, Faculty of Administration.

"Anti-Acne Drug Poses Dilemma for FDA

(Accutane and Birth Defects)." 1988. *Time*, 131, 63 (May 2).

Antonovsky, A. 1967. "Social Class, Life Expectancy and Overall Mortality," *Millbank Memorial Fund Quarterly*, 45: 31-73.

Antonovsky, A. 1979. *Health, Stress and Coping*. San Francisco: Jossey Bass.

Appleby, Timothy. 1989. "AIDS-infected prisoner's treatment cruel and unusual, judge rules," *Globe and Mail*, February 22: A11.

Armstrong, Liz, and Adrienne Scott. 1992. *Whitewash*. Toronto: Harper Collins.

Armstrong, Pat, Jacqueline Choinceri, and Claire Day. 1993. *Vital Signs: Nursing in Transition*. Toronto: Garamond Press.

Aronourtz, Robert A. 1988. "From Myalgic Encephalitis to Yuppie Flu: A History of Chronic Fatigue Syndrome," in C.E. Rosenbart and J. Golden, eds., *Framing Disease: Studies in Cultural History*. New Brunswick, N.J.: Rutgers University Press: 155-81.

Ashford, Nicholas A., and Claudia S. Miller. 1991. *Chemical Exposures*. New York: Van Nostrand Reinhold.

Ashworth, C.D., P. Williamson, and D. Montano. 1984. "A Scale to Measure Physician Beliefs about Psychosocial Aspects of Patient Care," *Social Science and Medicine*, 19: 1235-38.

Association of American Medical Colleges. 1984. *Physicians for the Twenty-first Century: Report of the Panel on the General Professional Education of the Physician and College Preparation for Medicine*. Washington: Anthon.

Bakwin, H. 1945. "Pseudoxia Pediatricia," *New England Journal of Medicine*, 232: 691-97.

Balshem, Martha. 1991. "Cancer, Control, and Causality: Talking about Cancer in a Working Class Community," *American Ethnologist*, 18: 152-72.

Bardossi, F. 1982. *Multiple Sclerosis: Grounds for Hope*. Toronto: Public Affairs Committee.

Barer, M.L., R.G. Evans, and G.L. Stoddart. 1979. "Controlling Health Care Costs by Direct Charges to Patients: Snare or Delusion?" Occasional Paper No. 10. Toronto: Ontario Economic Council.

Barker-Benfield, G.J. 1976. *The Horrors of the Half-Known Life*. New York: Harper Colophon Books.

Baron, S.H., and S. Fisher. 1962. "Use of Psychotropic Drug Prescriptions in a Prepaid Group Practice Plan," *Public Health Reports*, 77: 871-81.

Barrington, E. 1985. *Midwifery is Catching*. Toronto: NC Press.

Barsky, Arthur J. 1981. "Hidden Reasons Some Patients Visit Doctors," *Annals of Internal Medicine*, 94: 492-98.

Batt, Sharon. 1994. *Patient No More: The Politics of Breast Cancer*. Charlottetown, P.E.I.: Gynergy Books.

Battershell, Charles. 1994. "Social Dimensions in the Production and Practice of Canadian Health Care Professionals," in B. Singh Bolaria and Harley D. Dickinson, eds., *Health, Illness and Health Care in Canada*. Second Edition. Toronto: Harcourt Brace & Company: 135-57.

Baumgart, Alice J., and J. Larsen. 1988. *Canadian Nursing Faces the Future: Development and Change*. St. Louis: Mosby.

Baxter, J., J. Eyles, and D. Willms. 1992. "The Hagersville Tire Fire," *Qualitative Health Research*, 2: 208-37.

Beardshaw, Virginia. 1983. *Prescription for*

Change: Health Action International's Guide to National Health Products. The Hague: International Organization for Consumer Unions.

Becker, Gay, and Robert D. Nachtigall. 1992. "Eager for Medicalization: The Social Production of Infertility as a Disease," *Sociology of Health and Illness*, 14, 4: 456-71.

Becker, H. 1963. *The Outsiders: Studies in the Sociology of Deviance*. New York: Free Press.

Becker, Howard S., *et al*. 1961. *Boys in White: Student Culture in Medical School*. Chicago: University of Chicago Press.

Becker, H., *et al*. 1982. "Union Activity in Hospitals: Past, Present and Future," *Health Care Financing Review*, 3: 1-110.

Beckman, L.F. 1977. "Social Networks, Host Resistance and Mortality: A Follow-Up Study of Alameda County Residents," Ph.D. dissertation, Department of Epidemiology, School of Public Health, University of California, Berkeley.

Belkin, Lisa. 1990. "Seekers of Urban Living Head for Texas Hills," *New York Times*, December 2: A1-32.

Bell, C. 1971. "Occupational, Career, Family Cycle and Extended Family Relations," *Human Relations*, 24 (December): 463-75.

Bell, R.W., and J. Osterman. 1983. "The Compendium of Pharmaceuticals and Specialties: A Critical Analysis," *International Journal of Health Services*, 13: 107-18.

Bell, Susan E. 1989. "Technology in Medicine: Development, Diffusion, and Health Policy," in Howard E. Freeman and Sol Levine, eds., *Handbook of Medical Sociology*. Fourth Edition. Englewood Cliffs, N.J.: Prentice-Hall: 185-204.

Berger, Peter L., and Thomas Luckmann. 1966.

The Social Construction of Reality. Garden City, N.Y.: Doubleday.

Beyond Adjustment: Responding to the Health Crisis in Africa. 1993. Toronto: Inter-Church Coalition on Africa (ICCAF).

Bieliauskas, Linas A. 1982. *Stress and Its Relationship to Health and Illness*. Boulder, Colorado: Westview Press.

Biggs, C. Lesley. 1983. "The Case of the Missing Midwives: A History of Midwifery in Ontario from 1795-1900," *Ontario History*, 75: 21-35.

Biggs, C. Lesley. 1988. "The Professionalization of Chiropractic in Canada: Its Current Status and Future Prospects," in B. Singh Bolaria and Harley D. Dickinson, eds., *Sociology of Health Care in Canada*. Toronto: Harcourt Brace Jovanovich: 328-45.

Bilson, Geoffrey. 1980. *A Darkened House: Cholera in Nineteenth-Century Canada*. Toronto: University of Toronto Press.

Birnie, Lisa Hobbs, and Sue Rodriguez. 1994. *Uncommon Will: The Death and Life of Sue Rodriguez*. Toronto: Macmillan.

Black Report – DHSS Inequalities in Health: Report of Research Writing Group. 1982. London: Department of Health and Social Security.

Blane, D. 1985. "An Assessment of the Black Report's Explanation of Health Inequalities," *Sociology of Health and Illness*, 7 (November): 423-45.

Blaxter, Mildred. 1978. "Diagnosis as Category and Process: The Case of Alcoholism," *Social Science and Medicine*, 12: 9-77.

Blishen, Bernard R. 1969. *Doctors and Doctrines: The Ideology of Medical Care in Canada*. Toronto: University of Toronto Press.

Blishen, Bernard R. 1991. *Doctors in Canada*. Toronto: University of Toronto Press.

Bloom, Joan, and Larry Kessler. 1994.

"Emotional Support Following Cancer: A Test of the Stigma and Social Activity Hypothesis," *Journal of Health and Social Behavior*, 35 (June): 118-33.

Bluebond-Langer, Myra. 1978. *The Private Worlds of Dying Children*. Princeton, N.J.: Princeton University Press.

Bolaria, B. Singh, and Rosemary Bolaria. 1994. *Racial Minorities: Medicine and Health*. Halifax: Fernwood.

Bolaria, B. Singh, and Rosemary Bolaria, eds. 1994. *Women, Medicine and Health*. Halifax: Fernwood.

Bolaria, B. Singh, and Harley D. Dickinson, eds. 1994. *Health, Illness and Health Care in Canada*. Second Edition. Toronto: Harcourt Brace & Company.

Bosk, Charles. 1979. *Forgive and Remember: Managing Medical Failure*. Chicago: University of Chicago Press.

Boston Women's Health Collective. 1971. *Our Bodies, Our Selves*. New York: Simon & Schuster.

Boughy, Howard. 1978. *Insights of Sociology: An Introduction*. Boston: Allyn and Bacon.

"A Boycott over Infant Formula." 1979. *Business Week*, April 23: 137-40.

Brack, J., and Robert Collins. 1981. *One Thing for Tomorrow: A Woman's Personal Struggle with Multiple Sclerosis*. Saskatoon: Western Producer Prairie Books.

Braden, Charles Samuel. 1958. *Christian Science Today: Power, Policy, Practice*. Dallas: Southern Methodist University Press.

Bradshaw, York W., Rita Noonan, Laura Gash, and Claudia Buchmann Sershen. 1993. "Borrowing Against the Future: Children and Third World Indebtedness," *Social Forces*, 71, 3: 629-56.

Bransen, Els. 1992. "Has Menstruation Been Medicalized or Will It Never Happen?" *Sociology of Health and Illness*, 14, 1: 98-110.

Brook, R.H., C.J. Kanberg, A. Mayer Oakes, *et al*. 1989. "Appropriateness of Acute Medical Care for the Elderly." Santa Monica, Calif.: Rand Corporation: 1-58.

Brown, Phil. 1992. "Popular Epidemiology and Toxic Waste Contamination: Lay and Professional Ways of Knowing," *Journal of Health and Social Behavior*, 33 (September): 267-81.

Brown, Richard E. 1979. *Rockefeller Medical Men: Medicine and Capitalism in America*. Berkeley: University of California Press.

Brym, R.J., and B.J. Fox. 1989. *From Culture to Power: The Sociology of English Canada*. Toronto: Oxford University Press.

Bull, Angela. 1985. *Florence Nightingale*. London: Hamish Hamilton.

Bullard, Robert D. 1983. "Solid Waste Sites and the Black Houston Community," *Sociological Inquiry*, 53: 273-88.

Bullough, Bonnie, and Vern Bullough. 1972. "A Brief History of Medical Practice," in Judith Lorber and Eliot Freidson, eds., *Medical Men and Their Work: A Sociological Reader*. New York: Aldone Atherton: 86-101.

Bunker, J. 1970. "Surgical Manpower: A Comparison of Operations and Surgeons in the United States and in England and Wales," *New England Journal of Medicine*, 282, 3: 135-44.

Bunker, John P. 1985. "When Doctors Disagree," *New York Times Review of Books*, April 25: 7-12.

Bunker, J.P., V.C. Donahue, P. Cole, and M.P. Knotman. 1976. "Public Health Rounds at the Harvard School of Public Health. Elective Hysterectomy: Pro and Con," *New England Journal of Medicine*, 295, 5: 264-68.

Burke, Mike, and H. Michael Stevenson. 1993. "Fiscal Crises and Restructuring in Medicine: The Politics and Political Science of Health in Canada," *HCS/SSC*, 1, 1: 51-80.

Burkett, Gary, and Kathleen Knaft. 1974. "Judgment and Decision-Making in a Medical Specialty," *Sociology of Work and Occupations*, 1: 82-109.

Burnfield, A. 1977. "Multiple Sclerosis: A Doctor's Personal Experience," *British Medical Journal*, no. 6058 (February 12): 435-36.

Burstyn, Verna. 1992. "Making Babies," *Canadian Forum*, March: 12-17.

Burtch, Brian E. 1994. "Promoting Midwifery, Prosecuting Midwives: The State and the Midwifery Movement in Canada," in B. Singh Bolaria and Harley D. Dickinson, eds., *Health, Illness and Health Care in Canada*. Second Edition. Toronto: Harcourt Brace & Company: 504-23.

Bury, M.R. 1986. "Social Constructionism and the Development of Medical Sociology," *Sociology of Health and Illness*, 2: 137-69.

Butler, Irene. 1993. "Premature Adoption and Routinization of Medical Technology: Illustrations from Childbirth Technology," *Journal of Social Issues*, 49, 2: 11-34.

Cadman, D., *et al.* 1986. "Chronic Illness and Functional Limitation in Ontario Children: Findings of the Ontario Child Health Study," *Canadian Medical Association Journal*, 135 (October): 761-67.

Calnan, Michael, and Simon Williams. 1992. "Images of Scientific Medicine," *Sociology of Health and Illness*, 14, 2: 233-54.

Campbell, Marie. 1988. "The Structure of Stress in Nurses' Work," in B. Singh Bolaria and Harley D. Dickinson, eds., *Sociology of Health Care in Canada*. Toronto: Harcourt Brace Jovanovich: 393-406.

Canada Health Act Annual Report 1992-1993. 1993. Ottawa.

Canada's Green Plan. 1994. Ottawa: Minister of Supply and Services.

Canada's Health Promotion Survey. 1988. *Technical Report*. Ottawa: Health and Welfare Canada.

Canada Year Book. Various years. Ottawa: Statistics Canada.

Canadian Advisory Council on the Status of Women. 1987. *Recommendations*. Ottawa.

Canadian Cancer Society. 1994. *Protecting Health and Revenue: An Action Plan to Control Contraband and Tax-Exempt Tobacco*. Ottawa.

Canadian College of Naturopathic Medicine. 1995. *The Power to Heal*. Toronto.

The Canadian Encyclopedia. Second Edition. 1988. Edmonton: Hurtig.

Canadian Medical Association. 1989. *Code of Ethics*.

Canadian Social Trends. 1992. Ottawa: Statistics Canada, Cat. 11-008E, No. 24 (Spring).

Cannon, William B. 1932. *The Wisdom of the Body*. New York: Norton.

Capildeo, R., and A. Maxwell. 1982. *Progress in Rehabilitation: Multiple Sclerosis*. London: Macmillan.

Caplan, Elinor. 1989. "Speaking Out." Toronto: TVO, Winter.

Caplan, Ronald Lee. 1984. "Chiropractic," in *Alternative Medicines: Popular and Policy Perspectives*. New York: Tavistock: 80-113.

Carlson, Rick. 1975. *The End of Medicine*. Toronto: Wiley.

Carpenter, M. 1980. "Review Article: Medical Sociology and the Politics of Health," *Sociology of Health and Illness*, 3: 104-12.

Cartwright, A. 1967. *Patients and Their Doctors*. London: Routledge and Kegan Paul.

Cartwright, A., and R. Anderson. 1981. *General Practice Revisited: A Second Study of Patients and Their Doctors*. London: Tavistock.

Cassel, J. 1974. "Psychosocial Processes and 'Stress': Theoretical Formulation," *International Journal of Health Services*, 4: 471-82.

Castiglioni, Arturo. 1941. *A History of Medicine*. New York: Knopf.

Charles, Catherine A. 1976. "The Medical Profession and Health Insurance: An Ottawa Case Study," *Social Science and Medicine*, 10: 33-38.

Charmaz, Kathy. 1987. "Struggling for a Self: Identity Levels of the Chronically Ill," *Research in the Sociology of Health Care*, 6: 283-321.

Chivian, Eric, Michael McCally, Howard Hu, and Andrew Haines. 1993. *Critical Condition: Human Health and Environment*. Cambridge, Mass.: MIT Press.

Chopra, Deepak. 1987. *Creating Health*. Boston: Houghton Mifflin.

Chopra, Deepak. 1989. *Quantum Healing: Exploring the Frontiers of Body Medicine*. New York: Bantam Books.

Clark, Jack A., and Elliott G. Mishler. 1992. "Attending to Patients' Stories: Reframing the Clinical Task," *Sociology of Health and Illness*, 14, 3.

Clark, Jack A., Deborah A. Potter, and John B. McKinlay. 1991. "Bringing Social Structure Back into Clinical Decision-Making," *Social Science and Medicine*, 32, 8: 853-63.

Clarke, Juanne N. 1980. "Medicalization in the Past Century in the Province of Ontario: The Physician as Moral Entrepreneur," Ph.D. dissertation, University of Waterloo.

Clarke, Juanne N. 1981. "A Multiple Paradigm Approach to the Sociology of Medicine, Health and Illness," *Sociology of Health and Illness*, 3, 1 (March): 89-103.

Clarke, Juanne N. 1983. "Sexism, Feminism and Medicalism: A Decade Review of the Literature on Gender and Illness," *Sociology of Health and Illness*, 5 (March): 62-82.

Clarke, Juanne N. 1984. "Medicalization and Secularization in Selected English Canadian Fiction," *Social Science and Medicine*, 18, 3: 205-10.

Clarke, Juanne N. 1985. *It's Cancer: The Personal Experiences of Women Who Have Received a Cancer Diagnosis*. Toronto: IPI Publishing.

Clarke, Juanne N. 1987. "The Paradoxical Effects of Aging on Health," *Journal of Gerontological Social Work*, 10: 3-20.

Clarke, Juanne N. 1990. *Health, Illness and Medicine in Canada*. Toronto: McClelland & Stewart.

Clarke, Juanne N. 1992a. "Cancer, Heart Disease and AIDS: What Do the Media Tell Us About These Diseases?" *Health Communication*, 4, 2: 105-20.

Clarke, Juanne N. 1992b. "Feminist Methods in Health Promotion Research," *Canadian Journal of Public Health*, Supplement 1 (March/April): 554-57.

Clarke, Juanne N. 1995. "Breast Cancer in Mothers: Impact on Adolescent Daughters," *Family Perspectives*, 29, 3: 243-57.

Clements, F.E. 1932. "Primitive Concepts of Disease," *Publications – American Archeology and Ethnology*, 32, 2: 182-252.

Clendening, Logan, ed. 1960. *Source Book of Medical History*. New York: Dover.

"Clioquinol: Time to Act." 1977. *Lancet*, 1, 8022 (May 28): 1139.

Cobb, Sidney. 1976. "Social Support as a Moderator of Life Stress," *Psychosomatic Medicine*, 38: 301-14.

Coburn, David, and C. Lesley Biggs. 1986. "Chiropractic: Legitimation or Medicalization?" in David Coburn *et al.*, eds., *Health and Canadian Society: Sociological Perspectives*. Second Edition. Don Mills, Ont.: Fitzhenry & Whiteside: 366-84.

Coburn, D., and J. Eakin. 1993. "The Sociology of Health in Canada: First Impressions," *Health and Canadian Society*, 1, 1: 83-112.

Coburn, D., G.M. Torrance, and J.M. Kaufert. 1983. "Medical Dominance in Canada in Historical Perspective: The Rise and Fall of Medicine," *International Journal of Health Services*, 13, 3: 407-32.

Cole, Stephen, and Robert Lejeune. 1972. "Illness and the Legitimation of Failure," *American Sociological Review*, 37: 347-56.

Coleman, James, Elihu Katz, and Herbert Menzel. 1957. "The Diffusion of Innovation Among Physicians," *Sociometry*, 20: 253-69.

Colliers Encyclopedia. 1970, 1973. New York: Crowell-Collier.

Collishaw, N.E. 1982. "Disability Attributable to Smoking – Canada, 1978-79," *Chronic Diseases in Canada* (Ottawa: Health and Welfare), 3 (December): 61.

Colombo, John Robert. 1993. *The Canadian Global Almanac. 1993. A Book of Facts*. Toronto: Macmillan.

Colquitt, W., and C. Killian. 1991. "Students Who Consider Medicine but Decide Against It," *Academic Medicine*, 66, 5: 273-78.

Conger, Rand D., Frederick O. Lorenz, Glen Ceda, Jr., Ronald L. Simons, and Xiaojia Ge. 1993. "Husband and Wife Differences in Response to Undesirable Life Events," *Journal of Health and Social Behavior*, 32 (March): 71-88.

Conley, M.C., and H.O. Maukasch. 1988. "Registered Nurses, Gender and Commitment," in A. Statham, E.M. Miller, and H.O. Maukasch, eds., *The Worth of Women's Work: A Qualitative Synthesis*. Albany: State University of New York Press.

Conrad, Peter. 1975. "The Discovery of Hyperkinesis: Notes on the Medicalization of Deviant Behaviour," *Social Problems*, 23 (October): 12-21.

Conrad, P. 1987. "The Experience of Illness: Recent and New Directions," in J. Roth and P. Conrad, eds., *Research in the Sociology of Health Care*, vol. 6. Greenwich, Conn.: JAI Press.

Conrad, P., and R. Kern, eds. 1990. *The Sociology of Health and Illness*. Third Edition. New York: St. Martin's Press.

Conrad, Peter, and Joseph W. Schneider. 1980. *Deviance and Medicalization: From Badness to Sickness*. St. Louis: Mosby.

Cooperstock, Ruth, and Jessica Hill. 1982. *The Effects of Tranquilization: Benzodiazepine Use in Canada*. Ottawa: Health Promotion Directorate.

Cooperstock, Ruth, and Henry L. Lennard. 1987. "Role Strains and Tranquilizer Use," in D. Coburn *et al.*, eds., *Health and Canadian Society: Sociological Perspectives*. Second Edition. Toronto: Fitzhenry & Whiteside: 314-32.

Corbin, Juliet, and Anselm Strauss. 1987. "Accompaniments of Chronic Illness: Changes in Body, Self, Biography and Biographical Time," *Research in the Sociology of Health Care*, 6: 249-81.

Cornwell, Jocelyn. 1984. *Hard-Earned Lives:*

Accounts of Health and Illness from East London. London: Tavistock.

Coulter, Harris L. 1984. "Homeopathy," in *Alternative Medicines: Popular and Policy Perspectives.* New York: Tavistock: 57-59.

Cousins, Norman. 1979. *Anatomy of an Illness as Perceived by the Patient.* Toronto: Bantam Books.

Cousins, Norman. 1983. *The Healing Heart.* New York: Norton.

Cousins, Norman. 1989. *Head First: The Biology of Hope.* New York: E.P. Dutton.

Cowie, W. 1976. "The Cardiac Patient's Perception of His Heart Attack," *Social Science and Medicine,* 10: 87-96.

Crane, Diana. 1975. *The Sanctity of Social Life: Physicians' Treatment of Critically Ill Patients.* New York: Russell Sage Foundation.

Crawford, Robert. 1984. "A cultural account of health: control, release, and the social body," in John B. McKinlay, ed., *Issues in the Political Economy of Health Care.* London: Tavistock.

Culane, Dora Speck. 1987. *An Error in Judgement – The Politics of Medical Care in an Indian/White Community.* Vancouver: Talon Books.

Currie, Dawn. 1988a. "Starvation Amidst Abundance: Female Adolescence and Anorexia," in B. Singh Bolaria and Harley D. Dickinson, eds., *Sociology of Health Care in Canada.* Toronto: Harcourt Brace Jovanovich: 198-216.

Currie, Dawn. 1988b. "Rethinking What We Do and How We Do: A Study of Reproductive Decisions," *Canadian Review of Sociology and Anthropology,* 25, 2: 231-53.

Currie, Dawn, and Valerie Raoul, eds. 1992. *Anatomy of Gender: Women's Struggle for the Body.* Ottawa: Carleton University Press.

Daniels, Arlene Kaplan. 1975. "Advisory and Coercive Functions in Psychiatry," *Sociology of Work and Occupations,* 2, 1: 55-78.

D'Arcy, Carl. 1986. "Unemployment and Health: Data and Implications," *Canadian Journal of Public Health,* 77, Supplement 1: 124-31.

D'Arcy, Carl. 1987. "Social Inequalities in Health: Implications for Priority Setting," paper presented at the Second Biennial Conference on Health Care in Canada: Setting Priorities. Waterloo, Ont.: Wilfrid Laurier University, April-May.

Davis, D. 1979. "Equal Treatment and Unequal Benefits: The Medical Program," in G.L. Albrecht and P.C. Higgins, eds., *Health, Illness and Medicine.* Chicago: Rand McNally.

Davis, F. 1963. *Passage Through Crisis.* Indianapolis: Bobbs-Merrill.

Davison, Charlie, George Davey Smith, and Stephen Frankel. 1991. "Lay Epidemiology and the Prevention of Paradox: The Implication of Coronary Candidacy for Health Education," *Sociology of Health and Illness,* 13, 1: 1-19.

Devita, V.T., Jr., S. Hellman, and S.A. Rosenberg. 1985. *AIDS, Etiology, Diagnosis, Treatment and Prevention.* Philadelphia: J.B. Lippincott.

Devitt, Neil. 1977. "The Transition from Home to Hospital Birth in the U.S. 1930-1960," *Birth and Family Journal* (Summer): 45-58.

Dickin McGinnis, Janice P. 1977. "The Impact of Epidemic Influenza: Canada, 1918-1919," Canadian Historical Association, *Historical Papers.*

Dickinson, Harley D., and David A. Hay. 1988. "The Structure and Cost of Health Care in Canada," in B. Singh Bolaria and Harley D. Dickinson, eds., *Sociology of Health Care in*

Canada. Toronto: Harcourt Brace Jovanovich: 51-74.

Dickinson, Harley D., and Mark Stobbe. 1988. "Occupational Health and Safety in Canada," in B. Singh Bolaria and Harley D. Dickinson, eds., *Sociology of Health Care in Canada*. Toronto: Harcourt Brace Jovanovich: 426-38.

Dickson, Geri L. 1990. "A Feminist Poststructural Analysis of the Knowledge of Menopause," *Advances in Nursing Science*, April: 15-31.

Dimich-Ward, Helen, *et al.* 1988. "Occupational Mortality among Bartenders and Waiters," *Canadian Journal of Public Health*, 79 (May/June): 194-97.

Doll, R., and J. Peto. 1981. *The Causes of Cancer: Quantitative Estimates of Avoidable Risks of Cancer in the United States Today*. New York: Oxford University Press.

Doran, Chris. 1988. "Canadian Workers' Compensation: Political, Medical and Health Issues," in B. Singh Bolaria and Harley D. Dickinson, eds., *Sociology of Health Care in Canada*. Toronto: Harcourt Brace Jovanovich.

Dossey, Larry. 1982. *Space, Time and Medicine*. Boston: New Science Library.

Dossey, Larry. 1991. *Meaning and Medicine*. New York: Bantam Books.

Douglas, Mary. 1973. *Natural Symbols*. New York: Vintage Books.

Doyal, Lesley. 1979. *The Political Economy of Health*. London: Pluto Press.

Dubos, Rene. 1959. *The Mirage of Health*. Garden City, N.Y.: Doubleday.

Dunbar, R. 1943. *Psychosomatic Diagnosis*. New York: Hoeber Press.

Dunkel-Schetter, C., and C. Wortman. 1982. "Interpersonal Dynamics of Cancer: Problems in Social Relationships and Their Impact on the Patient," in Howard S.F. Friedman and M. Robin DeMatteo, eds., *Interpersonal Issues in Health Care*. New York: Academic Press: 69-117.

"Early Warnings: An Uproar over Accutane." 1988. *Science*, 240 (May 21): 714-15.

Eberts, Mary. 1987. *Report of the Task Force on the Implementation of Midwifery in Ontario*. Toronto: Ontario Ministry of Health.

Economic and Ecological Interdependence: A Report on Selected Environment and Resource Issues. 1982. Paris: Organization for Economic Co-operation and Development.

Eddy, Mary Baker. 1934. *Science and Health with a Key to the Scriptures*. Boston: Published by the Trustees under the Will of Mary Baker Eddy.

Edelstein, M.R. 1988. *Contaminated Communities*. Boulder: Westview Press.

Educating Future Physicians for Ontario (EFPO). 1990. *Annual Report*. Hamilton: McMaster University, Faculty of Health Sciences.

Educating Future Physicians for Ontario (EFPO). 1991. *Progress Report*. Hamilton: McMaster University, Faculty of Health Sciences, June.

Ehrenreich, B., M. Dowie, and S. Minkin. 1979. "The Charge Genocide: The Accused the U.S. Government," *Mother Jones* (November): 28-29.

Ehrenreich, Barbara, and Deidre English. 1973a. *Witches, Midwives and Nurses: A History of Women Healers*. Old Westbury, N.Y.: Feminist Press.

Ehrenreich, Barbara, and Deidre English. 1973b. *Complaints and Disorders: The Sexual Politics of Sickness*. Old Westbury, N.Y.: Feminist Press.

Ehrenreich, Barbara, and Deidre English. 1978. *For Her Own Good: 150 Years of the Experts' Advice to Women*. New York: Anchor Press/Doubleday.

Eichler, Margrit. 1988. *Families in Canada Today*. Toronto: Gage.

Eisenberg, David M., *et al.* 1993. "Unconventional Medicine in the United States – Prevalance, Costs, and Patterns of Use," *New England Journal of Medicine*, 328, 4: 246-52

Eisenberg, L. 1977. "Disease and Illness: Distinctions between Professional and Popular Ideas of Sickness," *Culture, Medicine and Psychiatry*, 1, 11: 9-23.

Elliott, S., *et al.* 1993. "Modelling Psychological Effects of Exposure to Solid Waste Facilities," *Social Science and Medicine*.

Encyclopedia Britannica. 1976. Chicago: William Brenton.

Engel, George L. 1971. "Sudden and Rapid Death during Psychological Stress: Folklore or Folk Wisdom?" *Annals of Internal Medicine*, 74: 771-82.

Engels, Frederick. [1845] 1985. *The Condition of the Working Class in England*. Stanford, Calif.: Stanford University Press.

Epp, Jake. 1986. *Achieving Health for All: A Framework for Health Promotion*. Ottawa: Minister of National Health and Welfare.

Epstein, Samuel S. 1979. *The Politics of Cancer*. Revised Edition. New York: Doubleday.

Epstein, Samuel S. 1993. "Evaluation of the National Cancer Program and Proposed Responses," *International Journal of Health Services*, 23, 1: 15-44.

Epstein, Samuel S. 1994. "Environmental and Occupational Pollutants are Avoidable Causes of Breast Cancer," *International Journal of Health Services*, 21, 1: 145-50.

Eyer, Joe. 1984. "Capitalism, Health and Illness," in John B. McKinlay, ed., *Issues in the Political Economy of Health Care*. New York: Tavistock: 23-59.

Eyles, John. 1993. "From Disease Ecology and Spatial Analysis to . . .?: The Challenges of Medical Geography in Canada," *Health and Canadian Society*, 1, 1: 113-46.

Eyles, J., D. Sider, J. Baxter, S.M. Taylor, and D. Willms. 1990. "The Impacts and Effects of the Hagersville Tire Fire," *Environment Ontario: The Challenge of a New Decade* (Toronto: Environment Ontario), 2.

Fabrega, Horatio. 1973. "Toward a Model of Illness Behavior," *Medical Care*, 11, 6: 470-84.

Fabrega, Horatio. 1974. *Disease and Social Behavior: An Interdisciplinary Perspective*. Cambridge, Mass.: MIT Press.

Facts about Chiropractic. n.d. Toronto: Ontario Chiropractic Association.

Ferguson, J.A. 1990. "Patient Age as a Factor in Drug Prescribing Practices," *Canadian Journal of Aging*, 9: 278-95.

Findlay, Deborah. 1993. "The Good, the Normal and the Healthy: The Social Construction of Medical Knowledge About Women," *Canadian Journal of Sociology*, 18, 2: 115-33.

Finlayson, Ann. 1988. "Blood on the Coal," *Maclean's* (October): N1-N4.

Fisher, Sue. 1986. *In the Patient's Best Interest: Women and the Politics of Medical Decisions*. New Brunswick, N.J.: Rutgers University Press.

Fisher, Sue, and Alexandra Dundas Todd, eds. 1963. *The Social Organization of Doctor-Patient Communication*. Washington, D.C.: Center for Applied Linguistics.

Fitzpatrick, Roy, *et al.* 1984. *The Experience of Illness*. London: Tavistock.

Flexner, Abraham. 1910. *Medical Education in the United States and Canada. A Report to the Carnegie Foundation for the Advancement of Teaching*. Bulletin No. 4. New York: Carnegie Foundation.

Foster, Michelle, and Elizabeth Huffman. 1995. "The Portrayal of the Elderly in Medical Journals," paper written for Qualitative Methods sociology course at Wilfrid Laurier University.

Foucault, Michel. 1973. *The Birth of Illness*, trans. A.M. Sheridan Smith. New York: Pantheon Books.

Foucault, Michel. 1975. *The Birth of the Clinic: An Archeology of Medical Perception*, trans. A.M. Sheridan Smith. New York: Vintage Books.

Fox, Nicholas J. 1993. "Discourse, Organization and the Surgical Ward Round," *Sociology of Health and Illness*, 15, 1.

Fox, Nicholas J. 1994a. "Anaesthetists, The Discourse on Patient Fitness and the Organization of Surgery," *Sociology of Health and Illness*, 16, 1: 1-18.

Fox, Nicholas J. 1994b. *Postmodernism, Sociology and Health*. Toronto: University of Toronto Press.

Fox, Renee C. 1957. "Training for Uncertainty," in Robert K. Merton, George G. Reader, and Patricia L. Kendall, eds., *The Student Physician*. Cambridge, Mass.: Harvard University Press: 207-18, 228-41.

Fox, Renee C. 1976. "Advanced Medical Technology: Social and Ethical Implications," *Annual Review of Sociology*, 2: 231-68.

Fox, Renee C. 1977. "The Medicalization and Demedicalization of American Society," in John H. Knowles, ed., *Doing Better and Feeling Worse: Health in the United States*. New York: Norton: 9-22.

Frank, Arthur. 1991. *At the Will of the Body*. Boston: Houghton Mifflin.

Frank, Arthur. 1993. "The Rhetoric of Self-change: Illness Experience as Narrative," *The Sociological Quarterly*, 32, 1: 39-52.

Frankel, V. 1965. *Man's Search for Meaning*, trans. I. Lasch. Boston: Beacon Press.

Frankenberg, Ronald. 1974. "Functionalism and After: Theory and Development in Science Applied to the Health Field," *International Journal of Health Services*, 4, 3: 411-27.

Freeland, M.S., and C.E. Schendler. 1981. "National Health Expenditures: Short-term Outlook and Long-term Projections," *Health Care Financing Review*, 2: 97-126.

Freidson, Eliot. 1970. *Professional Dominance: The Social Structure of Medical Care*. New York: Atherton Press.

Freidson, Eliot. 1975. *The Profession of Medicine: Study in the Sociology of Applied Knowledge*. New York: Dodd Mead.

Freund, Peter E., and Meredith B. McGuire. 1991. *Health, Illness and the Social Body*. Englewood Cliffs, N.J.: Prentice-Hall.

Friedman, Meyer, and Ray H. Rosenman. 1974. *Type A Behaviors and Your Heart*. New York: Alfred A. Knopf.

Friedman, Meyer, *et al.* 1982. "Feasibility of Altering Type A Behavior Pattern After Myocardial Infarction," *Circulation*, 66, 1: 83-92.

Fries, J.F. 1980. "Aging, Natural Death and Compression of Morbidity," *New England Journal of Medicine*, 303: 130-35.

Fry, H. 1987. "Ontario Task Force Disagrees with CMA About Need for Midwives," *Canadian Medical Association Journal*, 137, 11 (December 1): 1032.

Gabe, J., and M. Calnan. 1989. "The Limits of Medicine: Women's Perception of Medical

Technology," *Social Science and Medicine*, 28: 223-31.

Gallagher, Eugene B., and C. Maureen Searle. 1989. "Content and Context in Health Professional Education," in Howard E. Freeman and Sol Levine, eds., *Handbook of Medical Sociology*. Fourth Edition. Englewood Cliffs, N.J.: Prentice-Hall: 437-55.

Garland, L.H. 1959. "Studies in the Accuracy of Diagnostic Procedures," *American Journal of Roentgeneology*, 82: 25-38.

Gasner, Douglas. 1982. *The American Medical Association Book of Heart Disease*. New York: Random House.

General Social Survey Analysis Series. 1987. Ottawa: Statistics Canada.

Gerber, L.A. 1983. *Married to their Careers: Career and Family Dilemmas in Doctors' Lives*. New York: Tavistock.

Gibson, R.G., S.L.M. Gibson, A.D. MacNeill, W. Watson, and W. Buchanan. 1980. "Homeopathic Therapy in Rheumatoid Arthritis: Evaluation of Double-Blind Clinical Therapeutic Trial," *British Journal of Clinical Pharmacology*, 9: 453-59.

Gidney, R.D., and W.P.S. Millar. 1984. "Origins of Organized Medicine, Ontario, 1850-1869," in Charles G. Roland, ed., *Health, Disease and Medicine: Essays in Canadian History*. Toronto: The Hannah Institute for the History of Medicine: 72-95.

Gilman, Charlotte Perkins. 1973. *The Yellow Wallpaper*. Old Westbury, N.Y.: Feminist Press.

Glaser, Barney G., and Anselm L. Strauss. 1967. *The Discovery of Grounded Theory*. Chicago: Aldine.

Glaser, Barney G., and Anselm L. Strauss. 1968. *Time for Dying*. Chicago: Aldine.

Globe and Mail. June 12, 1970; February 5, 1989; February 14, 1989.

Godon, D., J.P. Thouez, and P. Lajoie. 1989. "Analyse géographique de l'incidence des concer au Québec en function de l'utilisation des pesticides en agriculture," *Canadian Geographer*, 33: 204-17.

Goffman, Erving. 1959. "The Moral Career of the Mental Patient," *Psychiatry*, 22: 123-35.

Goffman, Erving. 1961. *Asylums: Essays on the Situation of Mental Patients and Other Inmates*. New York: Anchor Press/Doubleday.

Goffman, Erving. 1963. *Stigma: Notes on the Management of Spoiled Identity*. Englewood Cliffs, N.J.: Prentice-Hall.

Goldscheider, C. 1971. *Population, Modernization and Social Structure*. Boston: Little Brown.

Good, Mary Jo, Byron J. Good, and Paul D. Cleary. 1987. "Do Patient Attitudes Influence Physician Recognition of Psychosocial Problems in Primary Care?" *Journal of Family Practice*, 25: 53-59.

Goode, William J. 1956. "Community within a Community: The Professions," *American Sociological Review*, 22 (April): 194-200.

Goode, William J. 1960. "Encroachment, Charlatanism and the Emerging Profession: Psychology, Sociology and Medicine," *American Sociological Review*, 25, 6: 902-14.

Goode, William J. 1969. "The Theoretical Limits of Professionalization," in Amitai Ezioni, ed., *The Semi-Professions and Their Organizations: Teachers, Nurses, Social Workers*. New York: Free Press.

Gordon, Deborah R. 1988. "Tenacious Assumptions in Western Medicine," in M. Lock and D.R. Gordon, eds., *Biomedicine Examined*. Dordrecht, Netherlands: Kluwer: 19-56.

Gordon, Sidney, and Ted Allan. 1952. *The Scalpel, The Sword*. Toronto: McClelland and Stewart.

Gort, Elaine. 1986. "A Social History of Naturopathy in Ontario: The Formation of an Occupation," M.A. thesis, University of Toronto.

Gouldner, Alvin W. 1970. *The Coming Crisis in Western Sociology*. New York: Basic Books.

Gove, N.R. 1973. "Sex, Marital Status and Mortality," *American Journal of Sociology*, 79: 45-67.

Graham, Hilary. 1984. *Women, Health and the Family*. Brighton, Sussex: Wheatsheaf Books.

Graham, Hilary. 1985. "Providers, Negotiators, and Mediators: Women as the Hidden Carers," in Ellen Lewis and Virginia Olesen, eds., *Women, Health, and Healing: Toward a New Perspective*. New York: Tavistock: 25-52.

Graham, Wendy. 1994. "Sexual Harassment of Physicians," *Women's Health Office Newsletter*, April 14.

Grimes, David A. 1993. "Editorial: Over-the-Counter Oral Contraceptives – An Immodest Proposal," *American Journal of Public Health*, 83, 8: 1092-93.

Grippando, Gloria M. 1986. *Nursing Perspectives and Issues*. Third Edition. Albany, N.Y.: Delmar.

Groce, Stephen B. 1991. "The Nature and Type of Doctors' Cultural Assumptions about Patients as Men," *Sociological Focus*, 24, 3: 211-23.

Gross, M.J., and C.W. Schwenger. 1981. *Health Care for the Elderly in Ontario*. Toronto: Ontario Economic Council.

Growe, S.J. 1991. "The Nature and Type of Doctors' Cultural Assumptions about Patients as Men," *Sociological Focus*, 24, 3: 211-23.

Grymonpre, R.E., P.A. Metenko, *et al.* 1988. "Drug Associated Hospital Admission in Older Medical Patients," *Journal of American Geriatric Society*, 36: 1092-98.

Hacker, Carlotta. 1974. *The Indomitable Lady Doctors*. Toronto: Clarke, Irwin.

Hall, Edward T. 1977. *Beyond Culture*. Garden City, N.Y.: Anchor Press/Doubleday.

Hall, Emmett M. 1980. *Canadian National-Provincial Health Program for the 1980's. A Commitment for Renewal*. Ottawa: National Health and Welfare.

Hall, Oswald. 1948. "The Stages of a Medical Career," *American Journal of Sociology*, 53 (March): 328-36.

Hall, R. 1946. "Some Organizational Considerations in Professional-Organizational Relationship," *Administrative Science Quarterly*, 12, 3: 461-78.

Hamilton, Vivian, and Barton Hamilton. 1993. "Does Universal Health Insurance Equalize Access to Care?: A Canadian-U.S. Comparison," paper presented at Northwestern University Fourth Annual Health Economics Workshop, August.

Hamilton, V. Lee, Clifford L. Broman, William S. Hoffman, and Deborah S. Renner. 1990. "Hard Times and Vulnerable People: Initial Effects of Plant Closings on Auto Workers' Mental Health," *Journal of Health and Social Behavior*, 31 (June): 123-40.

Hammer, Vicki. 1981. "So Many Like Her," *World Health*. Geneva: World Health Organization.

Hamowy, Ronald. 1984. *Canadian Medicine: A Study in Restricted Entry*. Vancouver: The Fraser Institute.

Harding, Jim. 1987. "The Pharmaceutical Industry as a Public Health Hazard and an

Institution of Social Control," in D. Coburn et al., eds., *Health and Canadian Society: Sociological Perspectives*. Second Edition. Toronto: Fitzhenry & Whiteside: 314-32.

Harding, Jim. 1994a. "Environmental Degradation and Rising Cancer Rates: Exploring the Links in Cancer," in B. Singh Bolaria and Harley D. Dickinson, eds., *Health, Illness and Health Care in Canada*. Second Edition. Toronto: Harcourt Brace & Company: 649-67.

Harding, Jim. 1994b. "Social Basis of the Over-Prescribing of Mood-Modifying Pharmaceuticals to Women," in Bolaria and Bolaria, eds., *Women, Medicine and Health*. Halifax: Fernwood: 157-81.

Harding, T.W., and W. Curran. 1978. *The Law and Mental Health: Harmonizing Objectives*. Geneva: World Health Organization.

Harpur, Tom. 1994. *Uncommon Touch*. Toronto: McClelland & Stewart.

Harrison, Michelle. 1982. *A Woman in Residence*. New York: Random House.

Heacock, Helen Jane, and Jason Keller Rivers. 1986. "Occupational Diseases of Hairdressers," *Canadian Journal of Public Health*, 77 (March/April): 109-13.

Heagerty, John J. 1928. *Four Centuries of Medical History in Canada*, 2 vols. Toronto: Macmillan.

Health and Social Support 1985. 1987. Series No. 1 of General Social Survey Analysis Series. Ottawa: Statistics Canada.

Health and the Status of Women. 1980. Geneva: World Health Organization.

Health and Welfare – Mental Health Division. 1984. *Alzheimer's Disease: A Family Information Handbook*. Ottawa: Published in Cooperation with the Alzheimer Society.

Health and Welfare Canada. 1982, 1984. *National Health Expenditures in Canada 1970-1982*. Ottawa.

Health and Welfare Canada. 1986. *Issues for Health Promotion in Family and Child Health: A Sourcebook*. Ottawa: Medical Service Branch, Indian and Inuit Services.

Health and Welfare Canada and Statistics Canada. 1981. *The Health of Canadians: Report of the Canada Health Survey*. Cat. 82-538E. Ottawa: Minister of Supply and Services and the Minister of National Health and Welfare.

Health, Health Care and Medicine: A Report to the National Council of Welfare. 1990. Ottawa.

Health Promotion. 1987, 1988. Ottawa: Health and Welfare Canada.

Hemminki, Elina. 1975. "Review of Literature on the Factors Affecting Drug Prescribing," *Social Science and Medicine*, 9: 111-15.

Hilfiker, David. 1985. *Healing the Wounds: A Physician Looks at His Work*. New York: Pantheon Books.

Holling, S.A. 1981. "Primitive Medicine among the Indians of Ontario," in Holling et al., eds., *Medicine for Heroes*. Mississauga: Mississauga Historical Society.

Holmes, T.H., and M. Masuda. 1974. "Life Change and Illness Susceptibility," in B.S. and B.P. Dohrenwend, eds., *Stressful Life Events: Their Nature and Effect*. New York: Wiley.

Holmes, T.H., and R.H. Rahe. 1967. "The Social Readjustment Rating Scale," *Journal of Psychosomatic Research*, 11: 213-18.

Horn, Joshua. 1969. *Away With All Pests: An English Surgeon in People's China: 1954-1969*. New York: Monthly Review Press.

Horowitz, Lawrence C. 1988. *Taking Charge of Your Medical Fate*. New York: Random House.

"Horror Stories Start with Pill, End in Murder." *Winnipeg Free Press*, December 20: A11.

House, J., K.R. Landis, and D. Umberson. 1988. "Social Relationships and Health," *Science*, 241: 540-45.

Hughes, C.C. 1967. "Ethnomedicine," in David Gills, ed., *International Encyclopedia of Social Sciences*, vol. 10. New York: Macmillan and Free Press: 87-92.

Hughes, David. 1989. "Paper and People: The Work of the Casual Reception Clerk," *Sociology of Health and Illness*, 11, 4.

Hughes, Everett C. 1971. *The Sociological Eye*. Chicago: Aldine-Atherton.

Hunt, Charles W. 1989. "Migrant Labour and Sexually Transmitted Disease: AIDS in Africa," *Journal of Health and Social Behavior*.

Hunt, L., B. Jordan, S. Irwin, and C.H. Browner. 1989. "Compliance and the Patient's Perspective: Controlling Symptoms in Everyday Life," *Culture, Medicine and Psychiatry*, 13: 315-34.

Ideas transcript. 1983. "We Know Best: Experts' Advice to Women." Transcript of *Ideas*, January 2-23. Toronto: Canadian Broadcasting Corporation.

Illich, Ivan. 1976. *Limits to Medicine: Medical Nemesis, the Expropriation of Health*. Toronto: McClelland and Stewart.

Imbert, Lorrie. 1993. "Ayurveda: Ancient Medicine for Modern Times," *Health Naturally*, October/November: 5-7.

Indian Conditions – A Survey. 1980. Ottawa: Minister of Indian Affairs and Northern Development.

James, W.J., and S. Lieberman. 1975. "What the American Public Knows About Cancer and Cancer Tests," in Patricia Hubbs, ed., *Public Education About Cancer*. Geneva: International Union Against Cancer.

Jarvis, G.K., and M. Boldt. 1982. "Death Styles Among Canada's Indians," *Social Science and Medicine*, 16: 1345-52.

Jennett, P.A., M. Cooper, S. Edworthy, *et al.* 1991. "Consumer Use of Official Health Care: Facts and Implications," *Proceedings of the 5th ACMC Conference on Physician Manpower*, April 28, Association of Canadian Medical Colleges, Ottawa.

Jensen, Phyllis Marie. 1988. "Nursing," in *The Canadian Encyclopedia*, vol. 3. Edmonton: Hurtig: 1546.

Jin, Robert L., Chandrakant P. Shah, and Tomislav J. Svoboda. 1994. "The Health Impact of Unemployment," unpublished paper, available from Dr. C.P. Shah, Faculty of Medicine, University of Toronto.

Johnson, M. 1975. "Medical Sociology and Sociological Theory," *Social Science and Medicine*, 9: 227-32.

Johnson, Robert J., and Frederic D. Wolinsky. 1990. "The Legacy of Stress Research: The Course and Impact of This Journal," *Journal of Health and Social Behavior*, 32 (September): 217-25.

Johnson, Terrence. 1972. *The Professions and Power*. London: Macmillan.

Johnson, Terrence. 1977. "Industrial Society: Class Change and Control," in R. Scase, ed., *The Professions in the Class Structure*. London: Allen and Unwin: 93-110.

Johnson, Terrence. 1982. "Social Class and the Division of Labour," in A. Giddens and G. Mackenzie, eds., *The State and the Professions: Peculiarities of the British*. Cambridge: Cambridge University Press: 182-208.

Johnstone, Tracey, and Julie Robinson. 1995. "*Shape* Versus *Men's Fitness*: Are Fitness Magazines Different for Each Sex?" paper for Qualitative Reasearch Methods course, Wilfrid Laurier University.

Jonas, H.A., and J. Lumley. 1993. "Triplets and Quadruplets Born in Victoria between 1982 and 1990: The Impact of IUF and GIFT in Raising Birthrates," *Medical Journal of Australia*, 17, 5: 158; 17, 10: 659-63.

Jones, K., and G. Moon. 1987. *Health, Disease and Society*. London: RKP.

Jones, W.H.S. 1943. *Hippocrates*, vol. II. London: Heinemann.

Jossa, Diana. 1985. *Smoking Behaviour of Canadians 1983*. Ottawa: National Health and Welfare.

Justice, Blair. 1987. *Who Gets Sick: Thinking and Health*. Houston: Peak Press.

Kanter, Rosabeth Moss. 1977. *Men and Women of the Corporation*. New York: Basic Books.

Kaplan, Howard B. 1991. "Social Psychology of the Immune System: A Conceptual Framework and Review of the Literature," *Social Science and Medicine*: 909-23.

Karliner, Joshua. 1994. "Toxin Town," *New Statesman and Society*, 7, 2 (December 2): 18.

Kaufert, P. 1988. "Through Women's Eyes: The Case for Feminist Epidemiology," *Healthsharing*: 10-13.

Kaufert, Patricia. 1992. "Mammography and the Misplacement of Faith," paper presented at the American Anthropological Association annual meeting, San Francisco.

Kaufert, P.A., and P. Gilbert. 1987. "Medicalization and the Menopause," in D. Coburn, C. D'Arcy, G. Torrance, and P. New, eds., *Health and Canadian Society*. Second Edition. Toronto: Fitzhenry & Whiteside.

Keating, M. 1986. *To the Last Drop: Canada and the World's Water Crisis*. Toronto: Macmillan.

Kelly, Ken. 1962. "Baby-Deforming Drug Sold First as Sleeping Pill," *Kitchener-Waterloo Record*, August 17: 27.

Kelly, O. 1979. *Until Tomorrow Comes*. New York: Everest House.

Kelner, Merrijoy, Oswald Hall, and Jan Coultner. 1980. *Chiropractors: Do They Help?* Toronto: Fitzhenry & Whiteside.

Kendall, P.P., and G.G. Reader. 1988. "Innovations in Medical Education of the 1950's contrasted with Those of the Early 1970's and 1980's," *Journal of Health and Social Behavior*, 29, 4: 279-93.

Kidder, Louise H. 1986. *Research Methods in Social Relations*. New York: Holt, Rinehart and Winston.

Kim, Kwang Kee, and Philip M. Moody. 1992. "More Resources, Better Health? A Cross-National Perspective," *Social Science and Medicine*, 34, 8: 837-42.

Kirk, Jo-Ann. 1994. "A Feminist Analysis of Women in Medical School," in B. Singh Bolaria and Harley D. Dickinson, eds., *Sociology of Health Care in Canada*. Second Edition. Toronto: Harcourt Brace & Company: 158-83.

Kirkby, Pat. 1986. *Education for Excellence 1986/1987*. Toronto: Registered Nurses Association of Ontario.

Kirkmayer, Laurence J. 1988. "Mind and Body as Metaphor: Hidden Values in Biomedicine," in M. Lock and E.R. Gordon, eds., *Biomedicine Examined*. Dordrecht, Netherlands: Kluwer: 57-93.

Klass, Alan. 1975. *There's Gold in Them Thar Pills*. London: Penguin Books.

Kleinman, Arthur. 1988. *The Illness Narratives: Suffering, Healing and the Human Condition*. New York: Basic Books.

Koblinsky, M., J. Timyan, and J. Gay. 1993. *The Health of Women: A Global Perspective*. Boulder, Colorado: Westview Press.

Kohler-Reissman, Catherine. 1989. "Women and Medicalization: A New Perspective," in Phil Brown, ed., *Perspectives in Medical Sociology*. Belmont, Calif.: Wadsworth.

Koss, Mary P., Lori Heise, and Nancy F. Russo. 1994. "The Global Health Burden of Rape," *Psychology of Women Quarterly*, 18: 509-37.

Kramer, Peter D. 1993. *Listening to Prozac*. New York: Penguin Books.

Krause, Elliott A. 1978. *Power and Illness: The Political Sociology of Health and Medical Care*. New York: Elsevier.

Kroll-Smith, Steve, and Anthony E. Ladd. 1993. "Environmental Illness and Biomedicine: Anomalies, Exemplars and the Politics of the Body," Visiting Fellow Centre for Disaster Management, University of New England, Armidale NSW, Australia.

Kuhn, Thomas. 1962. *The Structure of Scientific Revolutions*. Chicago: University of Chicago Press.

Lalonde, Marc. 1974. *A New Perspective on the Health of Canadians*. Ottawa: Information Canada.

Langone, John. 1982. *Chiropractors*. New York: Addison-Wesley.

Last, J. 1963. "The Iceberg: Completing the Clinical Picture in General Practice," *Lancet*, 2, 729: 28-31.

"Last Gasp." 1992. *Equinox*, May/June: 85-98.

LaVeist, Thomas A. 1992. "The Political Empowerment and Health Status of African-Americans: Mapping a New Territory," *American Journal of Sociology*, 97, 4: 1080-95.

Lazarus, R.S., and A. Delongis. 1983. "Psychological Stress and Coping in Aging," *American Psychologist*, 38: 245-54.

Lee, Charles. 1987. *Toxic Waste and Race in the U.S.* New York Commission for Racial Justice, United Church of Christ.

Lesage, J. 1991. "Polypharmacy in Geriatric Patients," *Nursing Clinics of North America*, 26: 273-90.

LeShan, Larry. 1978. *You Can Fight for Your Life*. New York: M. Evans.

Levin, Lowell S., and Ellen L. Idler. 1981. *The Hidden Health Care System*. Cambridge, Mass.: Battlinger.

Lewis, Charles E., and Mary Anne Lewis. 1977. "The Potential Impact of Sexual Inequality on Health," *New England Journal of Medicine*, 297 (October): 863-69.

Lexchin, Joel. 1984. *The Real Pushers: A Critical Analysis of the Canadian Drug Industry*. Vancouver: New Star Books.

Lexchin, Joel. 1988a. "Profits First: The Pharmaceutical Industry in Canada," in B.S. Bolaria and H.D. Dickinson, eds., *Sociology of Health Care in Canada*. Toronto: Harcourt Brace Jovanovich: 497-513.

Lexchin, Joel. 1988b. "Pushing Pills: Who's to Blame for so much Poor Prescribing?" *Globe and Mail*, December 13: A7.

Lexchin, Joel. 1988c. "Pharmaceutical Industry," in *The Canadian Encyclopedia*, vol. 3. Edmonton: Hurtig: 1653-54.

Lexchin, Joel. 1990. "Drug Makers and Drug Regulators: Too Close for Comfort: A Study of the Canadian Situation," *Social Science and Medicine*, 31, 11: 1257-63.

Lexchin, Joel. 1994a. "Profits First: The Pharmaceutical Industry in Canada," in B. Singh Bolaria and Harley D. Dickinson, eds., *Sociology of Health Care in Canada*. Second Edition. Toronto: Harcourt Brace & Company: 700-20.

Lexchin, Joel. 1994b. "Canadian Marketing

Codes: How Well Are They Controlling Pharmaceutical Promotion," *International Journal of Health Services*, 24, 1: 91-104.

Liang, M.H., *et al.* 1973. "Chinese Health Care: Determinants of the System," *American Journal of Public Health*, 63, 2: 102-10.

Light, Donald W., and Grace Budrys. "Health Care Technology: Social Constructions of Reality," unpublished paper, Rutgers University and DePaul University.

Litoff, J. 1978. *American Midwives: 1860 to the Present*. Westport, Conn.: Greenwood Press.

Lock, Margaret, and Gilles Bibeau. 1993. "Healthy Disputes: Some Reflections on the Practice of Medical Anthropology in Canada," *HCS/SSC*, 1, 6: 147-76.

Lorber, J. 1975. "Women and Medical Sociology: Invisible Professionals and Ubiquitous Patients," in M. Millman and R. Kanter, eds., *Another Voice*. New York: Anchor Books: 75-105.

Lorber, J. 1984. *Women Physicians, Careers, Statuses and Power*. New York: Tavistock.

Loring, Marti, and Brian Powell. 1988. "Gender, Race and DSM-III: A Study of the Objectivity of Psychiatric Diagnostic Behaviors," *Journal of Health and Social Behavior*, 29 (March): 1-22.

Lupton, Deborah. 1994. "Femininity, Responsibility, and the Technological Imperative: Discourses on Breast Cancer in the Australian Press," *International Journal of Health Services*, 24, 1: 73-89.

Lynch, James L. 1977. *The Broken Heart: The Medical Consequences of Loneliness*. New York: Basic Books.

MacIntyre, S., and D. Oldman. 1984. "Coping with Migraine," in N. Black *et al.*, eds., *Health and Disease, A Reader*. Milton Keynes: The Open University Press: 271-75.

Macionis, John T. 1995. *Sociology*. Fifth Edition. Englewood Cliffs, N.J.: Prentice-Hall.

Macionis, John T., *et al.* 1994. *Sociology. Canadian Edition*. Englewood Cliffs, N.J.: Prentice-Hall.

MacLeod, L. 1987. *Wife Battering in Canada: The Vicious Circle*. Report of the Canadian Advisory Council on the Status of Women. Ottawa: Ministry of Supply and Services.

Major, Ralph H. *A History of Medicine*, vols. I-II. Springfield, Ill.

Makdessian, Frances. 1987. *Occupational Health and Safety Management Book*. Don Mills, Ont.: Corpus Information Services.

Manga, Pran. 1993. "Health Economics and the Current Health Care Cost Crisis: Contributions and Controversies," *HCS/SSC*, 1, 1: 177-203.

Manning, Peter K., and Horatio Fabrega. 1973. "The Experience of Self and Body: Health and Illness in the Chiapas Highlands," in George Psathas, ed., *Phenomenological Sociology*. New York: John Wiley and Sons.

Mansour, Valerie. 1995. "Judge ends trial in N.S. mine deaths," *Toronto Star*, June 10: A3.

Mao, Y., H. Morrison, R. Semenciw, and D. Wigle. 1986. "Mortality on Canadian Indian Reserves, 1977-1982," *Canadian Journal of Public Health*, 77: 263-68.

Marchak, Patricia. 1975. *Ideological Perspectives on Canada*. Toronto: McGraw-Hill.

Marks, Geoffrey, and William K. Beatty. 1976. *Epidemics*. New York: Charles Scribner's.

Marshall, V.W. 1980. *Last Chapters: A Sociology of Aging and Dying*. Monterey, Calif.: Brooks/Cole Publishing.

Martin, Emily. 1987. *The Woman in the Body: A Cultural Analysis of Reproduction*. Boston: Beacon Press.

Martindale, Don. 1960. *The Nature and Types of*

Sociological Theory. Boston: Houghton Mifflin.

Marx, Karl. 1964. *The Economic and Philosophic Manuscripts of 1844*, trans. M. Milligan. New York: International Publishers.

May, J., and J. Wasserman. 1984. "Selected Results from an Evaluation of the New Jersey Diagnosis Related Group System," *Health Services Research*, 19, 5 (December): 548.

McCormack, Thelma. 1991. "Public Policies and Reproductive Technology: A Feminist Critique," *Research in the Sociology of Health Care*, 9: 105-24.

McCrea, F.B. 1983. "The Politics of Menopause: The Discovery of a Deficiency Disease," *Social Problems*, 31, 1: 111-23.

McDaniel, Susan. 1988. "Women's Roles, Reproduction and the New Reproductive Technologies: A New Stork Rising," in Nancy Mandell and Ann Duffy, eds., *Reconstructing the Canadian Family: Feminist Perspectives*. Toronto: Butterworths: 175-207.

McKeown, T. 1976. *The Role of Medicine: Dream, Mirage or Nemesis*. London: Nuffield Provincial Hospitals Trust.

McKeown, T., and R.G. Record. 1975. "An Interpretation of the Decline in Mortality in England and Wales during the Twentieth Century," *Population Studies*, 29: 391-422.

McKinlay, John B. 1982. "Toward the Proletarianization of Physicians," in C. Derber, ed., *Professionals as Workers*. Boston: Hall: 37-62.

McKinlay, John B., and Sonja M. McKinlay. 1987. "Medical Measures and the Decline of Mortality," in Howard D. Schwartz, ed., *Dominant Issues in Medical Sociology*. Second Edition. New York: Random House.

McKinlay, J.B., and S.M. McKinlay. 1977. "The Questionable Contribution of Medical Measures to the Decline of Mortality in the United States in the Twentieth Century," *Millbank Memorial Fund Quarterly* (Summer).

McKinlay, J.B., and S.M. McKinlay. 1981. "From Promising Report to Standard Procedure: Seven Stages in the Career of a Medical Innovation," *Millbank Memorial Fund Quarterly*, 59: 374-411.

McKinlay, S.M. 1971. "Some Approaches and Problems in the Study of the Use of Services: An Overview," *Journal of Health and Behaviour*, 13: 115-52.

McLeod, Thomas H., and Ian McLeod. 1987. *Tommy Douglas: The Road to Jerusalem*. Edmonton: Hurtig.

McNab, E. 1970. *A Legal History of Health Professions in Ontario*. Committee on the Healing Arts. Toronto: Queen's Printer.

Meador, Clifton. 1965. "The Art and Science of Non-Disease," *New England Journal of Medicine*, 235: 424-45.

Mechanic, David. 1978. *Medical Sociology: A Comprehensive Text*. New York: Free Press.

Mechanic, J. David. 1993. "Sociological Research in Health and the American Socio-political Context," *Social Science and Medicine*, 36, 2: 95-102.

Meddison, D., and W.L. Walker. 1967. "Factors affecting the outcome of conjugant bereavement," *British Journal of Psychiatry*, 113: 1057-67.

Meichenbaum, Donald. 1983. *Coping with Stress*. Toronto: Wiley.

Merton, Robert K., George Reader, and Patricia Kendall, eds. 1957. *The Student Physician: Introductory Studies in the Sociology of Medical Education*. Cambridge, Mass.: Harvard University Press.

Mhatra, Sharmila, and Raise B. Deber. 1992. "From Equal Access to Health Care to Equitable Access to Health: A Review of Canadian Provincial Health Commissions and Reports," *International Journal of Health Services*, 22, 4: 645-68.

Milgram, Stanley. 1974. *Obedience to Authority*. New York: Harper & Row.

Millar, Wayne. 1992. "A Trend to a Healthier Lifestyle," *Canadian Social Trends*. Ottawa: Statistics Canada Cat. 11-008E, no. 24 (Spring).

Millman, Marcia. 1977. *The Unkindest Cut*. New York: William Morrow.

Mills, C. Wright. 1959. *The Sociological Imagination*. New York: Oxford University Press.

Mills, Donald. 1964. *Royal Commission on Health Services Study of Chiropractors, Osteopaths, and Naturopaths in Canada*. Ottawa: Queen's Printer.

Mishler, Elliott. 1984. *The Discourse of Medicine: Dialectics of Medical Interviews*. Norwood, N.J.: Ablex.

Mitchinson, Wendy. 1987. "Medical Perceptions of Healthy Women: The Case of Nineteenth Century Canada," *Canadian Women Studies*, 8, 4: 42-43.

Montini, Theresa, and Kathleen Slobin. 1991. "Tensions Between Good Science and Good Practice: Lagging Behind and Leapfrogging Ahead Along the Cancer Care Continuum," *Research in the Sociology of Health Care*, 9: 127-40.

Morbidity and Mortality Weekly Report. 1981. Atlanta Centers for Disease Control. Atlanta, Georgia (June 5, July 3).

Morgan, P.P. 1984. "Pharmaceutical Advertising in Medical Journals," *Canadian Medical Association Journal*: 130-42.

Moyer, Anne, Susan Grenner, John Beavais, and Peter Salovey. 1994. "Accuracy of Health Research Reported in the Popular Press: Breast Cancer and Mammography," *Health Communication*, 7, 1.

Muller, M. 1982. *The Health of Nations*. London: Faber and Faber.

Mumford, Emily. 1983. *Medical Sociology: Patients, Providers and Policies*. New York: Random House.

Mustard, Fraser. 1987. "Health in a Post-Industrial Society," in J. Clarke *et al.*, eds., *Health Care in Canada: Looking Ahead*. Ottawa: Canadian Public Health Association.

Muzzin, Linda J., Gregory P. Brown, and Roy W. Hornosty. 1993. "Professional Ideology in Canadian Pharmacy," *Health and Canadian Society*, 1, 2: 319-46.

"NAFTA gives tobacco companies power to block plain packaging . . ." 1994. *CCPA Monitor* (Canadian Centre for Policy Alternatives), 1, 2 (June 3).

Nathanson, C.A. 1977. "Sex, Illness and Medical Care: A Review of Data, Theory and Method," *Social Science and Medicine*, 11: 13-25.

National Council of Welfare. 1991. *Funding Health and Higher Education: Danger Looming*. Ottawa: Minister of Supply and Services.

Navarro, Vincente. 1975a. "The Industrialization of Fetishism or the Fetishism of Industrialization: A Critique of Ivan Illich," *Social Science and Medicine*, 9, 7: 351-63.

Navarro, Vincente. 1975b. "Women in Health Care," *New England Journal of Medicine*, 202: 398-402.

Navarro, Vincente. 1976. "Social Class, Political Power and the State and Their Implications for Medicine," *Social Science and Medicine*, 10: 437-57.

Navarro, Vincente. 1992. "Has Socialism Failed? An Analysis of Health Indicators under Socialism," *International Journal of Health Services*, 22, 4: 586-601.

Naylor, C.D. 1982. "In Defense of Medicare: Are Canadian Doctors Threatening the Health Care System?" *Canadian Forum*, LXII (April): 12-16.

Neidhardt, J., M.S. Weinstein, and Robert R. Coury. 1985. *Managing Stress*. Vancouver: Self-Counsel Press.

Nicholson, G.W.L. 1967. *The White Cross in Canada*. Montreal: Harvest House.

Nicholson, Gerald W. 1975. *Canada's Nursing Sisters*. Toronto: Hakkert.

Nicholson, Malcolm, and Cathleen McLaughlin. 1987. "Social Constructionism and Medical Sociology: A Reply to M.R. Bury," *Sociology of Health and Illness*, 9, 2: 107-26.

Nikiforuk, Andrew. 1991. "The Great Fire," *Equinox*, 59 (September/October).

Noh, J., and R.J. Turner. 1983. "Class and Psychological Vulnerabilities Among Women: The Significance of Social Support and Personal Control," *Journal of Health and Social Behaviour*, 24: 2-15.

Oakley, A. 1984. *The Captured Womb: A History of the Medical Care of Pregnant Women*. Oxford: Basil Blackwell.

Oberg, Gary R. 1990. *An Overview of the Philosophy of the American Academy of Environmental Medicine*. Denver: American Academy of Environmental Medicine.

O'Brien, Patricia. 1987. "All a Woman's Life Can Bring: The Domestic Roots of Nursing in Philadelphia, 1830-1885," *Nursing Research*, 36, 1: 12-17.

Occupational Health and Safety: A Training Manual. 1982. Toronto: Copp Clark.

O'Connor, J. 1973. *The Fiscal Crisis of the State*. New York: St. Martin's Press.

Olson, E. 1984. *No Place to Hide*. Wheaton, Ill.: Tyndale House.

Omran, Abdel R. 1979. "Changing Patterns of Health and Disease During the Process of National Development," in G. Albrecht and P.C. Higgins, eds., *Health, Illness and Medicine*. Chicago: Rand McNally: 81-93.

Ontario Naturopathic Association. 1983. *A Brief Prepared by the ONA and the Board of Directors of Drugless Therapy for the Honorable Keith Norton, Minister of Health*. Toronto (July).

Oppenheimer, J. 1983. "Childbirth in Ontario: The Transition from Home to Hospital in the Early Twentieth Century," *Ontario History*, 75 (March): 36-60.

Orbach, Susie. 1986. *Hunger Strike: An Anorexic's Struggle as a Metaphor for Our Age*. New York: Norton.

Osler, W. 1910. "The Lumleian Lectures on Angina Pectoris: Delivered Before the Royal College of Physicians of London," *Lancet*: 839-44.

Paget, Marianne A. 1988. *The Unity of Mistakes: A Phenomenological Interpretation of Medical Work*. Philadelphia: Temple University Press.

Parsons, Evelyn, and Paul Atkinson. 1992. "Lay Constructions of Genetic Risk," *Sociology of Health and Illness*, 14, 4: 437-55.

Parsons, Talcott. 1951. *The Social System*. Glencoe, Ill.: Free Press.

Parsons, Talcott. 1954. "The Professions and the Social Structure," in *Essays in Sociological Theory*. Revised Edition. Glencoe, Ill.: Free Press: 428-47.

Patterns of Growth: Seventh Annual Review. 1970. Ottawa: Queen's Printer.

Payer, Lynn. 1988. "Borderline Cases," *Science*.

Peters-Golden, Holly. 1982. "Breast Cancer: Varied Perceptions of Social Support in the Illness Experience," *Social Science and Medicine*, 16: 483-91.

Pettigrew, Eileen. 1983. *The Silent Enemy: Canada and the Deadly Flu of 1918*. Saskatoon: Western Producer Prairie Books.

Phifer, James F., Z. Kryzsztof, and Fran H. Norris. 1988. "The Impact of Natural Disaster on the Health of Older Adults: A Multiwave Prospective Study," *Journal of Health and Social Behavior*, 29 (March): 65-78.

Phillips, D.P., and R.A. Feldman. 1973. "A Dip in Deaths Before Ceremonial Occasions: Some New Relationships Between Social Integration and Mortality," *American Sociological Review*, 38: 678-96.

Phlanz, Manfred. 1975a. "A Critique of Anglo-American Medical Sociology," *International Journal of Health Services*, 4, 3: 565-74.

Phlanz, Manfred. 1975b. "Relations Between Social Scientists, Physicians and Medical Organizations in Health Research," *Social Science and Medicine*, 9: 7-13.

Pirie, Marion. 1988. "Women and the Illness Role: Rethinking Feminist Theory," *Canadian Review of Sociology and Anthropology*, 25, 4: 628-48.

Pollard, John. 1994. "Attitudes toward Euthanasia in Metropolitan Toronto," *Institute for Social Research Newsletter*, 9, 2 (Summer): 3-4.

Porter, John. 1989. *Health for Sale*. New York: Manchester University Press.

Priest, Lisa. 1993. "Thalidomide survivors have tackled life with gusto," *Calgary Herald*, February 21: B8

Pulliam, C., J. Hanlon, and S. Moore. 1988.

"Medication and Geriatrics," in F. Abellah and S. Moore, eds., *Surgeon General's Workshop. Health Promotion and Aging. Background Papers*. California: Henry J. Kaiser Foundation.

Punnet, Laura. 1976. "Women-Controlled Medicine – Theory and Practice in 19th Century Boston," *Women and Health*, 1, 4 (July/August): 3-10.

Purvis, Trevor, and Alan Hunt. 1993. "Discourse, Ideology, Discourse, Ideology, Discourse Ideology," *British Journal of Sociology*, 44, 3: 499.

Quinn, K., M.J. Baker, and B. Evan. 1992. "A Population-wide Profile of Prescription Drug Use in Saskatchewan, 1989," *Canadian Medical Association Journal*, 146: 2177-86.

Rabe, Barry G. 1992. "When Citing Works, Canada-Style," *Journal of Health Politics, Policy and Law*, 17, 1 (Spring).

Rachlis, Michael, and Carol Kushner. 1989. *Second Opinion: What's Wrong with Canada's Health Care System*. Toronto: Harper Collins.

Rachlis, Michael, and Carol Kushner. 1994. *Strong Medicine: How to Save Canada's Health Care System*. Toronto: Harper Collins.

Raffel, S. 1979. *Matters of Fact: A Sociological Inquiry*. London: Routledge and Kegan Paul.

Rahe, R.H., J.J. Mahan, and R.J. Arthur. 1970. "Prediction of near-future health changes from subjects' preceding life changes," *Journal of Psychosomatic Research*, 14: 401-06.

Rahe, R.H., and J. Paasikivi. 1971. "Psychosocial factors and myocardial infarction, II: An Outpatient Study in Sweden," *Journal of Psychosomatic Research*, 15: 33-39.

Raven, Peter H., Linda R. Berg, and George B. Johnson. 1993. *Environment*. Toronto: Saunders.

Rawson, Nigel S.B., and Carl D'Arcy. 1991. "Sedative-Hypnotic Drug Use in Canada," *Health Report*, 3, 1: 33-57.

Raymond, Chris. 1989. "Distrust, Rage May Be 'Toxic Core' That Puts Type A At Personal Risk," *Journal of the American Medical Association*, 261, 6: 813.

Reasons, C.E., L.E. Ross, and Craig Patterson. 1981. *Assault on the Worker: Occupational Health and Safety in Canada*. Toronto: Butterworths.

Registered Nurses Association of Ontario. 1987. *The RNAO Responds: A Nursing Perspective on the Events at the Hospital for Sick Children and the Grange Inquiry*. Toronto: RNAO, April.

Regush, Nicholas. 1987. *Canada's Health Care System: Condition Critical*. Toronto: Macmillan.

Regush, Nicholas. 1993. *Safety Last: The Failure of the Consumer Health Protection System in Canada*. Toronto: Key Porter Books.

Reif, L. 1975. "Ulcerative Colitis: Strategies for Managing Life," in Anselm L. Strauss and Barney G. Glaser, eds., *Chronic Illness and the Quality of Life*. St. Louis: Mosby: 81-88.

Reinharz, Shulamit. 1992. *Feminist Methods in Social Research*. Oxford: Oxford University Press.

Reissman, Catherine Kohler. 1983. "Women and Medicalization: A New Perspective," in Howard D. Schwartz, ed., *Dominant Issues in Medical Sociology*. Second Edition. New York: Random House: 101-21.

Reitz, J.G. 1980. *The Survival of Ethnic Groups*. Toronto: McGraw-Hill Ryerson.

Relman, Arnold S. 1986. "The New Medical-Industrial Complex," in Howard D. Schwartz, ed., *Dominant Issues in Medical Sociology*. New York: Random House: 597-607.

Report to the OMA Board of Directors for the Ad Hoc Committee on Women's Health Issues. 1987. Toronto: Ontario Medical Association.

Reverby, Susan M. 1987. *Ordered to Care: The Dilemma of American Nursing*. Cambridge: Cambridge University Press.

Richardson, Diane, and Victoria Robinson. 1993. *Thinking Feminist: Key Concepts in Women's Studies*. New York: Guilford Press.

Rieff, P. 1966. *Triumph of the Therapeutic*. New York: Harper and Row.

Ritzer, George. 1975. "Sociology: A Multiple Paradigm Science," *American Sociologist*, 10: 156-67.

Roberts, Paul William. 1992. "Feud," *Saturday Night*, December: 52-57, 96-106.

Rochon-Ford, Anne. 1986. "In Poor Health," *Healthsharing*, 7 (Winter): 8-10.

Rochon-Ford, Anne. 1990. *Working Together for Women's Health: A Framework for the Development of Policies and Programs*. n.p.: Federal/Provincial/Territorial Working Group on Women's Health.

Roland, Charles. 1988. "Medicine, History of," in *The Canadian Encyclopedia*, vol. 2. Edmonton: Hurtig: 1330.

Romelis, Shelly. 1985. "Struggle between providers and recipients: The case of birth practices," in Ellen Lewin and Virginia Olesen, eds., *Women, Health and Healing*. London: Tavistock: 174-208.

Rosen, George. 1963. "The Evolution of Social Medicine," in Howard E. Freeman, Sol Levine, and Leo G. Reeder, eds., *Handbook of Medical Sociology*. Englewood Cliffs, N.J.: Prentice-Hall: 23-50.

Rosenberg, Charles E., and Janet Golden, eds. 1992. *Framing Disease: Studies in Cultural History*. New Brunswick, N.J.: Rutgers University Press.

Roth, Julius. 1962. "Management Bias in Social Science Research," *Human Organization*, 21: 47-50.

Roth, Julius. 1963. *Timetables: Structuring the Passage of Time in Hospital Treatment and Other Careers*. Indianapolis: Bobbs-Merrill.

Rotter, J.B. 1966. "Generalized Expectation for Internal Versus External Control of Reinforcement," *Psychological Monographs*, 80: 1-28.

Rutherford, R.D. 1975. "The Changing Sex Differential in Mortality," *International Population and Urban Research*, University of California, Berkeley, Studies in Population of Urban Demography No. 1. Westport, Conn.: Greenwood Press.

Salmon, J. Warren. 1984. *Alternative Medicines: Popular and Policy Perspectives*. New York: Tavistock.

Schacter, S. 1975. "Cognition and Peripheralist-Centralist Controversies in Motivation and Emotion," in M.S. Gazzaniga and C. Blakemore, eds., *Handbook of Psychobiology*. London: Academic Press: 529-62.

Scheff, Thomas J. 1963. "The Role of the Mentally Ill and the Dynamics of Mental Disorder," *Sociometry*, 26 (June): 463-83.

Schneider, Joseph, and Peter Conrad. 1980. "In the Closet with Illness: Epilepsy, Stigma Potential and Information Control," *Social Problems*, 28, 1 (October).

Schneider, Joseph W., and Peter Conrad. 1983. *Having Epilepsy: The Experience and Control of Illness*. Philadelphia: Temple University Press.

Schwendinger, Julia, and Herman Schwendinger. 1971. "Sociology's Founding Fathers: Sexist to a Man," *Journal of Marriage and the Family*, 334: 783-89.

Scientific American. 1995. 272, 2 (June): 16.

Scott, James C. 1985. *Weapons of the Weak: Everyday Forms of Peasant Resistance*. New Haven: Yale University Press.

Scrip. 1983. *World Pharmaceutical News*, 797 (May 16): 15.

Scully, Diana. 1980. *Men Who Control Women's Health: The Miseducation of Obstetrician-Gynecologists*. Boston: Houghton Mifflin.

Scutt, Jocelynne. 1983. *Even in the Best of Homes: Violence in the Family*. Sydney: Penguin Books Australia.

Sechzer, Jeri A., Anne Griffin, and Sheila M. Pfafflin. 1994. *Forging a Women's Health Research Agenda: Policy Issues for the 1990's*. New York: Annals of the New York Academy of Science.

Selye, H. 1956. *The Stress of Life*. New York: McGraw-Hill.

Shackleton, Doris French. 1975. *Tommy Douglas*. Toronto: McClelland and Stewart.

Shaffir, William B., Robert A. Stebbins, and Allan Tarowetz. 1980. *Fieldwork Experience: Qualitative Approaches to Social Research*. New York: St. Martin's Press.

Shannon, H., *et al*. 1988. "Lung Cancer and Air Pollution in an Industrial City: A Geographical Analysis," *Canadian Journal of Public Health*, 79: 255-59.

Shapiro, Martin. 1978. *Getting Doctored: Critical Reflections on Becoming a Physician*. Toronto: Between the Lines.

Shephard, D.A., ed. 1982. *Norman Bethune – his times and his legacy*. Ottawa: Canadian Public Health Association.

Shkilnyk, Anastasia. 1985. *A Poison Stronger than Love*. New Haven: Yale University Press.

Shorr, R.I., S.F. Bauwens, and C.S. Landefeld. 1990. "Failure to Limit Quantities of

Benzodiazepine Hypnotic Drugs for Outpatients: Placing the Elderly at Risk," *American Journal of Medicine*, 89: 725-32.

Shorter, Edward. 1992. *From Paralysis to Fatigue: A History of Psychosomatic Illness in the Modern Era*. New York: Free Press.

Siggner, A.J. 1979. *An Overview of Demographic, Social and Economic Conditions Among Canada's Registered Indian Population*. Ottawa: Research Branch, DIAND.

Siirla, Aarne. 1981. *The Voice of Illness: A Study in Therapy and Prophecy*. Second Edition. New York: The Edwin Mellen Press.

Simonton, O. Carl, Stephanie Mathews Simonton, and James L. Creighton. 1978. *Getting Well Again*. Toronto: Bantam Books.

Smith, Dorothy E. 1987. *The Everyday World as Problematic: A Feminist Sociology*. Toronto: University of Toronto Press.

Smith, Dorothy E. 1993. *Texts, Facts, and Femininity: Exploring the Relations of Ruling*. London: Routledge.

Smith, Marianne, and Kathleen Buckwalter. 1992. "Medication Management, Anti-depressant Drugs, and the Elderly: An Overview," *Journal of Psychosocial Nursing*, 30, 10: 30-36.

Smith, Murray E.G. 1992. "The Burznyski Controversy in the United States and in Canada: A Comparative Case Study in the Sociology of Alternative Medicine," *Canadian Journal of Sociology*, 17, 2: 133-60.

Soderstrom, Lee. 1978. *The Canadian Health System*. London: Croom Helm.

Sontag, Susan. 1978. *Illness as a Metaphor*. New York: Random House.

Sontag, Susan. 1989. *AIDS and Its Metaphors*. Markham: Penguin Books.

Speedling, E.J. 1982. *Heart Attack: The Family Response and the Hospital*. New York: Tavistock.

Spiegel, D., J. Bloom, H. Kraemer, and E. Gotheil. 1989. "Effects of Psychosocial Treatment on Survival of Patients with Metastatic Breast Cancer," *Lancet*, October: 15.

Squires, B.P. 1987. "In Whose Service?" *Canadian Medical Association Journal*, 137: 983.

Stanley, E.M.G., and M.P. Ramage. 1984. "Sexual Problems and Urological Symptoms," in S.L. Stanton, ed., *Clinical Gynecological Urology*. St. Louis: Mosby: 398-405.

Stanley, Liz, and Sue Wise. 1993. *Breaking Out Again: Feminist Ontology and Epistemology*. London: Routledge.

Starr, Paul. 1982. *The Social Transformation of American Medicine*. New York: Basic Books.

Statistics Canada. n.d. *Corporate Financial Statistics: Detailed Income and Retained Earnings Statistics for 182 Industries*. Ottawa.

Statistics Canada. Various years. *Census of Canada*. Ottawa.

Statistics Canada. 1977. *Vital Statistics 1977*, vol. III. Ottawa.

Statistics Canada. 1981. *Health of Canadians. Report of the Canada Health Survey*. Ottawa.

Statistics Canada. 1982. *Medicare: The Public Good and Private Practice*. Ottawa: National Council on the Welfare of Canada's Health Insurance System (May).

Statistics Canada. 1983. *Health of Canadians. Report of the Canada Health Survey*. Ottawa.

Statistics Canada. 1984. *Life Tables, Canada and the Provinces*. Ottawa (May).

Statistics Canada. 1985. *Health and Social Support*. General Social Survey Analysis Series. Ottawa.

Statistics Canada. 1986. *Mortality and Vital Statistics*, vol. III. Ottawa.

Statistics Canada. 1987. *Active Health Report: Perspectives on Canada's Health Promotion Survey*. Ottawa: Ministry of Supply and Services.

Statistics Canada. 1988. *Work Injuries, 1985-87*. Ottawa: Statistics Canada, Labour Division.

Statistics Canada. 1991. *Accidents in Canada*. Ottawa.

Statistics Canada. 1993a. "The Violence Against Women Survey," *The Daily*, November 18.

Statistics Canada. 1993b. *Work Injuries, 1990-92*. Cat. 72-208. Ottawa.

Steacy, Anne. 1989. "Facing the Future: Thalidomide Victims Seek Compensation," *Maclean's*, February 20.

Stein, Howard F. 1990. *American Medicine as Culture*. Boulder, Colorado: Westview Press.

Stein, L. 1987. "The Doctor-Nurse Game," in H.D. Schwartz, ed., *Dominant Issues in Medical Sociology*. Second Edition. New York: Random House.

Steinem, Gloria. 1983. *Outrageous Acts and Everyday Rebellions*. New York: Holt, Rinehart and Winston.

Stelling, Joan. 1994. "Staff Nurses' Perceptions of Nursing: Issues in a Woman's Occupation," in B. Singh Bolaria and Harley D. Dickinson, eds., *Health, Illness and Health Care in Canada*. Second Edition. Toronto: Harcourt Brace & Company: 609-26.

Sternglass, Ernest J., and Jay M. Gould. 1993. "Breast Cancer: Evidence for a Relation to Fission Products in the Diet," *International Journal of Health Sciences*, 23, 4: 783-804.

Stevenson, Lloyd. 1946. *Sir Frederick Banting*. Toronto: Ryerson Press.

Stewart, David C., and Thomas J. Sullivan.

1982. "Illness Behaviour and the Sick Role in Chronic Disease: The Case of Multiple Sclerosis," *Social Science and Medicine*, 16: 1307-1404.

Stewart, Roderick. 1977. *The Mind of Norman Bethune*. Westport, Conn.: Lawrence Hill.

Stoddart, Greg L., and Roberta J. Labelle. 1985. *Privatization in the Canadian Health Care System: Assertions, Evidence, Ideology and Options*. Ottawa: Health and Welfare Canada.

Stone, Deborah S. 1986. "The Resistible Rise of Preventive Medicine," *Journal of Health Politics, Policy and Law*, 11: 671-96.

Stones, Ilene. 1987. "Rotational Shiftwork: A Summary of the Adverse Effects and Improvement Strategies." Hamilton, Ont.: Canadian Centre for Occupational and Health Safety.

Stoppard, Janet M. 1992. "A Suitable Case for Treatment? Premenstrual Syndrome and the Medicalization of Women's Bodies," in Dawn H. Currie and Valerie Raoul, eds., *Anatomy of Gender*. Ottawa: Carleton University Press.

Strauss, Anselm L. 1975. *Chronic Illness and the Quality of Life*. St. Louis: Mosby.

Strauss, Arlene. 1987. "Alzheimer's Disease and the Family Care Provider," supervised research project, Sociology and Anthropology Department, Wilfrid Laurier University.

Strauss, Robert. 1957. "The Nature and Status of Medical Sociology," *American Sociological Review*, 22: 200-04.

Strauss, Steven. 1994. "Arctic Pollution," *Globe and Mail*, November 19.

Strong, P.M. 1979. "Sociological Imperialism and the Profession of Medicine: A Critical

Examination of the Thesis of Medical Imperialism," *Social Science and Medicine*, 13, 2: 194-215.

Sudnow, D. 1967. *Passing On: The Social Organization of Dying*. Englewood Cliffs, N.J.: Prentice-Hall.

"Sweet Dreams or Nightmare?" 1991. *Newsweek*, August 19.

Szasz, Thomas S. 1974. *The Myth of Mental Illness*. New York: Harper and Row.

Szasz, T.S., and M.H. Hollender. 1956. "A Contribution to the Philosophy of Medicine the Basic Models of Doctor-Patient Relationship," *Archives of International Medicine*, 97: 585-92.

Tamblyn, R.M., Peter J. McLeod, *et al.* 1994. "Questionable Prescribing for Elderly Patients in Quebec," *Canadian Medical Association Journal*, 151: 1801-09.

Task Force on Sexual Abuse of Patients. 1991. *Final Report*. Toronto College of Physicians and Surgeons of Ontario.

Tataryn, Lloyd. 1979. *Dying for a Living*. Ottawa: Deneau and Greenberg.

Taylor, Paul. 1993. "Why the Rich Live Longer, Healthier," *Globe and Mail*, October 16: 1-2.

Taylor, William. 1985. *Hormonal Manipulation: A New Era of Monstrous Athletes*. Jefferson, N.C.: McFarland & Company.

Taylor, William. 1991. *Macho Medicine: A History of the Anabolic Steroid Epidemic*. Jefferson, N.C.: McFarland & Company.

Terris, M. 1980. "Preventative Services and Medical Care: The Costs and Benefits of Basic Change," *Bulletin of the New York Academy of Medicine*, 56: 180-89.

Terry, Edith. 1991. "The Workplace as Killing Field," *Globe and Mail*, September 21: D2.

Tesh, Sylvia Noble. 1988. *Hidden Arguments,*

Political Ideology and Disease Prevention Policy. New Brunswick, N.J.: Rutgers University Press.

"Thalidomide's After-Effects Today." 1989. *Globe and Mail*, February 14.

"Thalidomide Tragedy, labelling snag linked." 1972. *Kitchener-Waterloo Record*, September 27.

Theorell, T., and R.H. Rahe. 1971. "Psychosocial factors and myocardial infarction, I: An Inpatient Study in Sweden," *Journal of Psychosomatic Research*, 15: 25-31.

Thoits, Peggy A. 1982. "Conceptual, Methodological and Theoretical Problems in Studying Social Support as a Buffer Against Life Stress," *Journal of Health and Social Behaviour*, 23: 145-59.

Thoits, Peggy A. 1983. "Dimensions of Life Events that Influence Psychological Distress: An Evaluation and Synthesis of the Literature," in H.B. Kaplan, ed., *Psychosocial Stress: Trends in Theory and Research*. New York: Academic Press.

Thoits, Peggy A. 1991. "On Merging Identity Theory and Stress Research," *Social Psychology Quarterly*, 54: 101-12.

Thoits, Peggy A. 1994. "Stressors and Problem Solving: The Individual as Psychological Activist," *Journal of Health and Social Behavior*, 35 (June): 145-59.

Thomas, D.J. 1982. *The Experience of Handicap*. London: Methuen.

Thomas, Lewis H., ed. 1982. *The Making of a Socialist: The Recollections of T.C. Douglas*. Edmonton: University of Alberta Press.

Thompson, Joanne Emily. 1974. *The Influence of Dr. Emily Howard Stowe on the Woman Suffrage Movement in Canada*. Waterloo, Ont.: Waterloo Lutheran University Press.

Tierney, D., P. Romita, and K. Messing. 1990. "She Ate Not the Bread of Idleness: Exhaustion is Related to Domestic and Salaried Working Conditions among 539 Quebec Hospital Workers," *Women and Health*, 16, 1: 21-42.

Torrance, G. 1987. "Socio-Historical Overview: The Development of the Canadian Health System," in D. Coburn *et al.*, eds., *Health and Canadian Society*. Second Edition. Toronto: Fitzhenry & Whiteside: 6-32.

Trovato, Frank, and Carl F. Grindstaff, eds. 1994. *Perspectives on Canada's Population: an introduction to concepts and issues*. Toronto: Oxford University Press.

Trussell, James, Felicia Stewart, Malcolm Potts, Felicia Guest, and Charlotte Ellertson. 1993. "Should Oral Contraceptives Be Available Without Prescription?" *American Journal of Public Health*, 83, 8: 1094-97.

Trypuc, Joann M. 1988. "Women's Health," in B. Singh Bolaria and Harley D. Dickinson, eds., *Sociology of Health Care in Canada*. Toronto: Harcourt Brace Jovanovich: 154-66.

Tuckett, David, ed. 1976. *An Introduction to Medical Sociology*. London: Tavistock.

Tuohy, Carolyn J. 1976. "Medical Politics after Medicare: The Ontario Case," *Canadian Public Policy*, 2 (Spring): 192-210.

Turner, B.S. 1987. *Medical Power and Social Knowledge*. London: Sage Publications.

Turner, R. Jay, and William R. Avison. 1992. "Innovations in the Measurement of Life Stress: Crisis Theory and the Significance of Event Resolution," *Journal of Health and Social Behavior*, 33, 1 (March): 36-50.

Turner, R. Jay, B.G. Frankel, and D. Levin. 1983. "Social Support, Conceptualization, Measurement and Implications for Mental Health," in J.R. Greenly, ed., *Research in Community Mental Health*, vol. III. Greenwich, Conn.: JAI: 27-67.

Turner, R.J., C.F. Grindstaff, and N. Phillips. 1990. "Social Support and Outcome in Teenage Pregnancy," *Journal of Health and Social Behavior*, 31, 1, (March): 43-57.

Twaddle, Andrew C. 1982. "From Medical Sociology to Sociology of Health: Some Changing Concerns in the Sociological Study of Sickness and Treatment," in T. Bottomore *et al.*, eds., *Sociology: The State of the Art*. London: Sage: 324-58.

Twaddle, A.C., and R.M. Hessler. 1977. *A Sociology of Health*. St. Louis: Mosby.

Tyre, Robert. 1982. *Douglas in Saskatchewan: The Story of a Socialist Experiment*. Vancouver: Mitchell Press.

United Nations. 1982. *Demographic Indicators of Countries: Estimates and Projections as assessed in 1980*. New York: UN.

Valpy, Michael. 1994. "A fine mythology up in smoke," *Globe and Mail*, February 11: A2.

Vavasour, M., and Y. Mennie. 1984. *For Health or Profit*. Ottawa: World Inter-Action Ottawa and Inter-Paris.

Vayda, Eugene, Robert G. Evans, and William R. Mindell. 1979. "Universal Health Insurance in Canada," *Journal of Community Health*, 4, 3 (Spring): 217-31.

Verbrugge, Lois M. 1985. "Gender and Health: An Update on Hypothesis and Evidence," *Journal of Health and Social Behavior*, 26: 156-82.

Verbrugge, Lois M. 1989. "The Twain Meet: Empirical Explanations of Sex Differences in Health and Mortality," *Journal of Health and Social Behavior*, 31, 3: 282-304.

Verbrugge, Lois, and Deborah Wingard. 1987.

"Sex Differentials in Health and Mortality," *Women and Health*, 12, 2: 103-45.

Wahn, Michael. 1987. "The Decline of Medical Dominance in Hospitals," in D. Coburn *et al.*, eds., *Health and Canadian Society: Sociological Perspectives*. Second Edition. Toronto: Fitzhenry & Whiteside: 422-41.

Waitzkin, Howard. 1989. "A Critical Theory on Medical Discourse: Ideology, Social Control, and the Processing of Social Context in Medical Encounters," *Journal of Health and Social Behavior*, 30 (June): 220-39.

Waitzkin, Howard, and Theron Brett. 1989. "Changing the Structure of Medical Discourse: Implications of Cross-National Comparisons," *Journal of Health and Social Behavior*, 30, 4 (December): 436-49.

Waldron, I. 1977. "Increased Prescribing of Valium, Librium and other Drugs – An Example of the Influence of Economic and Social Factors on the Practice of Medicine," *International Journal of Health Services*, 7: 37-62.

Waldron, Ingrid. 1981. "Why Do Women Live Longer than Men?" *Journal of Human Stress*, 2: 19-30.

Waldron, Ingrid, and Susan Johnston. 1981. "Why Do Women Live Longer than Men? Part II," *Journal of Human Stress*, 2.

Walters, Vivienne. 1982. "State, Capital and Labour: The Introduction of Federal-Provincial Insurance for Physician Care in Canada," *Canadian Review of Sociology and Anthropology*, 19: 157-72.

Walters, Vivienne. 1991. "Beyond Medical and Academic Agendas: Lay Perspectives and Priorities," *Atlantis*, 17, 1: 28-35.

Walters, Vivienne. 1992. "Women's Views of their Main Health Problem," *Canadian Journal of Public Health*, 83, 5: 371-74.

Walters, Vivienne. 1994a. "Women's Perceptions Regarding Health and Illness," in B. Singh Bolaria and Harley D. Dickinson, eds., *Health, Illness and Health Care in Canada*. Second Edition. Toronto: Harcourt Brace & Company: 317-25.

Walters, Vivienne. 1994b. "The Social Construction of Risk in Nursing: Nurses' Responses to Hazards in their Work," in B. Singh Bolaria and Harley D. Dickinson, eds., *Health, Illness and Health Care in Canada*. Second Edition. Toronto: Harcourt Brace & Company: 627-43.

Walters, Vivienne, and J. Haines. 1989. "Workload and Occupational Stress in Nursing," *Canadian Journal of Nursing Research*, 21, 3: 49-58.

Warburton, Rennie, and W. Carroll. 1988. "Class and Gender in Nursing," in B. Singh Bolaria and Harley D. Dickinson, eds., *Sociology of Health Care in Canada*. Toronto: Harcourt Brace Jovanovich: 364-75.

Wardwell, Walter. 1980a. "Limited and Marginal Practitioners," in H. Freeman, S. Levine, and L. Reeder, eds., *Handbook of Medical Sociology*. Third Edition. Englewood Cliffs, N.J.: Prentice-Hall.

Wardwell, Walter. 1980b. "The Present and Future Role of the Chiropractor," in S. Haldeman, ed., *Modern Developments in the Principles and Practice of Chiropractic*. Englewood Cliffs, N.J.: Prentice-Hall.

Wardwell, Walter. 1988. "Chiropractors: Evolution to Acceptance," in Norman Gentz, ed., *Other Healers: Unorthodox Medicine in America*. Baltimore: Johns Hopkins University Press: 174-84.

Weber, Max. 1947. *The Theory of Social and Economic Organization*, trans. A.M.

Henderson and Talcott Parsons. New York: Free Press.

Weber, Max. 1968. *Economy and Society: An Outline of Interpretive Sociology*, trans. Ephraim Fischoff; G. Roth and C. Witlich, eds. New York: Bedminster Press.

Weiger, Charles. 1995. "Our Bodies, Our Science," *The Sciences*, May/June.

Weigts, Wies, Hannecke Hontkoop, and Patricia Mullen. 1993. "Talking Delicately: Speaking About Sexuality During Gynecological Consultation," *Sociology of Health and Illness*, 15, 4.

Weil, Andrew. 1983. *Health and Healing: Understanding Conventional and Alternative Medicine*. Boston: Houghton Mifflin.

Weisman, D.K., and J.W. Worden. 1975. "Psychological Analysis of Cancer Deaths," *Omega*, 6: 61-75.

Weiss, Gregory L., and Lynne E. Lonnquist. 1992. "Dissecting the Medical Encounter: A New Model of the Physician-Patient Relationship," paper presented at the 87th meeting of the ASA, Pittsburgh.

Weller, G. 1986. "Health Care Delivery in the Canadian North: The Case of Northwestern Ontario," paper presented at the annual meeting of the Western Association of Sociology and Anthropology, Thunder Bay, February 13-15.

Wennberg, John E. 1984. "Dealing with Medical Practice Variations: A Proposal for Action," *Health Affairs*, 4 (Summer): 6-32.

Wennberg, John E., John P. Bunker, and Benjamin Barnes. 1980. "The Need for Assessing the Outcomes of Common Medical Practice," *Annual Review of Public Health*, 1: 277-95.

Wennemo, Irene. 1993. "Infant Mortality, Public Policy and Inequality – A Comparison of 18 Industrialized Countries, 1950-1985," *Sociology of Health and Illness*, 15, 4.

Wertz, Richard W., and Dorothy Wertz. 1977. *Lying-In: A History of Childbirth in America*. New York: Free Press.

Wertz, Richard W., and Dorothy C. Wertz. 1986. "Notes on the Decline of Midwives and the Rise of Medical Obstetricians," in Peter Conrad and Rochelle Kern, eds., *The Sociology of Health and Illness: Critical Perspectives*. New York: St. Martin's Press.

Weston, Marianne, and Bonnie Jeffrey. 1994. "AIDS: The Politicizing of a Public Health Issue," in B. Singh Bolaria and Harley D. Dickinson, eds., *Health, Illness and Health Care in Canada*. Second Edition. Toronto: Harcourt Brace & Company: 721-39.

"Westray probe's scope curbed." 1995. *Toronto Star*, November 17: A12.

White, Kevin. 1991. "Trend Report: The Sociology of Health and Illness," *Current Sociology*, 39, 2: 1-115.

Wilensky, Harold L. 1964. "The Professionalization of Everyone," *American Journal of Sociology*, 70: 137-58.

Wilkins, R., and O. Adams. 1983. *Healthfulness of Life: A Unified View of Mortality, Institutionalization and Non-Institutionalized Disability in Canada, 1978*. Montreal: Institute for Research on Public Policy.

Wilkinson, Richard G. 1990. "Income Distribution and Mortality: A 'Natural' Experiment," *Sociology of Health and Illness*, 12, 4: 391-412.

Williams, P., and D.R. Rush. 1986. "Geriatric Polypharmacy," *Hospital Practice*, 21: 109-20.

Willis, Evan. 1983. *Medical Dominance: The Division of Labour in Australian Health Care*. Sydney: George Allen and Unwin.

Wilson, Bryan R. 1961. *Sects and Society: A Sociological Study of Three Religious Groups in Britain*. London: Heinemann.

Wilson, C.W.M., J.A. Banks, R.E.A. Mapes, and S.M.T. Korte. 1963. "Influence of Different Sources of Therapeutic Information on Prescribing by General Practitioners," *British Medical Journal*, 3: 599.

Wilson, Edward O. 1991. "Biodiversity, Prosperity and Value," in Herbert F. Bohrmann and Stephen R. Kellert, eds., *Ecology, Economics and Ethics: The Broken Circle*, special issue of *Society*, 30, 1 (November/December): 90-93.

Wilson, Jane. 1987. "Why Nurses Leave Nursing," *The Canadian Nurse*, 83 (March): 20-23.

Wilson, S.J. 1982. *Women, the Family and the Economy*. Toronto: McGraw-Hill.

Winsor, Hugh. 1973. "Thalidomide," *Globe and Mail*, March 10.

Wnuk-Lipinski, Edmund, and Raymond Illsley. 1990. "International Comparative Analysis: Main Findings and Conclusions," *Social Science and Medicine*, 31, 8: 879-89.

Wohl, Stanley. 1984 *The Medical Industrial Complex*. New York: Harmony Books.

Wolfe, Naomi. 1990. *The Beauty Myth*. Toronto: Vintage Books.

Wolfe, Morris. 1993. "Dental Flaws," *Saturday Night* (November): 15-16, 20-24.

Women's Health Office Newsletter. 1994. Hamilton, Ont.: McMaster University, April.

Wood, S. 1980. *WHO International Classification of Impairments, Disabilities and Handicaps*. Geneva: World Health Organization.

Woolhandler, Steffie, David U. Himmelstein, Ralph Silba, Michael Bader, M. Narnley, and Alice A. Jones. 1985. "Medical Care and Mortality: Racial Differences in Preventable Deaths," *International Journal of Health Services*, 15, 1: 1-11.

World Development Report. 1992. *Development and the Environment*. Oxford: Oxford University Press.

World Health Organization. 1983. International Code (May, 1981, Article 1). Geneva: WHO.

Wysong, Peggy. 1986. "Health Profession Legislation Review," *RNAO News*, 9 (Spring): 18-19.

Yudkin, J.S. 1978. "Provisions of Medicines in a Developing Country," *Lancet* (April).

Zabolai-Csekme, Eva. 1983. *Women, Health and Development*. Geneva: World Health Organization.

Zborowski, Mark. 1952. "Cultural Components in Response to Pain," in E. Jaco, ed., *Patients, Physicians and Illness*. Glencoe, Ill.: Free Press: 256-68.

Zborowski, Mark. 1969. *People in Pain*. San Francisco: Jossey-Bass.

Zimmerman, Mark. 1983. "Methodological Issues in the Assessment of Life Events: A Review of Issues and Research," *Clinical Psychology Review*, 3: 339-70.

Zola, Irving K. 1972. "Medicine as an Institution of Social Control," *Sociological Review*, 20: 487-504.

Zola, Irving K. 1973. "Pathways to the Doctor: From Person to Patient," *Social Science and Medicine*, 7, 9: 677-89.

Zola, Irving K. 1975. "In the Name of Health and Illness: On Some Socio-Political Consequences of Medical Influence," *Social Science and Medicine*, 9: 83-87.

Index